Cartography in Antiquity and the Middle Ages

Technology and Change in History

VOLUME 10

Cartography in Antiquity and the Middle Ages

Fresh Perspectives, New Methods

Edited by

Richard J.A. Talbert and Richard W. Unger

BRILL

LEIDEN • BOSTON
2008

Cover illustration: Augustus Caesar commissioning a survey of the world, as imagined in the 1280s by the makers of the Hereford *mappa mundi*. From a nineteenth century facsimile edition, courtesy of Hereford Cathedral, Hereford, U.K.

This book is printed on acid-free paper.

Library of Congress Cataloging-in Publication data

Cartography in antiquity and the Middle Ages : fresh perspectives, new methods / edited by Richard J.A. Talbert and Richard W. Unger.
p. cm. — (Technology and change in history ; v. 10)
Includes bibliographical references and index.
ISBN 978-90-04-16663-9 (hardback : alk. paper) 1. Cartography—History. 2. Early maps. I. Talbert, Richard J. A., 1947- II. Unger, Richard W. III. Title. IV. Series.

GA205.C37 2008
526.093—dc22

2008009749

ISSN 1385-920X
ISBN 978 90 04 16663 9

PRINTED IN THE NETHERLANDS

To the Memory of
John Brian Harley and David Woodward

CONTENTS

LIST OF CONTRIBUTORS

Emily Albu is Associate Professor of Classics at the University of California, Davis. Her special interests are Late Antiquity, the twelfth century, and the classical tradition. She is the author of *The Normans in their Histories* (Boydell & Brewer, 2001), and "Imperial geography and the medieval Peutinger map" (*Imago Mundi* 2005).

Raymond Clemens is Associate Professor of History at Illinois State University. His research focuses on the archaeology of book and map production. With Timothy Graham, he is co-author of *Introduction to Manuscript Studies* (Cornell, 2007).

Lucy Donkin is a British Academy Postdoctoral Fellow at University College, Oxford. She specializes in perceptions of sacred space during the Middle Ages, and has published "Mosaici pavimentali medievali nell' Italia settentrionale e i loro rapporti con la liturgia" (AISCOM, Tivoli, 2005).

Evelyn Edson is Professor Emerita of History at Piedmont Virginia Community College, Charlottesville, Virginia. Her books include *Mapping Time and Space: How Medieval Mapmakers Viewed their World* (British Library, 1997), and (with Emilie Savage-Smith), *Medieval Views of the Cosmos: Picturing the Universe in the Christian and Islamic Middle Ages* (Bodleian Library, 2004).

Tom Elliott is Director of the Pleiades Project at the University of North Carolina, Chapel Hill. He specializes in Roman history, geography and epigraphy, and in humanities computing.

Patrick Gautier Dalché is Directeur de recherche (Centre national de la recherche scientifique, Institut de recherche et d'histoire des textes, Paris), and Directeur d'études (École pratique des hautes études, Sorbonne). He specializes in medieval representations of space (texts and images), and his many publications include: *La "Descriptio mappe mundi" de Hugues de Saint-Victor* (Études Augustiniennes, 1988); *Carte marine et portulan au XIIe siècle* (École française de Rome, 1995); *Géographie et culture. La représentation de l'espace du VIe au XIIe siècle* (Variorum, 1997); *Du Yorkshire à l'Inde* (Droz, 2005).

Benjamin Z. Kedar is Professor of History at the Hebrew University of Jerusalem. He specializes in the history of the Crusades, comparative history, and the use of aerial photographs as historical sources. Among his recent publications are: *The Changing Land between the Jordan and the Sea: Aerial Photographs from 1917 to the Present* (Jerusalem/Tel Aviv, 1999); *Franks, Muslims and Oriental Christians in the Latin Levant: Studies in Frontier Acculturation* (Variorum, 2006).

Maja Kominko is currently moving from a faculty position at Sabanci University, Istanbul, to one at Princeton University. She specializes in Byzantine art history. Her publications include "The map of Cosmas and its influence" (*Mediterranean Historical Review* 2005).

Natalia Lozovsky is a Visiting Scholar in the Office for History of Science and Technology, at the University of California, Berkeley. She specializes in the transmission and reception of classical texts on geography, ethnography and history in the Middle Ages. Her publications include *'The Earth is Our Book': Geographical Knowledge in the Latin West, 400-1000* (Michigan, 2000).

Yossef Rapoport is a Research Fellow at the Khalili Research Centre for the Art and Material Culture of the Middle East at the University of Oxford. Medieval Islamic history forms his principal research interest. He is the author of *Marriage, Money and Divorce in Medieval Islamic Society* (Cambridge, 2005).

Emilie Savage-Smith is Professor of the History of Islamic Science at the University of Oxford. Her recent publications include *Medieval Views of the Cosmos: Picturing the Universe in the Christian and Islamic Middle Ages* (with Evelyn Edson, Bodleian Library, 2004), and *Medieval Islamic Medicine* (with Peter Pormann, Georgetown, 2007).

Camille Serchuk is Professor of Art History at Southern Connecticut State University. She specializes in late medieval French art.

Among her publications is "Picturing France in the fifteenth century: the map in BNF Ms Fr 4991" (*Imago Mundi* 2006).

RICHARD TALBERT is Kenan Professor of History and Classics at the University of North Carolina, Chapel Hill. Roman maps, worldview and associated materials and ideas form his principal current research interest. He edited the *Barrington Atlas of the Greek and Roman World* (Princeton, 2000), and (with Kai Brodersen) *Space in the Roman World: its Perception and Presentation* (LIT, 2004).

JENNIFER TRIMBLE is Associate Professor of Classics at Stanford University, specializing in Roman art and archaeology. Her book *Replicating Women: Portrait, Place and Power in the Roman Empire* is forthcoming.

RICHARD UNGER is Professor of History at the University of British Columbia. He specializes in economic history and the history of technology. His books include *A History of Brewing in Holland, 900-1900: Economy, Technology and the State* (Brill, 2001), and *Beer in the Middle Ages and the Renaissance* (University of Pennsylvania, 2004).

LIST OF ILLUSTRATIONS

Tom Elliott

Emily Albu

Yossef Rapoport and Emilie Savage-Smith

Maja Kominko

Benjamin Kedar

EVELYN EDSON

Raymond Clemens

Camille Serchuk

ACKNOWLEDGMENTS

The present volume derives from a meeting originally conceived by Richard Talbert as a means to bring together historians who study classical and medieval cartography. The venue was the University of British Columbia, Vancouver, and the context a meeting in an annual series there which for the past thirty years and more has addressed current themes in medieval studies. We are keenly aware that neither this Thirty-Fifth Medieval Workshop in October 2005, nor the volume emerging from it, would have been possible without generous support furnished in a variety of ways. As joint organizers and editors, we are much indebted for funding contributed by the Office of the Vice-President, Research, at the University of British Columbia; the Dean of Arts; the President's Committee on Visiting Lecturers; the three Departments of History, English, and Classical, Near Eastern and Religious Studies; and the Committee for Medieval Studies. Our speakers unselfishly embraced the workshop character of the meeting, and used it accordingly as a long awaited opportunity to exchange ideas and practices across disciplines, continents and millennia. Members of the Committee for Medieval Studies rendered invaluable assistance in organizing the meeting, as did Professor Leanne Bablitz, co-organizer in the Department of Classical, Near Eastern and Religious Studies. To her, and to all, we offer our warmest thanks. Brill's reader, Prof. Paul Harvey, evaluated the complete set of contributions in a most positive and helpful manner. Last but not least, we greatly appreciate that Brill is publishing the volume in 2008, the year that marks the twenty-first anniversary of Harley and Woodward's inspiring *History of Cartography*, volume 1.

Richard Talbert, Chapel Hill, North Carolina
Richard Unger, Vancouver, British Columbia
September 2007

INTRODUCTION

Richard Talbert and Richard W. Unger

There has been growing recognition that the study of classical and medieval cartography is typically being pursued by two separate groups of scholars who feel nothing but mutual goodwill, yet in practice seldom cross paths to exchange results with one another. Hence the primary aim of the October 2005 meeting was to create such an opportunity, so that all who participated could gain a deeper sense of their colleagues' research and of how it might enrich their own. A gratifyingly broad and varied range of backgrounds, perspectives, materials and methods was represented across a vast arc of time and space. The diverse character of the participants' activity, so clearly evident from the papers they have contributed to this volume, well reflects the origins of their research and indeed of the history of cartography itself, with its roots in nationalism and in the history of science as much as of art, not to mention—among other sources—the passion of collectors.

Without doubt, the entire history of cartography has undergone a remarkable transformation during the past quarter-century. There was special concern, therefore, to mark the advances made above all since the 1987 publication of *Cartography in Prehistoric, Ancient, and Medieval Europe and the Mediterranean*, the opening volume in the path-breaking and still ongoing series *The History of Cartography* conceived and launched by Brian Harley and David Woodward. The intention was not so much to 'correct' this volume's scholarship, but rather to appreciate the value of the foundation which it lays, as well as to offer instructive illustration of how its fresh thinking has proven the springboard and inspiration for new departures in the field. The broader scope that Harley and Woodward's expanded vision of cartography and its significance opens up is not merely confined to the use of iconology and theory borrowed from postmodernism. In fact, those methods are now regarded with serious scepticism by some historians of cartography. Rather, the deeper and more durable impact of the 1987 volume lies in its vision of maps as sources with unrealized potential to advance understanding of past societies and their cultures, especially when an overly rigid definition of 'map' is abandoned and when such materials are interpreted with due recognition for their contemporary context. To persist in contemplating maps for little more than their accuracy or beauty from a modern Western perspective was an outmoded and disappointingly blinkered approach which Harley and Woodward had the creative boldness to supersede, to universal benefit. At the outset—as we now recognize in retrospect—even they did not anticipate the scale and productiveness of the seismic shift that they were setting in motion.

The volume's papers can naturally be divided into those with a classical focus, and those with a medieval one. The apparent separation between the two chronological periods is in many respects artificial, an arrangement inherited from the organizational structure of academic institutions over the last two hundred years. One of the goals of the meeting and of this volume, accomplished with visible success, was to diminish the significance of this divide. Such an attempt to appreciate the connections linking the Roman Empire and medieval Europe reinforces a widespread trend in recent historical scholarship. Indeed, for over twenty years now it has been a special concern of the Medieval Studies program at the University of British Columbia to equip medievalists with a profounder understanding of the classical past, while stimulating classicists to explore the implications of the Roman heritage through the centuries that culminate with the Renaissance.

The volume's opening pair of papers survey the study of classical and of medieval maps respectively. These papers reach both forward and back in time to illustrate connections across the entire period under consideration, while reflecting upon the current state of study and reporting some significant recent work. Richard Talbert lays out the role of *The History of Cartography* in encapsulating the lines of approach adopted for the history of Roman cartography up to 1987. He points to changed perspectives and emphases of recent years which hold out clear promise for the future. Patrick Gautier Dalché travels an even longer and more tortuous road, roaming from the early Middle Ages to the Renaissance, as well as reaching back to the intellectual roots of high medieval maps in late Antiquity. His analysis of shifts in cartographic outlook and purpose over many centuries draws from an exceptional grasp of scholarship. He points to literary sources as vital resources for historians of cartography, especially in their quest to understand *mappaemundi*. Those complex products of the high Middle Ages, with their varied images, have attracted intensified interest in recent years, aided by improved representation and more rigorous study. Not only has their reproduction in print become fuller, sharper and less costly, but digitization and dissemination through electronic media have also begun to make a marked impact.

The volume's papers about ancient maps all share a greater or lesser preoccupation with the reconstruction of that period's lost output, as well as with its original use and function. Here, recourse to the exploitation of digital technology has emphatically proven its worth. Jennifer Trimble addresses the monumental wall map of the city of Rome, only scattered segments of which survive. As she reveals, to achieve a keener sense of how it appeared to its viewers proves a fruitful, and rewarding, means of reassessing its purpose. Tom Elliott demonstrates the ways in which digital technology can advance our engagement with premodern cartography that manifestly departs from adherence to the norms taken for granted today, using as his example the peculiarly shaped Peutinger map of the Roman world and its land routes. In the following paper, Emily Albu contends that this map is a medieval creation, not—as has been widely accepted—an ancient one, and she seeks to establish where it may have been produced.

Focus on the early Middle Ages in particular bridges Roman and undeniably medieval cartography, and further serves to identify the linkages between the two. This said, there is no clear line of division between the two periods, a point superbly illustrated for the history of cartography by the next pair of papers. In the first, Emilie Savage-Smith and Yossef Rapoport together report on an extraordinary recent find with far reaching consequences. The *Book of Curiosities*, with its rich cartographic material, may have had naval or commercial purposes, or both. It is impossible to be sure of the author's goals or his sources from the remarkable text and illustrations. The maps in this book incorporate both classical learning and new knowledge, and form a priceless extension to the known body of Islamic cartography. In the second paper, Maja Kominko examines the maps associated with the work of an author notorious for his claim that the world is flat. As a result, for the past two centuries Cosmas Indicopleustes has been cited as (false) proof of the supposedly primitive understanding of geography prevalent in an era much swayed by religion. Kominko maintains that Cosmas' map of the world in fact derives from a separate stream in late classical Greek cartography. Ideas about the curvature of the earth or lack of it were not among its special concerns. Rather, Cosmas' map is merely one effort among others to represent the world in a way consistent with both classical knowledge and Christianity.

The diversity of maps of the high and late Middle Ages—creations based to a significant degree on those varied classical roots—is manifest from the maps next discussed by Benjamin Kedar, Natalia Lozovsky and Lucy Donkin. Kedar penetrates the so-called renaissance of the twelfth century, and through a different cartographic tradition. As he shows, in twelfth century western Europe there flourished mapping that was both non-Christian and non-Muslim, as well as reliant upon Biblical sources for both its questions and its answers. Specifically, Kedar discusses the cartography of Rashi, a Jewish mapmaker, whose distinctive and little known work draws upon both Christian and Muslim traditions. Lozovsky examines the relationship of words to maps, in particular how maps became vehicles for repeating and amplifying the written word. Recent scholarship has pointed to connections between maps and rulers. The latter grasped the value of maps for propaganda and the dissemination of information, as well as to record and interpret knowledge. Lozovsky underscores the persistence of ideas about the Roman Empire, even once such conceptions no longer reflected reality. Maps provided one useful means for the persistence of those ideas to be articulated. Even if her claims about the longevity of an imagined Roman hegemony do not prove to win full acceptance, her argument serves to emphasize the continuity of ideas about space, as well as mapmaking practices, across the first Christian millennium and beyond. Donkin focuses on floor mosaics, a context for maps in the Roman world and similarly, as she shows, in the medieval one too. Such placement has increased these maps' chances of survival, making them a valuable type of source even if the character of the medium gives rise to problems of interpretation. Donkin not only identifies a medieval connection with the classical tradition of siting maps on floors, but she also finds linkages between different medieval mosaic maps with methods and information transferred from one to another. Evelyn Edson brings to bear her encyclopedic knowledge of medieval maps in exposing various ways in which late classical ideas influenced the maps of the high Middle Ages. The formation of Christian identity in the late Roman Empire furnished cartographers with an agenda for some centuries. Over time, maps might appear to be different, and might expand in scope, size and complexity; even so, according to Edson, the agenda for mapmakers was already set in the early seventh century and before.

In the case of maps, as with many aspects of the intellectual life of medieval Europe, the influence of those writers who categorized, catalogued and organized knowledge for Christians may have waned in the high Middle Ages, but it did not disappear. While many signs of their sustained influence remained visible, by the fourteenth and fifteenth centuries new ideas about maps and new functions for them made an impact on cartographic products. In their papers, Raymond Clemens and Camille Serchuk both address a much more rapidly changing late medieval world. Clemens explores how maps could be integral tools in education, or at least how they were seen as such in fifteenth century Italy. The period when Renaissance ideas about art and learning were starting to influence mapmaking also saw a significant increase in the uses of maps as well as in the number of their users. The connection of maps to cosmology, as found by Clemens in the work of one Florentine writer, was normal. At the same time, there were signs that the embedding of theology, which was so much a part of many earlier maps, was waning. Hence the cartography of this period differed from maps of earlier traditions. Serchuk explores some new uses and users of maps along with the distinctive practical trends of the fifteenth century. The rise in the political functions of maps during the Renaissance is already well known. Less noticed is the increasing use of maps for legal disputes. Serchuk brings the eye of an art historian to questions of who made maps in fifteenth century France, how the artists gained their commissions, and how they were paid for their work. What appears in the maps reflects who made them and why. The new functions of the representations of lands were influential too, as Serchuk argues from certain fifteenth century images designed to illustrate historical works. Maps unquestionably had varying purposes, but these differences did not prevent practices from one type of map being transferred across a range of products. The same artists, it appears, created a range of map types.

By the fifteenth century, as Clemens and Serchuk show, mapmaking was without doubt taking on a new character.

As was only to be anticipated perhaps, the conference amply confirmed the increasing need to replace *The History of Cartography* volume 1 with a new edition. This recommendation seeks to cast no aspersions upon the volume. On the contrary, it underlines the inspiration which it has proved to be—and will continue to be—since its publication in 1987. Fortunately, Matthew Edney, current director of the ongoing project begun by Harley and Woodward, was able to attend the conference. Speaking about plans for the future, he readily endorsed the need to revisit the project's initial publication once the first edition is completed in its entirety.

Meantime the conference papers, as revised here for publication, show the study of maps to be at once varied and cohesive. The variety is both chronological and methodological. The expertise of contributors ranges from art historian to computer scientist. An open approach to the history of maps calls for a large set of skills, and these are ably deployed here both on a broad front and in detail. Especially striking, and very welcome, is the expanding range of methods represented. At the same time there is also a cohesion to the contributions, as there was to the discussions which formed a vital part of the conference itself. Participants shared the conviction of Harley and Woodward that maps, as cultural artefacts, represent the thinking of their time, and should therefore always be interpreted as products of specific periods and places. In consequence, the history of cartography must intersect with that of art, politics and the intellect, indeed with the whole history of culture, both high and popular.

The claim can fairly be made that the history of cartography offers a model guide for approaching the past. Not least in this volume, it decisively demonstrates that the separation of the mythical Middle Ages from a classical Antiquity crafted by Renaissance thinkers is false. To use the Renaissance as a marker for the end of an era in mapmaking nonetheless remains valid, insofar as it was then that thinkers conceived the idea of a unique 'middle age'. Humanists in the fifteenth and sixteenth centuries believed that they were ushering in a new era, and in cartography—to some degree—they did. Between the classical and medieval worlds there should be no such division, however, at least not for the history of cartography, nor perhaps elsewhere either. For scholars studying maps, as this volume attests, themes of continuity, connections and adaptation now promise to be among the most rewarding. New methods and technologies, together with fresh discoveries of materials, will allow the progress set in motion by Harley and Woodward to be sustained and expanded. Consequently, to an unprecedented degree, an awareness of cartography and its value will be integrated into the mainstream of historical study.

GREEK AND ROMAN MAPPING: TWENTY-FIRST CENTURY PERSPECTIVES

Richard Talbert

Surprising though it may seem to non-specialists,[1] the current academic preoccupation with Greek and Roman maps is a recent phenomenon that barely predates the 1980s. Moreover, it stems from the work of just a handful of individuals encapsulated in a few truly seminal books. Along with these books, three of the individuals will be singled out for special mention here: Dilke, Harley and Woodward.

Oswald Dilke (1915-1993)[2] was a classical scholar who in the course of his career held positions in several British universities, the last being the Chair of Latin at Leeds. He was a remarkable polymath who possessed the courage and originality to develop interests that were altogether neglected and unfashionable in his day. The first of note for present purposes was the Roman *agrimensores* or land surveyors. A notoriously dense body of technical writing by them survives, together with various documents and land-maps on stone, as well as remains of the checkerboard pattern that they imposed on the landscape detectable through archaeology or aerial photography. Their survey-maps were also exhibited in bronze on-site, and a copy was sent to Rome. Of all these bronzes from across the empire, however, just one tiny fragment has survived. It emerged from Spain about fifteen years ago.[3] [Fig. 1] Although Dilke neither edited nor translated the texts by the *agrimensores*, he did produce a pioneering book about their profession and its work, which appeared in 1971.[4]

[1] The attempt to make this keynote address accessible to them is maintained in the present revised version.

[2] Obituary by W.G. Arnott in *The Independent* (London), July 21, 1993.

[3] For a drawing and discussion, see most readily K. Brodersen, *Terra Cognita: Studien zur römischen Raumerfassung* (Hildesheim: Olms, 1995), 221-24.

[4] *The Roman Land Surveyors: an Introduction to the Agrimensores* (Newton Abbot: David and Charles, 1971).

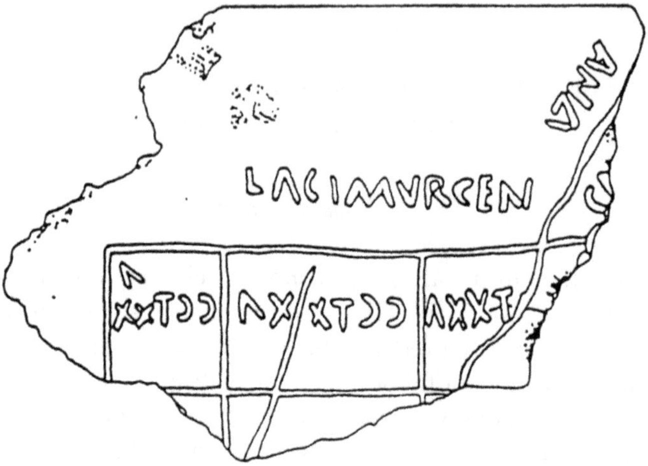

Fig. 1. Fragment of bronze land survey map from Spain (8.6 x 5.6 cm).

This achievement inspired him to widen his scope and to attempt a second book encompassing not just land-survey maps, but also other types of Greek and Roman maps and cartographic writings. The book formed part of the Thames & Hudson *Aspects of Greek and Roman Life* series, which called for a modest 22 x 14 cm trim-size, kept notes to a minimum, and preferred a length of no more than 200-250 pages. This highly original work, *Greek and Roman Maps*—for which his wife Margaret, herself a geographer, assisted him—appeared in 1985.

Meantime, however, Dilke had been recruited to contribute to the literally larger and far more ambitious enterprise which Brian Harley and David Woodward had launched as the History of Cartography Project on receiving their first grant from the National Endowment for the Humanities in 1981.[5] Harley, who died in 1991, and Woodward, who died in August 2004, both professors of geography and cartography, were visionaries who had identified the history of cartography as a potentially rich field woefully lacking in framework or focus. Up to the 1970s, historians of cartography typically limited themselves to elucidating the content of individual maps for fellow historians and collectors. Otherwise they had no organi-

[5] The project is ongoing: visit www.geography.wisc.edu/histcart

zational principle, no philosophy, no impetus to range further. The relationship between form and content embedded in maps was not being appreciated or probed. Woodward argued—to quote the obituary by Matthew Edney—[6] "that the history of cartography should be 'the study of maps and mapmaking in their human context through time.' He saw detailed studies in map production and design as the foundation for an autonomous, open, outward looking and necessarily interdisciplinary discipline."

Hence there occurred the development of the History of Cartography Project, with Harley and Woodward as its joint leaders and editors. As initially proposed, the work would comprise a total of four volumes to be completed by 1992, with individual chapters commissioned from specialists. It would, to quote Edney again, "give shape to the still inchoate field by unifying its literature, by providing it with a secure foundation for future growth and by demonstrating the worth of analyzing maps on their own terms rather than simply as sources of evidence for other historical disciplines."[7] For the purpose, moreover, a new definition of "map" was to be adopted, which marked a clear break with the traditional narrower understanding of maps as factual documents, namely: "Maps are graphic representations that facilitate a spatial understanding of things, concepts, conditions, processes, or events in the human world."[8]

As Edney reminds us,[9] in the early stages it was unavoidably a struggle for Harley and Woodward to achieve the openness and comprehensiveness, the interpretation and contextualization, that they were so eager to introduce. These difficulties were only compounded by their desire to devote the first volume to prehistoric, ancient, and medieval Europe and the Mediterranean, a bold and deliberate choice that gave unprecedented prominence to societies whose engagement with cartography had never formed the center of scholarly attention. Within the volume's vast arc of time and space, several chapters would plainly have to be devoted to ancient Greece and Rome, but what established experts were there to write them in this neglected, unshaped, field? A French scholar, Germaine

[6] *Imago Mundi* 57.1 (2005), 78.

[7] Ibid., 78.

[8] J.B. Harley and D. Woodward (eds.), "Introduction," in *The History of Cartography*, vol. 1 (Chicago: University of Chicago Press, 1987), xvi.

[9] Edney, "Obituary," 79.

Aujac, proved willing to supply materials which the two editors developed into three chapters on classical and Hellenistic Greek cartography.[10] Dilke's willingness to contribute was clearly a godsend. He committed to providing as many as five chapters that covered Roman cartography in its entirety, plus Greek cartography under the Roman Empire and on through the Byzantine period, about 100 printed pages in all. These chapters both draw from, and expand upon, what he wrote for his smaller, more comprehensive 1985 book *Greek and Roman Maps*. Ever since the appearance of *The History of Cartography* volume 1 in 1987, his chapters have represented a uniquely powerful, learned exposition of their themes.

It was Harley and Woodward's specific intention that an authoritative synthesis of the type offered by Dilke's contribution would regenerate, and perhaps re-orient, attention to Greek and Roman cartography and their related contexts. There is no question that this intention has succeeded in a variety of productive ways, so much so by now that one might say the time is ripe for a fresh synthesis to supersede that of Dilke. At the present date, to be sure, only a provisional update on "work in progress" is feasible,[11] because the concerns which Dilke provoked have by no means been fully explored or resolved as yet.

Dilke's assessment of Roman cartography in particular was quickly called into question by several scholars,[12] who found him prone to be credulous in identifying "maps". Neither ancient Greek nor Latin has any word that unequivocally signifies "map", and Dilke was hasty in assuming that in certain instances a term used in an ancient source which could mean "map" does mean that there, even though it might really be referring to something quite different. His most unfortunate such misstep was to assume that the official in the Late Roman Empire with the title Count (*Comes*) of *Formae* at Rome was "director of maps.... an official working for what must have been a civil service

[10] See further in this connection M.H. Edney, *The Origins and Development of J.B. Harley's Cartographic Theories*, *Cartographica* 40,1-2 [2005] = Monograph 54 (Toronto: University of Toronto Press), 54-55 with n. 7.

[11] Compare the review article by K. Brodersen, "Mapping (in) the ancient world," *Journal of Roman Studies* 94 (2004), 183-90.

[12] Note my review of his *Greek and Roman Maps*, in *Journal of Roman Studies* 77 (1987), 210-12.

maps and plans department."[13] *Forma* is indeed among the ambiguous Latin nouns that can signify "map". But it has several other meanings too, and in this instance further testimony leaves no doubt that *formae* signifies "water channels" or "conduits", so that the *Comes Formarum* was in fact the official in charge of the city of Rome's aqueducts, without any connection to cartography.

Other claims by Dilke are likewise dubious. When there are surely grounds for greater caution, he presents as a Roman map a block of sandstone said to have been found by chance on the site of a Roman military camp in France in which he recognizes the shape of ancient Gaul.[14] With similar lack of hesitation he accepts as truth the legend, also reflected on the Hereford *mappamundi*, that Julius Caesar commissioned four men to survey the different quarters of the world, and that each did so in periods ranging from, allegedly, 21 years 5 months 9 days to 32 years 1 month 20 days.[15]

I imagine that Dilke's credulity may be accounted for in part by a sense of frustration (whether conscious or not) that the number of surviving maps, and even the amount of testimony about use of maps, both seem so limited. Hence his reaction is to give questionable instances the benefit of the doubt too readily. More skeptical evaluators of his claims proceeded to draw the overall conclusion that he shied away from, perhaps because the very basis of his approach is subverted in consequence. These evaluators concluded that, while cartography had its role in classical antiquity, this remained a limited one, and that most Greeks and Romans even at the highest levels of civil or military authority found little use for maps.

Such an attitude may seem a counter-intuitive one to attribute to Romans of all people, who created a far-flung empire and maintained it for centuries. By chance, however, a book published in 1984 by an Italian scholar, Pietro Janni, had made a similar argument, urging that Romans preferred to envisage "space", not two-dimen-

---

[13] O. A. W. Dilke "Itineraries and Geographical Maps in the Early and Late Roman Empires," in *The History of Cartography*, vol. 1, 244.

[14] O. A. W. Dilke, "Maps in the Service of the State: Roman Cartography to the End of the Augustan Era," in *The History of Cartography*, vol. 1, 206-207 with Fig. 12.5.

[15] O. A. W. Dilke, "Maps in the Service of the State," 205-206 with Fig. 12.4.

sionally with maps, but rather just one-dimensionally through lines.[16] Janni observed that while there is scant testimony to Roman use of maps, ample evidence exists for itineraries which record routes by land or sea in linear fashion. Indeed, the one great Roman world-map, surviving in a medieval copy, the so-called Peutinger Map, presents land routes horizontally in this way.

The contrasting positions taken by Janni and Dilke quickly provoked a debate about the degree of Romans' map consciousness, as well as about the means by which Romans conceptualized their surroundings at every level from the local to the global. This debate is far from over yet. Janni's adherents, with their bold claim that the Roman concept of space was no more than linear, have made a deep impression. Perhaps the single most important contribution to the debate so far—Kai Brodersen's book *Terra Cognita: Studien zur römischen Raumerfassung* (1995)—reinforces this claim. Brodersen presents a very skeptical appraisal of all testimony to Roman maps and their use. He even goes so far as to propose[17] that the famous (but lost) world-map publicly displayed in Rome that had been begun by Marcus Agrippa, the close associate of the first Roman emperor Augustus, was never an image at all, but only an extensive text like Augustus' own *Res Gestae*, a posthumously published record of his own achievements. Another strong affirmation of Janni's position was offered by C.R. Whittaker in 2002.[18]

To the best of my knowledge, Dilke nowhere published a reaction to Janni's view. This silence apart, a fundamentally unsatisfying feature of Dilke's approach is its failure to conceptualize any distinctive Roman worldview. Rather, he remains content with the unspoken assumption that Romans produced maps and used them in ways regarded as normal today. His stance must surely have disappointed Harley and Woodward, since his chapters in *The History of Cartography* vol. 1 fall short of reflecting their wish for detailed exploration of

[16] Pietro Janni, *La Mappa e il Periplo: Cartografia Antica e Spazio Odologico* (Rome: Bretschneider, 1984).

[17] K. Brodersen, *Terra Cognita,* 268-87. This view is supported by S. Carey, *Pliny's Catalogue of Culture: Art and Empire in the* Natural History (Oxford: Oxford University Press, 2003), chap. 3.

[18] "Mental maps: seeing like a Roman," in P. McKechnie (ed.), *Thinking Like a Lawyer: Essays on Legal History and General History for John Crook on his Eightieth Birthday* (Leiden: Brill, 2002) 81-112; reprinted in C.R. Whittaker, *Rome and its Frontiers: the Dynamics of Empire* (London and New York: Routledge, 2004), chap. 4.

map design and production, and for analysis of maps and related materials, from a contemporary—in this case ancient—perspective rather than a modern one.

However, Dilke can only have been recruited by Harley and Woodward when his own work on Greek and Roman maps was already well advanced, and in practice a radical change of approach may have been too much to expect of him. Even so, as a result his chapters tend more to description than analysis; map design and production are seldom among his primary concerns; and it is characteristic of him to accept established wisdom with a minimum of discussion. Thus he envisages the large-scale, so-called Marble Plan of Rome as assisting the efficiency of the city's administration,[19] and treats the Peutinger Map as little more than a road map.[20] In similar vein, to him the titles of the so-called *Antonine Itineraries* by land and by sea credibly imply that the routes recorded are journeys completed by, or planned for, an emperor, most probably Caracalla.[21]

To many interested scholars today these views and others are liable to be seen as reflecting a measure of naiveté on Dilke's part, but the difference of opinion is itself a striking measure of how far scholarship has advanced since the late 1980s, and of the stimulus that Dilke's courageous synthesis gave to further research and evaluation. Harley and Woodward hoped that their *History of Cartography* would act as such a springboard, and in this vital respect Dilke's chapters should be rated an outstanding success.

One remarkable feature of the progress made is the sheer array of tools, materials and scholarship either already available now, or known to be in preparation, that were not at Dilke's disposal. Everyone now benefits from digital technology to a degree barely envisaged twenty years ago. Quite apart from the databases of bibliography, epigraphy, papyrology and much else which digital technology has made possible, it hugely assisted the production of my *Barrington Atlas of the Greek and Roman World*.[22] A digital version of this atlas that gives its printed maps unprecedented longterm value and versatility is now

[19] O. A. W. Dilke, "Roman Large-Scale Mapping in the Early Empire," in *The History of Cartography*, vol. 1, 227.

[20] O. A. W. Dilke, "Itineraries and Geographical Maps in the Early and Late Roman Empires," 238.

[21] O. A. W. Dilke, "Itineraries and Geographical Maps," 235.

[22] Princeton and Oxford: Princeton University Press, 2000.

a practical prospect. Moreover, before the appearance of the *Barrington Atlas* in 2000, there simply did not exist a set of scholarly modern maps that presented the entire classical world at a useful scale. In principle it is true that the investigation and appraisal of, say, the Antonine Itineraries and the Peutinger Map from an ancient perspective ought not to require recourse to modern maps of the classical world. In practice, however, it is hard to proceed without them as a frame of reference. Modern maps of the classical world, for example, can reliably demonstrate how many routes in these Roman sources are circuitous, untraceable, incomprehensible, repeated in whole or in part, and so on. At the same time, recourse to this modern frame of reference brings the distinctive characteristics of the ancient material into focus.

Even more fundamentally, presentation of the Peutinger Map as a digital image—my own current goal—makes it possible to view it and to study it in ways that have not been practicable by conventional means. The Map can be viewed in color, full-size, or even enlarged. It is sobering to recall that no full-size color photographs of it were published before the 1970s, and that even today the reduced-size 1887 lithograph issued by Konrad Miller—in fact a corrected version of an engraving dating back to 1753—remains widely used.[23] For digital presentation, the Map's eleven consecutive segments, each measuring approximately 60 cm long by 33 cm high, can be joined to form the single piece 674 cm long that its designer meant it to be. By contrast, the 1976 album of color photographs presents each segment separately.[24] A digital format, moreover, permits the creation of layers that may readily be superimposed on the Map image in any combination. Hence it becomes feasible to establish a clear reference grid for the first time, as well to assign individual identification numbers or the like to the great quantity of unnamed features. With Tom Elliott's invaluable assistance,[25] I have taken these steps and many more. Elementary the particular ones just cited may be, but they are indispensable for detailed analysis, which otherwise can prove almost impossible to report efficiently in

[23] It was reprinted as recently as 2003, forming the endpaper to F. Prontera (ed.), *Tabula Peutingeriana: Le Antiche Vie del Mondo* (Florence: Olschki).

[24] E. Weber, *Tabula Peutingeriana, Codex Vindobonensis 324* (Graz: Akademische Druck-u. Verlaganstalt, 1976).

[25] See his contribution to this volume.

writing, because innumerable features on the Map have to be referred to in clumsy *ad hoc* fashion.

The Stanford Forma Urbis Project[26] harnesses digital technology to transform access to all 1,200 or so Marble Plan of Rome fragments, and to provide an image of exceptional quality for each. As with the Peutinger Map, the limitations of the way in which the fragments were presented previously in the 1960 *Pianta Marmorea* volume[27] must surely have served to discourage researchers; with a print-run of no more than 400 copies, it has in any case always been a rare work. Thus in this instance too, like that of the Plan, the migration to digital presentation has acted as a stimulus to engage a wider audience, to re-assess old assumptions and to open up fresh perspectives. Were the Plan or the Map really intended to serve a useful, practical purpose? If not, then what was the primary aim in each instance? In the case of the Map, what reliable criteria are there (if any) by which to date it? Was it made in antiquity at all rather than later? What is the significance of its extreme shape, which can hardly be an accidental or random choice, not least because it creates so many challenges for the mapmaker? How much is lost at the lefthand end?

To summarize my own current view of the Peutinger Map, I see it as the project of a single designer, who was already an experienced cartographer, but who was sufficiently creative and ambitious to experiment here both with the map's shape and with featuring land routes. He aimed above all to convey how civilized, peaceful and united an appearance the entire *orbis terrarum* presented under Roman sway, with the city of Rome as its center and focal point. His map was to form only one part of a larger artwork—such as a globe image, divided horizontally into zones—for prominent display in a designated spot in a public building of the Tetrarchic period (c. A.D. 300). He needed to employ a team to assist him, and predictably they found their task more taxing than anticipated. They laid out the map from left to right, with more now lost at the left than is commonly assumed, and they proceeded with care. Towards the end, however, they found themselves pressed for space, confused

[26] See the contribution to this volume by J. Trimble.

[27] G. Carettoni *et al.* (eds.), *La Pianta Marmorea di Roma Antica. Forma Urbis Romae* (Rome: Arti grafiche M. Danesi, 1960).

about the geography of the Sassanid empire and India, and so harassed to meet the deadline fixed for completion that in their haste they perpetrated egregious slips and duplications in the final stages. But these shortcomings towards the map's righthand end pale into insignificance when compared to the fine overall impression that it makes. No one was ever meant to consult it seriously as a guide to land routes. Multiple other components, too—islands, peoples, regions, rivers, spas, and more—are equally intended to engage viewers and arouse their admiration.

It must be acknowledged that this view of the Peutinger Map departs considerably from established opinion. Fundamental to this departure is a close examination of cartographic design and technique, an approach that seems never to have been contemplated previously for the Peutinger Map, let alone attempted. The Stanford project has been attempting a similar one, and there is even reason to claim that the Marble Plan's design and technique resemble those of the Map to a surprising degree. Conceivably, the makers of both masterpieces were trained in the same tradition.[28]

Attention to cartographic design and technique has been encouraged by a remarkable discovery first partially disclosed in 1998, but yet to be published in full.[29] This is a papyrus roll from Egypt containing, among other items, a Greek text thought to have been copied around the mid-first century B.C. Its description of the Iberian peninsula can be identified as part of the lost *Geographica* written by Artemidorus of Ephesus around 100 B.C. A space approximately one metre long is left within the text for a map that the copyist has begun, but for whatever reason never finished. A considerable amount of linework has been drawn, and many symbols have been placed; but there is no lettering and no use of color. Clues to the scale or orientation of the map seem lacking, and hence its geographical scope remains to be identified. Even so, as the first surviving example of a classical text laid out to accommodate an

[28] See further in this connection my discussion "Rome's Marble Plan and Peutinger's Map: continuity in cartographic design," *Althistorisch-Epigraphische Studien* (Österreichischen Gesellschaft für Archäologie) 5 (2005), 627-34.

[29] See B. Kramer, "The earliest known map of Spain (?) and the Geography of Artemidorus of Ephesus on papyrus," *Imago Mundi* 53 (2001), 115-20. Further information (but not full publication) has since become available in the exhibition catalogue, C. Gallazzi and S. Settis (eds.), *Le Tre Vite del Papiro di Artemidoro: Voci e Sguardi dall' Egitto Greco-Romano* (Milan: Electa, 2006).

accompanying map, this is an important discovery. The map's very incompleteness compels attention to the process of how it was drawn.

Another, very different, but also important, discovery deserving mention is a monument from Patara in Lycia, now southern Turkey. It came to light in 1993 after a bushfire on an archaeological site revealed blocks from the monument reused in a corner tower of the city's Byzantine walls. Reassembled, the blocks form a rectangular monument approximately 6 m tall, 1.6 m wide and 2.3 m long, overlooking the ancient harbor. [Fig. 2] The top of the monument, probably erected in A.D. 45, was in all likelihood occupied by a statue of the Roman emperor Claudius on horseback. On three of the four sides were inscribed two Greek texts. The first is a dedication of thanks to Claudius by the Lycians for having restored order and stability to their country by annexing it. The second, not integral to the original design but added only a couple of years later, is a commemoration in Claudius' own name of how he has made roads throughout Lycia, followed by a detailed listing of the distances between at least fifty different places there. In this second text Claudius flaunts the remarkable title "emperor of the world" (*autokrator tes oikoumenes*).[30]

These texts have not just shed welcome new light on the history and geography of Lycia. They have also rekindled interest in the itinerary lists that Pietro Janni and his adherents regard as such an influential determinant of Roman conceptions of space. During the 1990s, Pascal Arnaud[31] and Mauro Calzolari[32] contributed valuable analyses of the extensive collection of land routes preserved in the so-called *Antonine Itinerary*.[33] Arnaud in particular has convincingly demonstrated that this work dates to around A.D. 300, and that it assembles individual itineraries of distinctly different character, origin, and perhaps even date, with no secure connection to travel by

[30] On the monument in general, and its texts in particular, see R.W.B. Salway, "Claudius 'emperor of the world' at Patara," forthcoming.

[31] "L' *Itinéraire d'Antonin*: un témoin de la littérature itinéraire du Bas-Empire," *Geographia Antiqua* 2 (1993), 33-47.

[32] *Introduzione allo Studio della Rete Stradale dell' Italia Romana: L'*Itinerarium Antonini, *Atti della Accademia Nazionale dei Lincei* (Classe di Scienze Morali, Storiche e Filologiche) *Memorie* Ser. IX vol. VII fasc. 4 (1996), 369-520.

[33] In full, *Imperatoris Antonini Augusti Itineraria Provinciarum et Maritimum.* The standard edition is by O. Cuntz (Leipzig: B. G. Teubner, 1929).

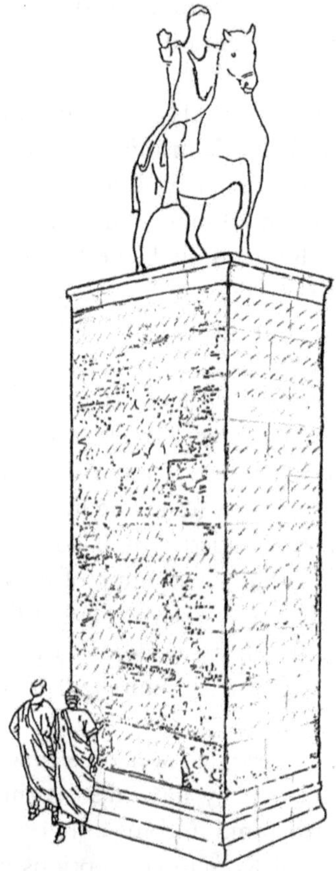

Fig. 2. Likely appearance of the monument erected in honor of the emperor Claudius at Patara, Lycia, probably in A.D. 45.

emperors. Moreover the collection is patchy in its coverage, loose in organization, confusingly repetitive, and uninformative where it offers a choice of routes. Most recently, Benet Salway has published discussions of the land and the sea parts of the collection, which are the most important contributions in English about both for a very long while.[34]

[34] "Travel, *itineraria* and *tabellaria*," in C. Adams and R. Laurence (eds.), *Travel and Geography in the Roman Empire* 22-66 (London and New York: Routledge, 2001) 23-25, 39-43; "Sea and river travel in the Roman itinerary literature," in R. Talbert and K. Brodersen (eds.), *Space in the Roman World: its Perception and Presentation* (Münster: LIT, 2004), 43-96.

In a forthcoming contribution,[35] I have gone further by probing some fundamental features of the land part which seem overdue for reappraisal. In particular, given that this collection is of very low quality and far from accommodating to potential users, why do scholars treat it so respectfully and assume that it was useful and put to use? Should it not be regarded instead as material compiled by an unsophisticated soldier or clerk in a desultory way, as a hobby, for no more than personal satisfaction? Perhaps it was unusual to compile an itinerary-collection, and maybe that is why this one miraculously survived to be copied as a curiosity. It is easy to imagine how by the Middle Ages it was valued as a kind of *Mirabilia Imperii Romani*, a precious evocation of the amazing long-distance journeys that had once had been feasible long ago when the Empire was thriving.[36] But this more respectful later assessment, which has persisted to the present day, does not justify the assumption either that this particular collection was similarly valued in antiquity, or even that collections of itineraries were common items then. That they may have been far from common is a key issue. Janni, after all, attaches primary importance to the use of itineraries in advancing his influential view that the Roman concept of space is only linear and one-dimensional.

Despite being convinced by Janni at first, I have increasingly come to doubt his view, at least as one that is exclusively applicable. I have no difficulty in believing that lines were one means by which Romans conceived space. But there have to be some further, two-dimensional means as well. If there are none, then what origin is there for the classical cartographic heritage that is to be perceived in medieval work?[37] Equally, how is the Peutinger Map to be accounted for? I have yet to be convinced either by Emily Albu's redating of it to the Carolingian period,[38] or by Kai Brodersen's claim that it is not a "map" at all, strictly speaking, but only a set of itineraries marked

[35] "Author, audience and the Roman empire in the *Antonine Itinerary*," in R. Haensch and J. Heinrichs (eds.), *Herrschen und Verwalten: Der Alltag der römischen Administration in der Hohen Kaiserzeit* (Köln, Weimar, Wien: Böhlau, 2007), 256-70.

[36] A mid-twelfth century work commonly referred to as *Mirabilia Urbis Romae* brought ancient Rome back to life for visitors to the city's ruins; see J.B. Friedman and K.M. Figg (eds.), *Trade, Travel, and Exploration in the Middle Ages: an Encyclopedia* (New York and London: Garland, 2000), s.v. *Marvels of Rome* (by C. Ho).

[37] See the contribution to this volume by P. Gautier Dalché.

[38] See her contribution to this volume.

out graphically.[39] Such a basis for the design is in my view demonstrably false.[40]

These controversies aside, plainer testimony is to be found in the Antonine Itinerary itself—not in the route listings, but in some of their headings, such as "*de Pannoniis in Gallias*" (231.8) or "*a Durrachio per Macedoniam et Trachiam*" (317.3-4). These make no sense to someone who lacks even an outline two-dimensional vision of the Roman Empire that relates the location of provinces and principal regions to one another. It is my contention, argued elsewhere, [41] that to envisage the Empire as a cluster of provinces was a widespread Roman form of worldview, in time reinforced ironically by Christianity, because Church leadership was organized from the start according to Roman provinces. This form of worldview is well known to have been widespread during the Middle Ages,[42] but an origin for it far back into the Roman period seems not to have been proposed previously. I propose it myself as only one form of Roman two-dimensional worldview, among no doubt others, as yet unidentified.

Given my reluctance to favor single or unduly sweeping explanations, I have no hesitation in advocating closer engagement with certain less well known materials both drawn and written. In the challenging case of the *agrimensores*, scholars everywhere are indebted to Brian Campbell, who dared to undertake the Herculean task which daunted even Oswald Dilke, namely, a full edition, English translation (the first into any language) and commentary on the entire corpus.[43] It is cause for regret that there is still no text of most of Ptolemy's *Geography* prepared with *apparatus criticus* according to modern scholarly standards; but fortunately one is now well on the way to completion by a team under the leadership of Alfred Stückelberger.[44] Meantime, Lennart Berggren and Alexander Jones have

[39] K. Brodersen, *Terra Cognita*, 186-87.

[40] See, in brief, my "Cartography and taste in Peutinger's Roman map," in *Space in the Roman World: its Perception and Presentation* (Münster: LIT, 2004), 124-25.

[41] "Rome's provinces as framework for world-view," in Luuc De Ligt et al. (eds.), *Roman Rule and Civic Life: Local and Regional Perspectives* (Amsterdam: Gieben, 2004), 21-37.

[42] See recently A. Hiatt, "Mapping the ends of empire," in A.J. Kabir and D. Williams (eds.), *Postcolonial Approaches to the European Middle Ages: Translating Cultures* (Cambridge: Cambridge University Press, 2005), chap. 3.

[43] *The Writings of the Roman Land Surveyors: Introduction, Text, Translation and Commentary* (London: Society for the Promotion of Roman Studies, 2000).

[44] Visit http://www.ptolemaios.unibe.ch

rendered the invaluable service of publishing the first sound translation of the theoretical chapters of the *Geography* (Book 1 and part of Book 2).[45] Today's non-scientific reader seeks explanation and guidance in addition, and Berggren and Jones skillfully provide these in a fifty-page Introduction.

Within the past twenty years, the Budé series of texts with commentary and French translation has been notably productive in publishing works of geographical importance that have had few, if any, such previous editions. They include, for example, Ampelius, *Liber Memorialis*,[46] and Pomponius Mela, *De Chorographia*, in Latin,[47] and in Greek, Arrian's *Periplous Euxeinou Pontou*,[48] Ctesias,[49] and Pseudo-Scymnus as the first volume in a set for the so-called Minor Greek Geographers.[50] Elsewhere, English translations with commentary have appeared of Agatharchides, *On the Erythraean Sea* by Stanley Burstein,[51] of the anonymous *Periplus Maris Erythraei* by Lionel Casson,[52] and of Arrian, *Periplus Ponti Euxini* by Aidan Little.[53] In 1998 the Kingdom of Jordan commissioned a volume to commemorate the centenary of the Madaba mosaic map's discovery by scholars, and this includes a superb presentation of the map itself.[54] Michael Lewis published a definitive treatment of Roman surveying instruments in 2001.[55] John Matthews is currently completing a full study of the papyrus archive[56] that records in detail the journey

[45] *Ptolemy's* Geography: *an Annotated Translation of the Theoretical Chapters* (Princeton: Princeton University Press, 2000).

[46] Ed. M.-P. Arnaud-Lindet (Paris: Belles Lettres, 1993).

[47] Ed. A. Silberman (Paris: Belles Lettres, 1988).

[48] Ed. A. Silberman (Paris: Belles Lettres, 1995).

[49] Ed. D. Lenfant (Paris: Belles Lettres, 2004).

[50] D. Marcotte (ed.), *Géographes grecs* 1 (Paris: Belles Lettres, 2000), incorporating an important introduction to the project as a whole.

[51] London: Hakluyt Society, 1989, 2nd series no. 172.

[52] Princeton: Princeton University Press, 1989.

[53] London: Bristol Classical Press, 2003.

[54] M. Piccirillo and E. Alliata (eds.), *The Madaba Map Centenary 1897-1997. Travelling through the Byzantine Umayyad Period* (Jerusalem: Studium Biblicum Franciscanum, 1998).

[55] *Surveying Instruments of Greece and Rome* (Cambridge: Cambridge University Press, 2001).

[56] C.H. Roberts and E.G. Turner (eds.), *Catalogue of the Greek and Latin Papyri in the John Rylands Library Manchester* IV (Manchester: Manchester University Press, 1952), nos. 627-38.

undertaken by the high official Theophanes from Egypt to Antioch in Syria and back around 320.

Alongside these new editions of existing materials, there are new discoveries to be taken into account too. In addition to those already mentioned, there is another mosaic "map", very different in character from the Madaba one, and much earlier, perhaps from the late third or early fourth century A.D.[57] It was found at Ammaedara in Tunisia, in a building whose function is uncertain. It measures about 30 square metres, covering the entire floor of a room, and representing sites connected with the goddess Aphrodite on Mediterranean islands from Sicily eastwards. Another notable new find, made by metal detector in February 2004, comes from Britain. This is a souvenir bowl of a type already known which commemorates forts along the western end of Hadrian's Wall or, as the wording phrases it, *RIGORE VALLI*, using the Roman surveyor's term for a straight line, *rigor*.[58] [Fig. 3]

The appearance of new discoveries and new editions should stimulate the quest for better understanding, and re-interpretation, of existing materials. Much more remains to be done in the area of Hellenistic geography, for example.[59] For my own part, looking to the end of antiquity, once I have finished with my current work on the Peutinger Map, there would be good reason to turn to the text by the so-called Cosmographer of Ravenna.[60] This work offers the largest single assemblage of toponyms in Latin to survive, but these have never been mapped out. Moreover, this text has never been translated into English, nor introduced into the recent discussions of Roman worldview.

The search for testimony about maps continues, and is not fruitless. James Scott has argued persuasively that an epigram by Philip

---

[57] F. Bejaoui, "Iles et villes de la Méditerranée sur une mosaïque d' Ammaedara," *Comptes Rendus des séances de l' Académie des Inscriptions et Belles-Lettres* (1997), 827-60.

[58] Visit http://www.roman-britain.org/frontiers/hw_souvenirs.htm (where the meaning of RIGORE VALLI is misunderstood).

[59] See the overviews by K. Geus in A. Erskine (ed.), *A Companion to the Hellenistic World* (Oxford: Blackwell, 2003), chap. 14, and by W.A. Koelsch, "Squinting back at Strabo," *The Geographical Review* 94 (2004), 502-18. Several of the texts in the Minor Greek Geographers collection mentioned above in note 50 await translation into English or indeed any modern language.

[60] Ed. J. Schnetz in *Itineraria Romana*, vol. 2 (Leipzig: B. G. Teubner, 1940). A 1990 reissue (Stuttgart : Teubner) added an index by M. Zumschlinge.

Fig. 3. Bowl (*patera*) from Britain commemorating forts along the western end of Hadrian's Wall (approx. 9 cm diameter x 4.5).

of Thessalonica[61] preserves mention of a tapestry world-map presented to a Roman emperor no later than Nero. In a forthcoming paper I dare to suggest that Claudius was displaying a map when he delivered his famous speech[62] in the Roman senate about the admission of Gallic notables to that august body. This is an obvious instance of seeing a familiar ancient text from a novel perspective that has attracted no attention to date.

Equally typical, albeit irksome, has been my recent experience of stumbling across known, but unfamiliar, testimony, only to realize too late that it could support a case already made and published. The particular case is my argument about provinces forming the basis of a two-dimensional Roman worldview. Ironically, this testimony is to be found in, of all places, Dilke's chapters for the *History of Cartography* volume 1: it is his two illustrations of a disassembled portable sundial,[63] a most accessible object in fact, since it hangs in an Oxford museum's display cabinet.[64] [Fig. 4] Dilke hardly discusses such sundials here,[65] nor does he cite any publication about this

[61] *Anth. Pal.* 9.778, discussed by J.M. Scott, *Geography in Early Judaism and Christianity:* The Book of Jubilees (Cambridge: Cambridge University Press, 2002), chapter 1.

[62] H. Dessau (ed.), *Inscriptiones Latinae Selectae* (Berlin: Weidmann, 1892-1916), = S. Riccobono (ed.), *Fontes Iuris Romani Antejustiniani* (Florence: S. a. G. Barbèra, 1941), I. 43.

[63] O. A. W. Dilke, "Roman Large-Scale Mapping," 215, figs. 13.4 and 5.

[64] Museum of the History of Science, Inventory no. 51358.

[65] See only O. A. W. Dilke, "Roman Large-Scale Mapping," 214 with n. 9, and his claim that a portable sundial was among the well informed traveler's principal aids (235, 254).

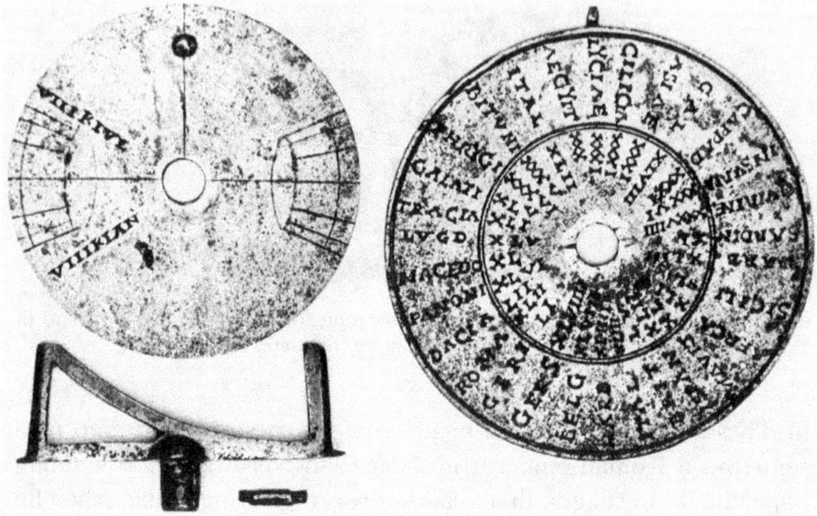

Fig. 4. Portable Roman sundial disassembled (6.1 cm diameter).

specific example, for which a fundamental discussion turns out to be in none other than the *Journal of the Royal Astronomical Society of Canada*.[66] But the striking use of province-names as the indicators for various latitudes is instructive, and further preliminary investigation confirms that other portable sundials preserved elsewhere follow the same practice. To tap this type of testimony did not occur to me earlier. The serendipitous discovery is a salutary reminder of how easily relevant material can be overlooked when it is so scattered and so varied in nature.

Almost twenty years on from the *History of Cartography* volume 1, there is plainly much more to be done, therefore. By nature, progress in scholarship is prone to be faltering, uneven, untidy, disputed. Beyond question, however, between them Dilke, Harley and Woodward in their different ways courageously established a secure foundation from which study of Greek and Roman cartography and worldview could proceed in a new, more open environment inviting fresh lines of enquiry. That a fuller, richer, more nuanced understanding is now developing, and will continue to advance, would surely please them all, and is for certain a compliment to them.

[66] F.A. Stebbins, "A Roman sun-dial," 52 (1958), 250-54, cited by Dilke in his earlier, fuller discussion, *The Roman Land Surveyors*, 70-73, 238.

I think they would be no less pleased that those of us whose focus is antiquity are eager to share our recent findings with medievalists, and to learn how they currently view the cartographic heritage from antiquity in their period.

L'HÉRITAGE ANTIQUE DE LA CARTOGRAPHIE MÉDIÉVALE: LES PROBLÈMES ET LES ACQUIS

Patrick Gautier Dalché

Les rapports de la cartographie antique et de la cartographie médiévale : voilà une question de l'histoire culturelle et intellectuelle qui n'est pas encore clairement résolue. La raison en est double, documentaire et idéologique. Les documents antiques sont rarissimes et leur date controversée. L'*ostrakon* de Soleto paraît d'une authenticité qui reste à confirmer.[1] Le papyrus d'Artémidore transmet l'oeuvre d'un cartographe hellénistique de la fin du Ier siècle avant notre ère; la signification intellectuelle de l'entreprise consistant à associer un prologue sur la valeur générale de la géographie, un schéma supposé de la péninsule Ibérique et un extrait de l'oeuvre du géographe d'Éphèse, demande à être précisée.[2] Que les cartes de Ptolémée aient été élaborées dans l'Antiquité est probable, sans être certain; mais, quoi qu'il en soit, elles sont une exception culturelle et scientifique dont la réception antique a été limitée.[3] Comme héritage de provenance antique, il ne reste que quelques rares et très schématiques figurations sur papyrus et la Table de Peutinger, à coup sûr une création romaine. En outre, les mentions de cartes dans la littérature latine sont difficiles à interpréter. Du côté du Moyen Age d'autre part, on ne dispose pas de la possibilité d'établir un arbre généalogique des *mappae mundi* du type de ce que les philologues réalisent à partir des variantes manuscrites des textes; les premiers exemples conservés sont des VIIIe-IXe siècle, et leur véritable essor

---

1 Un colloque lui a été consacré récemment: *La mappa di Soleto. Échanges de culture en Méditerranée ancienne.* Montpellier, 10-12 mars 2005 (Centre d'étude et de recherche sur les civilisations anciennes de la Méditerranée, Université Paul Valéry Montpellier III).

2 C. Gallazzi, B. Krämer, "Artemidor in Zeichensaal. Eine Papyrusrolle mit Text, Landkarte und Skizzenbüchern aus späthellenistischer Zeit," *Archiv für Papyrusforschung* 44 (1998), 189-208; B. Kramer, "The earliest known map of Spain (?) and the Geography of Artemidorus of Ephesus on papyrus," *Imago Mundi* 53 (2001), 115-20.

3 Comme je le montrerai dans un ouvrage à paraître sur la réception de la Géographie de l'Antiquité au XVIe siècles.

ne se produit qu'au XIIe, au moment où leur nombre s'accroît de façon importante.

Une deuxième raison ne tient pas aux données elles-mêmes, mais aux a priori idéologiques qui encombrent encore ce qu'il est convenu d'appeler l' "histoire de la cartographie." Notons tout d'abord un parallélisme paradoxal entre certains jugements d'ensemble formulés sur la cartographie romaine et sur la médiévale. Selon une opinion courante, le Moyen Age presque tout entier se caractériserait par la stagnation et l'absence de véritable créativité culturelle et scientifique. Il n'y aurait donc pas, à proprement parler, de cartographie médiévale. L'origine de ce jugement est à chercher dans une comparaison implicite des réalisations de l'*aetas media* avec celles de l'Antiquité et des Temps modernes, selon une problématique progressiste qui prend comme critère exclusif la science moderne. D'autre part, certains spécialistes de l'Antiquité ont développé des vues systématiques partant du même a priori épistémologique. Pour eux, des schémas imprécis et non fondés sur la localisation exacte des êtres géographiques à l'échelle ne sont pas des cartes: *ergo*, il n'a pas existé de cartes romaines.[4] Cette vision essentialiste et anachronique de la carte reproduit comme pérenne notre propre définition de la carte. Il est évident que la notion de représentation figurée de l'espace peut recouvrir plusieurs espèces, selon les cultures où elle est réalisée. On emploiera ici les termes "carte" et "cartographie" par commodité, sans leur attacher une signification moderne.

Il peut paraître donc inconcevable et presque injustifié d'étudier le problème qui m'occupe ici. Il y a pourtant de bons arguments pour penser que les représentations médiévales dépendent de prototypes antiques. Pourquoi la cartographie serait-elle différente des autres domaines de la pensée ? Mais comme pour toutes les discipli-

[4] On aura reconnu les thèses de Kai Brodersen, *Terra cognita. Studien zur römischen Raumerfassung* (Hildesheim: Georg Olms, 1995; éd. 2, 2003); plus récemment, "The presentation of geographical knowledge for travel and transport in the Roman world. *Itineraria non tantum adnotata sed etiam picta*," in C. Adams and R. Laurence (eds.), *Travel and Geography in the Roman Empire* (Londres et New York: Routledge, 2001), 7-21. R. Talbert a montré que l'aspect de la *Tabula Peutingeriana* résulte de choix opérés consciemment dans une perspective cartographique excédant la simple représentation diagrammatique de données itinéraires écrites: "Cartography and taste in Peutinger's Roman map," in R. Talbert and K. Brodersen (eds.), *Space in the Roman World. Its Perception and Representation* (Münster: LIT, 2004), 113-41.

nes dont l'origine et les textes fondateurs remontent à l'Antiquité, le Moyen Age a interprété, modifié, corrigé les éléments de l'héritage. Comment mesurer ces interprétations, ces modifications, ces corrections, si nous ne disposons ni du point de départ, ni des étapes intermédiaires entre les témoins évanouis de l'Antiquité et ceux, abondants, du XIIe siècle ? Les travaux sur ce sujet émanant de médiévistes sont extrêmement rares. Il n'y eut guère qu'un seul savant qui ait tenté de l'affronter, non pas en partant d'idées préconçues mais en restant au plus près des documents: Gerald Crone, ancien *map curator* de la Royal Geographical Society, dont les travaux sont aujourd'hui négligés, probablement parce qu'ils résultent d'une érudition qui n'est plus trop dans les habitudes des historiens. En suivant son exemple, en abandonnant tout a priori dans l'analyse des documents et en formulant quelques hypothèses, je pense qu'il est possible, sinon d'aboutir à un ensemble bien construit de certitudes, du moins d'ouvrir quelques pistes.

Dans les remarques qui vont suivre, je ne m'occuperai, d'un point de vue de médiéviste, que de cartographie à petite échelle, c'est-à-dire des images de l'oecumène ou de la sphère terrestre dans son ensemble, et je laisse de côté les cartes régionales.[5] De plus, je n'aborderai que la cartographie latine, c'est pourquoi je ne m'attarderai pas sur le problème de l'œuvre géographique et cartographique de Ptolémée qui, jamais oubliée, vécut d'une existence fantômatique en Occident pendant une dizaine de siècles avant de réapparaître à Florence, au début du XVe siècle.[6] Je m'attacherai d'abord à rejeter quelques vieilles idées qui encombrent encore, consciemment ou non, certaines études savantes et nombre d'ouvrages de vulgarisation. Je montrerai ensuite que les *mappae mundi*, qu'elles soient schématiques ou détaillées, ont pour prototypes des cartes de l'Antiquité tardive, c'est-à-dire élaborées, en gros, entre le IIIe et le VIIe siècle. Cette conclusion n'a pas grand intérêt en elle-même si l'on ne s'efforce pas de relier la production de tels objets à des pratiques intel-

[5] La tentative pour reconstruire conjecturalement une carte de Terre sainte attribuée à Théodose par Y. Tsafrir est peu convaincante: "The maps used by Theodosius: On the Pilgrim Maps of the Holy Land and Jerusalem in the Sixth Century," *Dumbarton Oaks Papers* 40 (1986), 129-45.

[6] P. Gautier Dalché, "Le souvenir de la *Géographie* de Ptolémée dans le monde latin médiéval (VIe-XIVe siècles)," *Euphrosyne* 27 (1999), 79-106.

lectuelles et des milieux culturels précis, ce qui me conduira nécessairement à examiner le rôle joué dans ce processus par la christianisation des savoirs qui se produisit dans les premiers siècles de notre ère. Enfin, dans un troisième temps, je m'efforcerai de définir à grands traits les apports proprement médiévaux qui ont modifié cet héritage: non pas tant des détails topographiques présents sur les *mappae mundi* que des constructions idéologiques élaborées durant le haut Moyen Age, associant de façon nouvelle des systèmes d'idées d'origine antique et acquérant ainsi des significations et une efficacité sociale nouvelles.

Ce qu'il faut rejeter

J'ai scrupule à donner place ici à des idées anciennes aujourd'hui abandonnées, pour deux raisons. D'une part, ces idées se retrouvent couramment dans des travaux de vulgarisation ou dans des travaux d'amateurs, qui sont, comme on sait, en grande abondance en matière d'histoire de la cartographie: il n'est que de constater combien les sites disponibles sur l'internet sont remplis d'affirmations ou d'hypothèses datant des siècles passés, pêchées dans des travaux de seconde main et présentées comme des vérités évidentes. Or ces idées anciennes appartiennent à des systèmes idéologiques lourds, notamment l'idée du progrès linéaire et indéfini de la science aujourd'hui revivifiée par la domination des utopies néotechnologiques. Leur efficacité passée leur assure un avenir alors même que les raisons sociales objectives de cette efficacité ont disparu; ils continuent ainsi à imprégner nos esprits, parfois à notre insu.

Une grande partie de la recherche sur la cartographie médiévale procède des travaux d'un infatigable historien indépendant, Konrad Miller (1844-1933). Son grand œuvre paru à la fin du XIXe siècle, *Mappaemundi. Die ältesten Weltkarten* (1895-1898), s'achève par un sixième volume dont le titre résume la problématique d'ensemble: *Rekonstruierte Karten.* Tel était en effet le but de Miller en cet âge d'or de la philologie et de la *Quellenforschung*: reconstruire l'un des pans de la civilisation antique en utilisant la méthode stemmatique élaborée depuis un demi-siècle afin d'approcher au plus près de l'original des grands textes. En cet âge de croyance aux progrès de la science, mus par l'*hubris* philologique, certains savants en vinrent à penser qu'il était possible, en recueillant les citations faites par des auteurs postérieurs et en les soumettant à la critique, de reconstituer des œuvres majeures disparues, comme par exemple les *Prata* de Suétone. Dans le domaine de recherche qu'il s'était choisi, où il éprouvait d'ailleurs le besoin de prendre une revanche sur l'Université à cause de son manque de légitimité institutionnelle,[7] Konrad Miller se livra à un minutieux travail de comparaisons et de recoupements qui devait aboutir à reconstruire l'image archétypique de la carte romaine dont, selon lui, les témoins médiévaux procédaient. Nul besoin de dire que cette entreprise supposait précisément résolu le principe qui était à démontrer. Il est peut-être plus utile de souligner combien profondément la mentalité dont procédait l'entreprise était imprégnée d'esprit progressiste: pour Miller, de même que les manuscrits médiévaux remplis de fautes de copie, les cartes médiévales ne pouvaient qu'avoir adultéré la carte romaine qui devait en être l'origine ultime. Il convient toutefois d'apporter une nuance: Konrad Miller ne se livra jamais à la dévalorisation systématique de la cartographie médiévale par rapport à la moderne.

Les savants du XIXe siècle se sont souvent livrés aux délices de la "reconstruction" de cartes anciennes supposées. Leurs oeuvres sont remplies d'images des cartes d'Hérodote, d'Eratosthène, de Strabon ou bien, du côté latin, de Pomponius Mela ou d'Orose, que l'on retrouve dans des ouvrages de vulgarisation—alors que nous avons de bonnes raisons de penser que ces auteurs, à part Eratosthène, n'ont jamais élaboré de cartes ! Une variante de cet étrange état d'esprit consiste à retrouver, en tel ou tel exemplaire médiéval, une carte de l'Antiquité qui aurait ainsi miraculeusement subsisté, à peu près inchangée. Ces démonstrations ont assez peu de titre à la scientificité. Dans cette lignée, Richard Uhden proposa de voir dans telle mappemonde médiévale la carte émanant de tel auteur antique. C'est ainsi que, pour lui, l'une des représentations élaborées par Lambert de Saint-Omer dans son encyclopédie *Liber floridus*, au début du XIIe siècle, était, sous réserve de quelques modifications, la carte de Martianus Capella, auteur, probablement au Ve siècle, d'une encyclopédie intitulée *Les noces de Philologie et de Mercure*. De même,

[7] Sur sa carrière, voir G. Husslein, "Konrad Miller," *Orbis Terrarum* 1 (1995), 213-33.

un manuscrit de la fin du VIIIe siècle aurait contenu la carte d'Isidore de Séville.[8] A partir des développements de K. Miller,[9] une semblable fiction historiographique a été bâtie à propos des deux célèbres cartes d'Asie mineure et de Palestine qui accompagnent le *Liber de situ et nominibus locorum Hebraicorum* de saint Jérôme dans un manuscrit du XIIe siècle (British Library, Additional 10049), couramment désignées comme "les cartes de saint Jérôme." Certains sont allés jusqu'à croire justifier cette attribution en prenant pour un avertissement de Jérôme lui-même une préface du XVIIe siècle qui mentionne une carte élaborée par les éditeurs modernes.[10]

Il faut enfin ranger parmi les illusions scientifiques d'un autre âge les tentatives de classification minutieuse des *mappae mundi* médiévales qui sont apparues dans les premières années du XXe siècle.[11] Le raffinement illusoire dans la distinction des genres et des espèces avait pour objectif de mieux réaliser le programme fallacieux de K. Miller, c'est-à-dire de reconstruire le ou les prototypes romains disparus.[12] Mais, selon les a priori de ces temps, les cartes du Moyen Age ne pouvaient que refléter de façon très imparfaite la "science antique." Il convenait donc de distinguer celles qui exprimaient l'idée "ancienne" d'une terre plate, et celles, "scientifiques," qui reflétaient

[8] R. Uhden, "Die Weltkarte des Martianus Capella," *Mnemosyne. Bibliotheca classica Batava* 3 (1936), 97-124; id., "Die Weltkarte des Isidorus von Sevilla," *Mnemosyne. Bibliotheca classica Batava* 3 (1936), 1-28. Sur cette dernière, voir E. Edson, "World maps and Easter tables: medieval maps in context," *Imago Mundi* 48 (1996), 25-42 (sp. 30-32), et L. Chekin, "Easter tables and the Pseudo-Isidorean Vatican map," *Imago Mundi* 51 (1999), 13-23.

[9] *Mappaemundi. Die ältesten Weltkarten*, t. 3, *Die kleineren Weltkarten* (Stuttgart: Roth, 1896), 1-21.

[10] A.-D. von den Brincken, *Fines terrae. Die Enden der Erde und der vierte Kontinent auf mittelalterlichen Weltkarten* (Hanovre: Hahn, 1992), 25-27; ead., "Weltbild der lateinischen Universalhistoriker und -kartographen," dans *Popoli e paesi nella cultura altomedievale* I, Settimane di studio del Centro italiano di studi sull'alto medioevo 29 (Spolète, 1983), 377-408. Le texte allégué a été écrit par les éditeurs Maffei et Vallarsi en vue d'accompagner une carte dressée au XVIIe siècle.

[11] J'ai présenté une critique détaillée de ces conceptions dans "*Mappae mundi* antérieures au XIIIe siècle dans les manuscrits de la Bibliothèque nationale de France," *Scriptorium* 52 (1998), 102-62 (sp. 103-107).

[12] Voir M. Destombes, ed., *Mappemondes A.D. 1200-1500. Catalogue préparé par la Commission des cartes anciennes de l'Union géographique internationale* (Amsterdam: N. Israel, 1964), Introduction, notamment p. xv. Les formulations contradictoires de cet auteur ne laissent pas d'étonner. Après avoir refusé l'image d'un Moyen Age "chaotique" entre deux périodes d'ordre rationnel, il note que "les lumières de la science ont été voilées," mais que l' "obscurité n'a pas été totale."

sa sphéricité. A cette aune, tous les schémas médiévaux de l'oecumène étaient considérés comme exprimant la première notion. Nous savons aujourd'hui qu'une telle opposition n'a jamais existé, puisque, hormis chez un ou deux auteurs chrétiens des premiers siècles, grecs pour l'essentiel, jamais la terre n'a été considérée comme plate. Il n'empêche que de telles banalités continuent de hanter la littérature de vulgarisation et même encore, parfois, des travaux à visées scientifiques.

Ces tentatives sont fondées sur de fausses prémisses. Mais gardons-nous de toute fierté illégitime en considérant que nous, historiens du début du XXIe siècle, serions entièrement indemnes de ces incompréhensions d'origine idéologique. Derrière les jugements très généraux et souvent répétés sur le poids particulièrement important que joueraient en ces matières les aspects " imaginaires," derrière les "déconstructions" de toute sorte, il y a toujours un système d'idées implicite qui attribue au Moyen Age une étrangeté plus ou moins radicale, suivant les cas, par rapport à notre notion de rationalité. Bien entendu, c'est nous qui apprécions cette étrangeté. Mais qui appréciera la nôtre, et selon quels critères ? Poser cette question conduit à rabattre de nos prétentions et à nous garder de tout jugement qui partirait d'idéologies modernes, même critiques: il faut interpréter le passé avec tout l'acquis positif et critique de l'histoire, mais en tentant d'observer avec les yeux des hommes du passé—autant qu'il est possible.

C'est ce que je vais tenter de faire maintenant, en partant d'une considération fort simple et d'une question de même nature. L'ensemble de la culture médiévale dépend, dans ses fondements mêmes, d'un héritage antique, conservé, travaillé et modelé par des hommes qui vécurent en dehors des cadres de vie et des systèmes de pensée de l'Antiquité. Puisque nous avons conservé des documents en grand nombre qui attestent l'intérêt de l'Occident médiéval pour les cartes, peut-on distinguer les éléments, de nature factuelle ou idéelle, qui, dans ces cartes, sont clairement "antiques" et ceux qui sont clairement "médiévaux" ? Peut-être cette problématique semble-t-elle un peu simple. J'espère montrer qu'il n'en est rien.

L'origine tardo-antique des mappae mundi

Deux catégories de figures cartographiques doivent ici être distinguées. Dans la première, se trouvent les très nombreuses cartes schématiques qui représentent soit l'oecumène divisée en trois parties, Asie, Europe, Afrique (ce que l'on appelle les cartes T-O), soit les zones climatiques de la sphère. La seconde catégorie est formée des cartes détaillées de la terre habitée, souvent de grandes dimensions et nettement moins nombreuses, mais dont le contenu toponymique permet des comparaisons plus précises.

Les cartes schématiques apparaissent de façon privilégiée—mais non exclusive—dans les manuscrits de quelques auteurs antiques. Des passages de l'historien Salluste (*Bellum Jugurthinum*) et du poète Lucain (*De bello civili*, ou *Pharsalia*) sont illustrés de cartes TO. La plupart des cartes à zones, quant à elles, se voient dans le texte du *Commentaire sur le Songe de Scipion* de Cicéron, dû à Macrobe, que l'on situe au Ve siècle, ou de néo-platoniciens médiévaux suivant sa tradition. La question qui se pose est évidemment de savoir si ces dessins remontent aux auteurs eux-mêmes. Dans le cas de Macrobe et des cartes à zones, il n'y a aucun doute. Il déclare qu'il illustre ses développements cosmographiques à l'aide de schémas explicatifs; les mappemondes sont annoncées par des formules *ad hoc* et, dans les manuscrits, elles sont toujours situées au même passage du texte. L'auteur a voulu que des représentations diagrammatiques de la sphère, comportant des détails topographiques, soient étroitement associées à ses propos.[13] Observons que Macrobe est un excellent représentant de cette attitude culturelle qui a été désignée du terme d'*eruditio* et qui caractérise la culture scolaire et littéraire de l'Antiquité tardive.[14]

Les choses sont moins claires dans le cas de Salluste et de Lucain. Les schémas TO inscrits dans les marges ont pour fonction d'expliquer le texte historique ou poétique sur un point qui nous semble scurrile, mais qui revêtait une certaine importance pour le lecteur

[13] Par exemple: "*Omnia haec ante oculos locare potest descriptio substituta, ex qua et nostri maris originem, quae totius una est, et Rubri atque Indici ortum videbis, Caspiumque mare unde oriatur inuenies...* " (II, 9, 7). On notera la précision sur la localisation des *ortus* de la mer Rouge, de la mer Indienne et de la mer Caspienne.

[14] H.-I. Marrou, *Saint-Augustin et la fin de la culture antique* (Paris: Boccard, 1958), 85-124.

antique et médiéval: la partition de l'oecumène en deux ou bien en trois parties. Le terme "continent," bien que d'emploi courant, est à proscrire. C'est en effet une notion moderne incompatible avec la perception antique et médiévale d'une masse unique de terre émergée, qui occulte les interrogations antiques et médiévales au sujet de la structure de l'*orbis terrarum* et empêche par là de les prendre en considération comme des objets légitimes d'enquête.

A partir d'une dizaine de manuscrits du *Jugurtha*, K. Miller avait prudemment conclu que leur modèle devait remonter à une date ancienne, sans préciser davantage—prudence qui n'a pas toujours été observée par les historiens postérieurs.[15] Dans la *Pharsale*, la fonction explicative et mnémotechnique du diagramme est claire: il met en rapport la tripartition de la terre, les points cardinaux et les vents selon leurs dénominations grecque et latine. Il est admis, sans plus, que la carte est "ancienne" et qu' "elle inspira peut-être celle d'Isidorus,"[16] mais on attend toujours la démonstration de cette dernière affirmation. En réalité, il est impossible de considérer que ces schémas émanent de Salluste ou de Lucain, ni de leurs contemporains, ni même des premiers temps de l'Empire. Ils n'apparaissent que dans des manuscrits tardifs, pour l'essentiel à partir des Xe-XIe siècles, si bien que l'on pourrait conclure, comme il m'est déjà arrivé de le faire à tort, qu'ils ont été élaborés durant le haut Moyen Age.[17] Ils émanent plutôt des IIIe-VIIe siècles, et ont été ajoutés aux textes dans le cadre des pratiques de l'école, le *grammaticus* expliquant les auctores par un commentaire continu, ou de l'*eruditio* qui caractérise les intérêts des lettrés de cette période et qui est particulièrement

---

[15] K. Miller, *Mappaemundi. Die ältesten Weltkarten*, t. 3. *Die kleineren Weltkarten*, 113. Selon M. Destombes, *Mappemondes A.D. 1200-1500*, 65, K. Miller démontre que l'origine commune de toutes les cartes des mss de Salluste remonte à une époque antérieure à la fin de l'empire romain, et peut-être même antérieure à la conversion de l'empire au christianisme. Mais Miller ne démontre en réalité rien de tel; le seul argument avancé par Destombes est très faible: certaines de ces cartes sont centrées sur Rome et non sur Jérusalem. Le jugement de Miller et de Destombes est repris par P. Arnaud, "*Plurima orbis imago*. Lectures conventionnelles des cartes au Moyen Age," *Médiévales* 18 (1990), 80.

[16] M. Destombes, *Mappemondes A.D. 1200-1500*, 74.

[17] P. Gautier Dalché, "De la glose à la contemplation. Place et fonction de la carte dans les manuscrits du haut Moyen Age," dans *Mappemondes A.D. 1200-1500. Testo e immagine nel alto medioevo*, Settimane di studio del Centro italiano di studi sull'alto medioevo, 41, 2 (Spolète, 1994), 693-771 (reproduit dans *Géographie et culture. La représentation de l'espace du VIe au XIIe siècle*, Variorum, Collected Studies Series (Aldershot: Ashgate, 1997), VIII).

visible dans l'oeuvre de Macrobe. Ces schémas s'apparentent à des gloses. Dans les manuscrits de la *Pharsale*, ils sont intégrés aux gloses marginales qui transmettent des éléments de commentaires scolaires issus de l'Antiquité, tels les *Commenta Bernensia* ou les *Adnotationes super Lucanum*.

Quelques exemples de *mappae mundi* diagrammatiques dans un environnnement textuel différent, probablement issues de modèles antiques, sont par ailleurs conservés. Le manuscrit Milan, Bibl. Ambrosiana C 264 inf. (XIVe siècle), est considéré comme une copie d'un exemplaire ancien produit entre V^e^ et VIIIe siècles.[18] Il contient deux textes: les *Collectanea* de Solin illustrés, entre autres, de dessins topographiques[19] dont certains rappellent les vignettes de la *Notitia dignitatum*, et de mirabilia; la *Cosmographia* du Pseudo-Aethicus accompagnée d'une mappemonde tripartite (f. 71v).[20] La même association de textes, accompagnés d'une *mappa mundi* identique, apparaît dans un manuscrit copié au XIIe ou au XIIIe siècle dans le conservatoire de traditions antiques qu'est le Mont Cassin (Bibl. del Mon. Naz. 391, p. 170).

On dispose de témoignages rares, mais significatifs, sur l'enseignement de la géographie dans l'Antiquité tardive à l'aide de représentations figurées. Un ensemble de papyrus égyptiens s'échelonnant du IIIe siècle de notre ère au VIe siècle (je ne prends pas en considération la période antérieure), contient essentiellement des listes de noms de peuples ou de lieux, souvent issus de textes classiques. Ils répondent à des préoccupations littéraires aussi bien que scientifiques. Malgré leur aspect de simple nomenclature, la structure et le contenu

[18] P. Revelli, "Figurazione cartografiche dell'età imperiale in un codice ambrosiano di Solino," dans *Raccolta di studi in onore di F. Ramorino* (Milan: Vita e Pensiero, 1927), 615-26. L'auteur attribue hardiment l'original antique à l' "école palatine de Constantinople" sous Théodose II, 625; A. et M. Levi, "The medieval map of Rome in the Ambrosian library manuscript of Solinus (C.246 inf)," *Proceedings of the American Philosophical Society* 118 (1974), 573, 576.

[19] P. Gautier Dalché, "Les diagrammes topographiques dans les manuscrits des classiques latin (Lucain, Solin, Salluste)," dans *La tradition vive. Mélanges d'histoire des textes en l'honneur de Louis Holtz* (Turnhout: Brepols, 2002), 291-306.

[20] La *Cosmographia* du Pseudo-Aethicus, sur laquelle, à cause de sa désignation pseudépigraphique, de nombreuses confusions sont faites, est une compilation chrétienne du VIIe siècle qui associe le chapitre géographique des *Histoires* d'Orose et la *Cosmographie* de Julius Honorius: A. Riese (ed.), *Geographi Latini Minores* (Heilbronn, 1878), 71-103; elle n'a rien à voir, sauf son attribution erronée de certains manuscrits, avec un autre texte du VIIe siècle, écrit dans un milieu franc, la *Cosmographie* d'Aethicus Ister, récit de voyage attribué à un "philosophe scythe."

de ces listes révèlent des intérêts idéologiques différents et successifs: propagande lagide au service d'un espace impérial marqué par l'entreprise d'Alexandre; résistance à la romanisation; perpétuation de l'empire romain à l'époque byzantine.[21] L'usage de cartes ou de schémas cartographiques dans un contexte scolaire est attesté par deux textes. La fonction de propagande de ces objets se retrouve dans le discours prononcé en 298 par le rhéteur Eumène lors de la restauration des écoles d'Autun. Il y décrit un *orbis depictus* situé sur un portique où la jeunesse pourrait localiser les conquêtes impériales.[22] On a supposé sans raison que cet *orbis depictus* était une copie de la carte d'Agrippa, elle aussi instrument de propagande impériale.[23]

Le lien est énoncé de façon plus nette dans la *Cosmographia* de Julius Honorius. C'est une liste de noms de lieux classés par grandes catégories géographiques, analogue aux listes des papyrus égyptiens, mais plus complète et plus systématique. Elle a été diversement datée, mais elle est certainement antérieure au milieu du VIe siècle, époque où Cassiodore puis Jordanès l'utilisèrent dans leurs oeuvres historiques. L'une des recensions donne pour auteur un "Iulius Honorius magister peritus et sine ulla dubitatione doctissimus" qui recommande, en vue d'éviter les erreurs, de ne pas séparer le livre d'une *sphaera*.[24] La critique a généralement considéré que ces listes avaient été transposées d'une carte, quelle qu'aient été la forme et l'aspect de celle-ci. Au milieu du VIe siècle, conscient de son utilité, Cassiodore recommanda le texte du *magister peritus* Julius Honorius aux moines de Vivarium dans ses *Institutiones diuinarum litterarum*, afin qu'ils s'initient aux complexités de la géographie.[25] Il n'évoque pas la

[21] B. Legras, "L'horizon géographique de la jeunesse grecque d'Égypte (IIIe s. avant notre ère-VIe siècle de notre ère)," dans A. Bülow-Jacobsen (ed.), *Proceedings of the 20th International Congress of Papyrologists* (Copenhague: Museum Tusculanum Press, 1994), 165-76.

[22] *Pro scholis*, 20-21, éd. L. Galletier, *Panégyriques latins*, t. 1 (Paris: Belles Lettres, 1949), 137-38.

[23] Nombreuses occurrences de cette opinion, par exemple J. O. Thomson, *History of Ancient Geography* (Cambridge: Cambridge University Press, 1948), 378.

[24] *Geographi Latini Minores*, 24-55. Voir sur ce texte P. Gautier Dalché, "Les 'quatre sages' de Jules César et la 'mesure du monde' selon Julius Honorius: réalité antique et tradition médiévale, II: la tradition médiévale," *Journal des Savants* (1987), 184-209, notamment 184-92 (reproduit dans *Géographie et culture. La représentation de l'espace du VIe au XIIe siècle*, I).

[25] *Institutiones*, I, 25.

"sphaera," ce qui semble montrer que le conseil du maître n'avait pas été respecté en cette occasion.

Il nous importe peu que, dans ces exemples, les représentations figurées aient été composées de simples listes de lieux regroupés par proximité topographique, ou qu'elles aient offert des contours plus "géographiques" correspondant à la "réalité." Il s'agit dans tous les cas d'images à visée mimétique. Il est aussi sans conséquence que les textes géographiques n'aient pas toujours été accompagnés de figures et, pour cette raison, il est exclu d'affirmer que cette pratique était générale. Mais, comme le montre l'ensemble de la documentation disponible, les cartes diagrammatiques ont été créées et utilisées dans les écoles et correspondent au type de culture promu par l'école.

L'étude des conditions matérielles et intellectuelles dans lesquelles furent réalisées les prototypes des grandes mappemondes en est encore à ses débuts. Les raisons de ce retard, alors que de nombreux travaux leur ont été consacrés depuis le début du XIXe siècle, tiennent à plusieurs circonstances. Comme il arrive souvent, par son apparence exhaustive et totalisante, le grand oeuvre de K. Miller a exercé une influence par certains aspects négative sur le développement de la recherche. La philologie triomphante du XIXe siècle connaît aujourd'hui un recul peu propice aux études minutieuses et souvent incertaines qu'une telle question rendait nécessaires. A partir des années 60 du siècle dernier les problématiques se sont modifiées, notamment grâce aux travaux pionniers d'A.-D. von den Brincken, dans le sens d'une plus grande attention aux aspects proprement médiévaux des *mappae mundi.* Mais, par l'insistance sur leur nature et leurs fonctions proprement médiévales, le lien entre Antiquité et Moyen Age s'est trouvé occulté. Fort peu de savants s'y sont intéressés, exception faite de Gerald Crone.

Je voudrais présenter un faisceau de données qui permettent de situer les modèles ultimes des *mappae mundi* entre le IIIe et le VIIe siècle, sans qu'il soit possible de préciser la chronologie. Ces arguments sont relatifs à leur structure et au matériau toponymique des cartes.

a) *Structure.* Des détails du contenu ne sont compréhensibles que s'ils proviennent d'un contexte antique, païen ou chrétien. Les Cyclades, autour d'une grande île qui doit être Délos, sont au centre de la carte de l'abbaye de Sawley (Cambridge, Corpus Christi College

66, fin du XIIe siècle).[26] Sur celle qui accompagne les *Étymologies* d'Isidore de Séville dans un manuscrit du XIIe siècle (Munich, Bayerische Staatsbibliothek, Clm 10058), c'est la ville de Tyr qui occupe le centre, ce qui s'explique sans doute par réminiscence plus ou moins bien comprise d'un passage du prophète Ezéchiel où la ville est dite située "in corde maris," c'est-à-dire au cœur ou au centre de la mer.[27]

La plupart des grandes mappemondes présentent des limites linéaires de pays ou de provinces qui n'ont de sens que dans le cadre de l'Empire. Supposons en effet que la carte d'un ms. daté entre 762 et 777 (Vatican, B.A.V., Vat. Lat. 6018)—identifiée à tort comme la carte d'Isidore de Séville—ait été élaborée au moment où le manuscrit fut copié: quel intérêt le dessinateur aurait-il trouvé à faire figurer ces lignes divisant, par exemple, les provinces romaines de l'*Hispania ulterior* et de l'*Hispania inferior* ? La carte conservée dans la collection Cotton de la British Library (Cotton Tiberius B.v, XIe siècle), la carte de Hereford, celle dite du Psautier de Londres (Addit. 28681, XIIIe siècle) montrent, elles aussi, des lignes de séparation, correspondant au souci de délimiter les provinces qui est une constante de la description géographique dans l'Antiquité romaine. Il serait utile d'étudier la forme et la distribution de ces limites en vue de déterminer si leur présence sur plusieurs *mappae mundi* est un phénomène erratique ou si l'on peut y discerner des régularités. Dans le même domaine des formes de la représentation, il conviendrait de comparer l'aspect général des contours terrestres apparaissant sur les *mappae mundi*, par exemple celui des différentes péninsules méditerranéennes; de tels travaux gagneraient sans doute en facilité et en objectivité s'ils étaient effectués par des moyens électroniques de reconnaissance des formes.

Des structures récurrentes apparaissent, dont je donnerai un seul exemple. Sur un grand nombre de *mappae mundi* le golfe Persique et la mer Caspienne, tous deux considérés comme des golfes de l'océan extérieur, sont alignés. Ce fait est très ancien: sur sa carte élaborée au IIIe/IIe siècle à partir de données empiriques recueillies auprès des voyageurs et organisée selon un réseau de coordonnées, Ératos-

[26] P. D. A. Harvey, "The Sawley map and other world maps in twelfth-century England," *Imago Mundi* 49 (1997), 33-42.

[27] A.-D. von den Brincken, *Fines terrae*, 183.

thène avait situé les deux mers sur le même alignement.[28] On reconnait cet alignement aussi bien sur ce qui est sans doute la plus ancienne carte dessinée au Moyen Age, la carte dite d'Albi (Bibl. mun. 29, VIIIe siècle), que sur l'*Atlas catalan* du XIVe siècle et sur des cartes marines. Une telle structure ne peut avoir été conçue au Moyen Age, alors que la carte d'Eratosthène était inconnue et qu'aucun texte ne permettait de se faire la moindre idée de ses caractéristiques. L'alignement présent sur de nombreux documents cartographiques est donc un héritage de la carte antique, transmis à travers de nombreuses copies.

b) *Toponymie.* Les indications les plus claires d'une origine tardo-antique des grandes *mappae mundi* proviennent des légendes et de la toponymie. Il ne s'agit pas là d'une méthode qui prendrait en compte la date de fondation ou de disparition de telle ou telle cité portée sur la carte: étant donné le conservatisme général de la toponymie de ces objets, il serait naïf et illusoire de penser que de telles données permettent d'établir une date absolue. En revanche, les toponymes et les légendes des *mappae mundi* ont des parallèles précis avec des sources textuelles antiques, traités de géographie au sens large et itinéraires.

Un tel travail de comparaison minutieuse avait été entrepris à propos de la carte de Hereford par G. Crone. J'ai suivi ses traces en étudiant deux textes, qui sont aussi des monuments de la cartographie médiévale: la *Descriptio mappe mundi*, transcription d'une *mappa mundi* due à la plume du maître parisien Hugues de Saint-Victor, dans les années 30 du XIIe siècle, et l'*Expositio mappe mundi*, qui enregistre le contenu d'une carte identique à celle de Hereford (XIIIe siècle), mais antérieure d'un siècle. Le choix de textes décrivant des cartes plutôt que les cartes elles-mêmes est dû à une simple raison d'opportunité: il est plus facile d'opérer des comparaisons entre textes. Une fois posé que les auteurs de ces descriptions ont respecté le contenu toponymique et les légendes de leurs modèles, ce qu'il est facile d'établir, la méthode ne soulève pas de difficultés.

---

[28] La ligne portes Caspiennes—embouchure du golfe Persique signe la limite entre la IIe et la IIIe *sphragìs* (voir la reconstruction de C. Müller reproduite in G. Aujac (ed.), *Optima Hereditas. Sapienza giuridica romana e conoscenza dell'ecumene* (Milan: Scheiwiller, 1992, 293); cf. H. Berger, *Die geographischen Fragmente des Eratosthenes* (Leipzig: Teubner, 1880; repr. Amsterdam, 1964), 200-203.

Sans entrer dans un détail qui serait fastidieux, notamment parce que les sources textuelles des légendes, traitant des mêmes êtres cartographiques, se répètent et se recoupent souvent, ce qui entraîne parfois des incertitudes quant à la véritable origine de tel ou tel élément textuel, on se limitera ici aux conclusions les plus sûres.

i. *Descriptio mappe mundi.*[29] Les correspondances les plus précises et les plus nombreuses proviennent d'Orose, auteur au Ve siècle des *Historiae adversus paganos*, dont l'objet, sous l'influence de saint Augustin, est d'exposer les calamités affectant constamment les païens du fait de leurs fausses croyances. Orose plaça en ouverture de son œuvre un tableau de l'*orbis terrarum* qui eut une influence décisive et profonde sur les représentations géographiques durant tout le durée du Moyen Age. On retrouve sur de nombreuses *mappae mundi*, dont celle de la *Descriptio mappae mundi*, des structures provenant directement de son exposé géographique, notamment la grande chaîne du Caucase qui s'étend sur toute l'Asie et dont chaque partie porte un nom différent. L'influence directe et immédiate exercée par le texte d'Orose sur le contenu toponymique et la structure de la carte est énorme. On peut conjecturer sans grand risque d'erreur que le tableau géographique de l'historien a été "mis en carte" à une date quelconque après le début du Ve siècle. Je dis bien "mis en carte," car il est douteux, et en tous cas indémontrable, qu'Orose ait lui-même réalisé une carte, malgré les efforts de certains savants pour appuyer cette thèse renouvelée de K. Miller.[30]

Il y a aussi des parallèles précis avec les source suivantes:

- Pomponius Mela, *De chorographia* (Ier s.)
- Ampelius, *Liber memorialis* (IIe s.)
- Ammien Marcellin, *Res gestae* (IVe s.)
- Martianus Capella, *De nuptiis Philologiae et Mercurii* (Ve s. probablement)
- *Epistola de rebus in oriente mirabilibus* (avant le VIIe-VIIIe s.).

Aucun de ces textes n'était disponible par un grand nombre de copies dans le plein Moyen Age. Le *De chorographia* de Pomponius Mela ne

[29] Cf. P. Gautier Dalché, *La Descriptio mappe mundi de Hugues de Saint-Victor* (Paris: Études Augustiniennes, 1988), 62-77.

[30] *Mappaemundi. Die ältesten Weltkarten*, t. 6 *Rekonstruierte Karten* (Stuttgart: Roth, 1898), 61-68.

fut lu qu'au IXe siècle, dans un cercle très restreint proche de Loup de Ferrières: son disciple Heiric d'Auxerre est probablement responsable du seul manuscrit qui transmet le texte et dont descendent les témoins beaucoup plus tardifs, multipliés notamment après que Pétrarque eut découvert l'une de ses très rares copies du XIIe siècle.[31] Le *Liber memorialis* d'Ampelius n'est transmis que par une copie du XVIIe siècle d'un manuscrit du Xe siècle perdu.[32] La fortune médiévale d'Ammien Marcellin est elle aussi très limitée: deux manuscrits de ses *Histoires* datant du IXe siècle reposèrent à Fulda et à Hersfeld, en Saxe, sans exercer d'influence repérable.[33] Avant l'époque carolingienne où elle fut utilisée dans l'enseignement et glosée, l'encyclopédie de Martianus Capella vécut souterrainement, sans doute dans la zone périphérique qu'était l'Irlande, quoique Grégoire de Tours, au VIe siècle, témoigne de sa connaissance.[34] L'*Epistola de rebus mirabilibus* est transmise à partir du VIIIe-IXe siècle en des versions toutes différentes, qui remontent à un original ancien prenant la forme d'un périple.

Les points de contact entre ces textes et les légendes de la carte de Hugues de Saint-Victor portent sur un très petit nombre de toponymes ou de légendes. Ils sont en outre exclusifs, c'est-à-dire qu'on ne les trouve pas ailleurs, dans quelque autre texte antique que ce soit. L'introduction de ces toponymes et de ces légendes dans une carte, ou encore dans un texte géographique mis par la suite sous forme cartographique, n'a pas pu être faite en des temps où ces textes rares ne circulaient pas, ou n'exerçaient aucune influence.

ii. L'*Expositio mappe mundi* est la description exacte d'une carte semblable à la mappemonde de Hereford;[35] les conclusions que nous tirerons de son examen seront donc valables aussi bien pour cette

[31] C. Gormley, M. A. et R. H. Rouse, "The medieval circulation of the *De chorographia* of Pomponius Mela," *Medieval Studies* 46 (1984), 266-320.

[32] L. D. Reynolds (éd.), *Texts and Transmission. A Survey of Latin Classics* (Oxford: Clarendon Press, 1983), 8.

[33] Ch. U. Clark, *The Text Tradition of Ammianus Marcellinus* (New Haven: the Author, 1904); W. Seyfarth, *Der Codex Fuldensis und der Codex E des Ammianus Marcellinus* (Berlin: Akademie-Verlag, 1962).

[34] C. Leonardi, "I codici di Marziano Capella," *Aevum* 33 (1959), 459-62; C. Leonardi, "Nota introduttiva per un indagine sulla fortuna di Marziano Capella nel Medioevo," *Bulletino dell'Istituto storico italiano per il medio evo e Archivio Muratoriano* 67 (1955), 265-88.

[35] Cf. P. Gautier Dalché, *Du Yorkshire à l'Inde. Une "géographie" urbaine et maritime de la fin du XIIe siècle (Roger de Howden ?)* (Genève: Droz, 2005), 70-74.

dernière. Contrairement à la *Descriptio mappe mundi*, ce n'est pas Orose qui joue le rôle essentiel, mais Solin, auteur, vraisemblablement au IIIe siècle, d'un recueil de *mirabilia* organisé selon un ordre géographique et copié en grande partie sur la *Naturalis historia* de Pline l'Ancien.[36] K. Miller, qui avait bien noté la présence massive de Solin dans les légendes et les noms de lieux de la mappemonde de Hereford, en avait tiré une théorie inutilement compliquée selon laquelle Solin n'avait fait que transcrire une carte, les cartographes du Moyen Age, tel l'auteur du prototype de la carte de Hereford, s'étant bornés à imiter à leur tour l'imitateur.[37] Une comparaison détaillée montre que, dans les légendes de la carte, le texte de Solin a été remanié pour être adapté aux nécessités propres à l'objet cartographique. Il n'a été reproduit exactement que dans les cas, rares, de courtes légendes. Dans la majorité des occurrences, le texte solinien a été simplifié quant au style ou résumé. Ces trois types de modifications révèlent, au rebours de ce que pensait K. Miller, qu'à une époque donnée après le IIIe siècle, le texte solinien a été travaillé de façon à être mis en carte. Cette opération fut double: on fit des extraits de passages signifiants, puis on les plaça sur la carte en fonction des localisations relatives explicitement formulées par Solin.

La carte de l'*Expositio* montre aussi des points de contact très précis avec l'*Itinéraire d'Antonin* (daté du IIIe siècle) notamment par une succession identique de toponymes sur la côte africaine, correspondant à des lieux d'importance diverse: colonies, mais aussi simples stations, ce que G. Crone avait relevé sur la carte de Hereford;[38] dans l'*Expositio*, le nombre de ces emprunts à l'*Itinéraire* est encore plus grand que sur la carte conservée. Enfin d'autres sources, telles que la *Cosmographia* d'Aethicus Ister (VIIe siècle), sont parfois associées aux légendes provenant de Solin.[39]

---

[36] Sur les sources de la carte de Hereford, voir en dernier lieu S. Westrem, *The Hereford Map: A Transcription and Translation of the Legends with Commentary* (Turnhout: Brepols, 2001), xxvii-xxxvii.

[37] *Mappaemundi. Die ältesten Weltkarten*, t. 3, 48.

[38] *The world map by Richard of Haldingham in Hereford Cathedral, circa A.D. 1285* (Londres: Royal Geographical Society, 1954), 20-23; "New light on the Hereford map," *The Geographical Journal* 131 (1965), 455.

[39] Voir par example l'*Expositio mappae mundi* I, 20, dans P. Gautier Dalché, *Du Yorkshire à l'Inde*, 145. L'attribution de la *Cosmographia* à Virgile, évêque de Salzbourg († 784), est une fiction savante aujourd'hui presque totalement abandonnée.

Il est donc difficile de dater précisément la fabrication du prototype. Le texte solinien a pu être "mis en carte" à n'importe quel moment postérieur à l'époque de son auteur, soit à partir du IIIe siècle. Si l'on tient compte de l'ensemble des sources tardives en considérant que les légendes ont été composées en une seule opération, celle-ci prendrait place au VIIe siècle. Mais plus vraisemblablement, elle fut effectuée en plusieurs étapes, la première étant antique, par des remaniements et des additions textuelles successives à partir de sources plus récentes intervenant lors de copies effectuées sur des exemplaires antérieurs.

La carte dite cottonienne paraît être un produit typique d'une opération de même nature. Dans le manuscrit du début du XIe siècle qui la transmet,[40] elle accompagne le texte de la *Cosmographie* du grammairien latin Priscien (vers 500), traduction de la *Périégèse* de Denys. L'incipit annonce la carte en l'attribuant, elle aussi, à Priscien: encore une association entre le monde scolaire et la cartographie. Elle paraissait légitime au rédacteur de l'incipit, quelle que soit la vérité de cette attribution.[41]

Il reste à examiner le rôle éventuel d'un document que l'on considère habituellement comme d'origine antique indubitable. La *Tabula Peutingeriana,* carte itinéraire du monde romain, est datée avec plus ou moins de précision et de certitude entre le IIIe et le Ve siècle.[42] La *Descriptio mappe mundi* comporte des noms de lieux qui ont leurs correspondants exacts sur la *Table de Peutinger.* Les traces de l'*Itinéraire d'Antonin* dans l'*Expositio mappae mundi* et sur la carte de Hereford pourraient aussi bien provenir de la *Tabula,* qui transpose sous forme cartographique des listes de noms de lieux se succédant au long des routes. D'autres *mappae mundi* portent des légendes plus développées semblables à celles de la *Tabula.* Ainsi, sur la plus complète des cartes accompagnant le *Commentaire sur l'Apocalypse* du moine hispanique Beatus de Liébana (776), on peut lire les inscriptions "In his locis elefanti nascuntur," "Saline inmense que cum luna crescunt et

[40] British Library, Cotton Tiberius B.v.

[41] "Incipit liber Pergesis id est de situ terrae Prisciani grammatici urbis Rome Caesariensis doctoris quem de priscorum dictis excerpsit ormistarum sed et huic operi de tribus partibus, uidelicet Asia, Africa, Europa mappam depinxerat aptam, in qua nationum, promontoriorum, fluminum insularumque situs atque monstrorum formatur honeste" (f. 59r).

[42] Voir E. Weber, *Tabula Peutingeriana. Codex Vindobonensis 324. Kommentar* (Graz: Akademische Druck- u. Verlagsanstalt, 1976), 20-23.

decrescunt."[43] En eux-mêmes, ces parallèles sont d'interprétation délicate. De telles légendes communes—dont certaines ont un contenu chrétien—pourraient, certes, avoir été ajoutées à la *Tabula* dans une étape de sa transmission médiévale. Mais pour quelle raison aurait-on fait entrer dans plusieurs *mappae mundi* différentes non seulement des légendes complexes, mais encore des noms de lieux sans importance particulière ? Il est généralement admis que Beatus ou un modèle antérieur, *donc au plus tard en 776*, emprunta quelques légendes et toponymes d'une carte semblable à la *Tabula* par son contenu.[44] Cette conclusion vaut aussi pour les prototypes de la carte de Hugues de Saint-Victor et de celle de Hereford qui résultent d'un semblable processus plus probablement tardo-antique que médiéval.

Mais selon une théorie récente, la *Tabula Peutingeriana*, loin d'être une création antique, aurait été entièrement élaborée à l'époque carolingienne à l'aide d'itinéraires écrits, plus nombreux que ceux qui nous sont parvenus.[45] L'hypothèse n'est pas invraisemblable si on la replace dans le contexte de la culture carolingienne, capable d'innovations significatives. Sans aller jusqu'à dire qu'à cette époque, la cartographie était "an advanced art," pour quoi les indices factuels sont absents,[46] il est certain—on y reviendra plus loin—que l'impul-

[43] K. Miller, *Mappaemundi. Die ältesten Weltkarten*, t. 1 *Die Weltkarte des Beatus (776 n. Chr.)* (Stuttgart: Roth, 1895), 54, 56.

[44] E. Schweder, "Über eine Weltkarte des achten Jahrhunderts," *Hermes* 24 (1889), 586-604 (594-603 sur les rapports avec la *Tabula*); H. Gross, *Zur Entstehungsgeschichte der Tabula Peutingeriana* (diss. Berlin, 1913), 68, 90-91. L'opinion de K. Miller, selon qui la principale source de Beatus aurait été une seule et même carte de l'Empire romain dont par ailleurs Orose et Isidore auraient conservé des extraits et qui aurait été à la base de la Tabula, est évidemment une pure construction de l'esprit: *Mappaemundi. Die ältesten Weltkarten*, t. 1 *Die Weltkarte des Beatus (776 n. Chr.)*, 65.

[45] E. Albu, "Imperial geography and the medieval Peutinger map," *Imago Mundi* 57 (2005), 136-48.

[46] Ibid. 138. On ne conserve pas de cartes d'époque carolingienne, sauf les exemplaires d'Albi (Bibl. mun. 29) et du Vatican (Vat. lat. 6018)—ce qui ne veut évidemment pas dire qu'il n'en a pas existé en bien plus grand nombre; il n'y a pas de moyen de savoir *précisément et en détail* l'aspect qu'elles revêtaient. Il ne va pas de soi que la "table d'argent" du testament de Charlemagne ait porté une "carte du monde" ("totius orbis descriptio" peut aussi bien renvoyer au cosmos dans son ensemble, d'autant qu'elle était "ex tribus orbibus conexa", ainsi que la décrit Eginhard). Pour une critique de l'attribution traditionnelle à Théodulf d'Orléans de la mappemonde contenue dans le ms. Vatican, BAV, Reg. Lat. 123, cf. P. Gautier Dalché, "Notes sur la 'carte de Théodose II' et sur la 'mappemonde de

sion donnée par l'idéologie de la *renovatio imperii* a joué un rôle sans doute très important dans la constitution et l'expansion du modèle médiéval de la *mappa mundi.* Toujours pour tenter de soutenir la validité de cette hypothèse, on pourrait aussi considérer que tous les indices de "mise en carte" de textes antiques renvoient plutôt aux temps carolingiens plutôt qu'à l'Antiquité: les textes rares au Moyen Age tels que ceux de Mela ou Ammien Marcellin étaient certes disponibles, mais par un ou deux manuscrits seulement. Enfin, une analyse de type philologique permet de penser avec un assez haut degré de probabilité qu'il y a eu en effet une étape carolingienne dans la transmission de la *Tabula.*[47] Les parallèles avec les *mappae mundi* pourraient ainsi s'expliquer en supposant l'existence, aux VIIIe-IXe siècles, d'un véritable chantier cartographique, assez important pour avoir produit plusieurs types de cartes, différentes dans certains détails mais très semblables dans d'autres, et surtout fort proches dans leurs structures.

Mais cette solution ne paraît pas vraisemblable. Il conviendrait d'abord de montrer de façon précise, et non pas par une simple accumulation d'hypothèses, comment la *Tabula* aurait pu concrètement être ainsi réalisée aux temps carolingiens et selon quelles motivations. Il est insuffisant de noter à ce sujet que "it is reasonable to expect Carolingian scribes to have known many more itineraries than we do":[48] quelles en sont les preuves précises ?[49] Où les créateurs carolingiens auraient-ils trouvé le type de vignette représentant les villes—qui a clairement son origine dans l'iconographie romaine[50]—si

Théodulf d'Orléans'," *Geographia Antiqua* 3 (1994-95), 91-108 (reproduit dans *Géographie et culture. La représentation de l'espace du VIe au XIIe siècle*, IX).

[47] Indépendamment même de ce que l'on peut déduire de la mention d'une "mappa mundi in rotulis II" dans le catalogue de la bibliothèque de Reichenau de 821-822, la présence de noms de régions (Francia, Suevia, Alamannia) à la place d'ethniques est nécessairement "médiévale".

[48] E. Albu, "Imperial geography and the medieval Peutinger map," 138.

[49] L'intérêt pour l'*Itinéraire d'Antonin*, associé à la *Cosmographie* du Pseudo-Aethicus, est attesté par plusieurs manuscrits de la fin du IXe siècle (L. Bieler, "The text tradition of Dicuil's Liber de mensura orbis terrae," *Proceedings of the Royal Irish Academy*, 64 C1 (1965), 1-3), dont le célèbre *Spirensis* perdu copié dans la deuxième moitié du IXe siècle ou au début du suivant dans un milieu proche de la cour (P. Lehmann, *Die mittelalterliche Dombibliothek zu Speyer*, in *Erforschung des Mittelalters* t. II (Stuttgart: Hiersemann, 1959), 201). Il ne s'agit que de l'*Itinéraire d'Antonin*; et l'on est loin de Charlemagne.

[50] A. et M. Levi, *Itineraria picta. Contributo allo studio della Tabula Peutingeriana* (Rome: Bretschneider, 1967), Parte Seconda.

ce n'est sur un document cartographique antérieur ?[51] Comment le canevas général aurait-il pu être établi ex nihilo, sans indications précises sur les rapports de contiguïté topographique entre les régions que pouvait seule fournir une représentation figurée totalisante—et donc une carte antérieure ? Du point de vue des motivations, enfin, on ne peut affirmer sans preuves que les "cartes" du testament de Charlemagne (la table d'argent représentant le "monde," plus une *effigies* de Rome et une *descriptio* de Constantinople) "répondait aux revendications du pape à la domination universelle." Il conviendrait de prouver que l'empereur eut l'ambition de présenter une carte de propagande, désir d'où découlerait l'aspect antiquisant de la *Tabula*, y compris le format en rouleau "associé avec l'Antiquité païenne," ce qui est une généralité abusive: saint Jérôme et ses contemporains chrétiens n'utilisaient-ils pas des *volumina* ?[52] Une formulation comme celle qui est affectée à Jérusalem "antea dicta Hierusalem modo Helya Capitolina" ne peut être interprétée comme une référence volontaire à l'Antiquité païenne choisie de préférence à une référence chrétienne—évident anachronisme. A ces remarques s'ajoutent plusieurs difficultés qu'il faudrait résoudre en évitant l'emploi d'arguments *ad hoc*:

• comment expliquer, dans cette hypothèse, que la carte, fabriquée par Beatus au plus tard en 776 dans une région périphérique par rapport aux centres de pouvoir carolingiens, comporte déjà des éléments communs avec la *Tabula* ?

• comment expliquer les relations entre la *Tabula* et la *Cosmographie* de l'Anonyme de Ravenne (VIIIe siècle), qui utilisa une carte analogue comportant des caractères grecs ?[53] Et comment expliquer l'existence très probable à Padoue, à la fin du XVe siècle, d'un exemplaire "vetustissimus" comportant lui aussi des caractères grecs—donc probablement d'origine antique ?[54]

---

[51] Il n'y a toutefois rien qui permette d'affirmer que la *Tabula* descend de la carte d'Agrippa, contrairement à l'affirmation d'E. Weber, *Tabula Peutingeriana. Codex Vindobonensis 324. Kommentar*, 2.

[52] E. Albu, "Imperial geography and the medieval Peutinger map," 139.

[53] P. Arnaud, "L'origine, la date de rédaction et la diffusion de l'archétype de la table de Peutinger," *Bulletin de la Société nationale des antiquaires de France* (1988), 316 et n. 42; et surtout L. Dillemann, *La Cosmographie du Ravennate*, Collection Latomus, t. 235 (Bruxelles: Latomus, Revue d'Etudes Latines, 1997), 38-40, 52-53.

[54] P. Gautier Dalché, "Du nouveau sur la transmission et la découverte de la

• comment expliquer, enfin, si un vaste "chantier cartographique" motivé par une politique consciente d'opposition à la papauté avait existé au IXe siècle, que l'on n'en ait pas conservé de traces moins ténues ?

Je concluerais donc volontiers dans le sens d'un scepticisme à la David Hume: nous avons des preuves raisonnables de l'existence d'une cartographie romaine, malgré les vues systématiques formulées çà et là; nous n'avons, en revanche, aucune preuve des intentions de Charlemagne en matière de "propagande," et les exemples de l' "art cartographique" à l'époque sont inexistants ou douteux. Il est donc rationnel, jusqu'à plus ample informé et jusqu'à ce que des preuves ou des indices plus nombreux et plus conséquents aient été apportés, de douter que la *Tabula Peutingeriana* soit une création carolingienne, et de continuer à la tenir pour une création antique.

Revenons aux *mappae mundi.* De tels objets existaient déjà au moment où Jonas de Bobbio rédigeait sa *Vie* de saint Colomban, au VIIe siècle. Il y a dans ce texte la première attestation de l'existence d'une véritable *mappa mundi*—sans qu'elle porte encore ce nom. Une représentation du monde apparaît à Colomban, en réponse à son désir de partir évangéliser le peuple slave des Wendes: "...un ange du Seigneur lui apparut dans une vision et lui montra l'assemblage du monde enfermé dans un circuit restreint, comme on a coutume de représenter à la plume la terre habitée en traçant un cercle sur une page."[55] Peu nous importe l'apparence "réelle" de cet objet. Il peut s'agir, dans l'esprit de Jonas, d'un diagramme analogue à celui des manuscrits de Salluste et de Lucain, ou d'une carte détaillée déployant des contours et de nombreux toponymes. Je précise "dans l'esprit de l'auteur," car il est évident que ce témoignage littéraire nous renseigne sur la valeur fonctionnelle attribuée par son auteur à la *mappa mundi*, non pas sur son aspect, qui est pour lui secondaire. Ce qui est important, c'est que l'ostension de cet objet est considéré comme naturelle par Jonas (et donc son public), et qu'elle intervient dans un processus démonstratif de nature véritablement géographique: elle rend évident que le monde est vaste et que l'espace à évan-

---

Tabula Peutingeriana: la "*cosmographia vetustissima*" de Pellegrino Prisciani (†1518)," *Geographia Antiqua* 13 (2004), 71-81.

[55] " ...*paruoque ambitu, uelut paginali solent stilo discribere circulum, mundi conpagem monstrauit,*" *Vita Columbani*, I, 27 (56), G. Krush (éd.), *M. G. H., Scriptores rerum Germ. in usum schol.* (Hanovre: Hahn, 1905), 217.

géliser excède largement le territoire limité des Slaves sur lequel le saint, dans un premier temps, voulut se rendre en obéissant à son propre désir qui est ainsi critiqué.

Une "mise en carte" des textes géographiques s'est donc produite entre le IIIe et le VIIe siècle.[56] Au contraire des hypothèses qui postulent la dégénérescence de modèles romains exacts, l'existence des témoins médiévaux présentant les caractéristiques évoquées ci-dessus ne s'explique que par la réalisation dans cette période de différentes figurations cartographiques apparentées par leurs buts idéologiques, leurs formes et leur contenu. Ces cartes et ces schémas, nés dans des milieux distincts influencés par les méthodes de l'école étaient des représentations du monde qui ne correspondaient ni à notre idée anachronique de la carte romaine "scientifique," ni à notre notion moderne de la carte, la première s'expliquant par la seconde. Il est sans valeur heuristique de penser ce processus sous le mode d'une dégénerescence qui aurait fait disparaître les aspects scientifiques et techniques d'une cartographie romaine supposée "correcte."[57] Cette opinion résulte d'un raisonnement circulaire où ce qui est à démontrer est déjà présent dans les prémisses: les conditions de réalisation de cartes "exactes"—notion d'échelle, calcul des coordonnées, idée de la forme "réelle" du monde étant supposées disparues, les témoins médiévaux ne peuvent être que des copies fausses de cartes correctes. Pour justifier cela, il faudrait prouver que les prototypes des *mappae*

[56] Étudiant les sources classiques de trois *mappae mundi* du VIIIe siècle, E. Edson a bien repéré l'importance de l'étape de "mise en carte" de textes. A propos des diagrammes, elle cite sans prendre parti l'opinion de K. Miller, selon qui ils furent établis et utilisés sous le haut Empire; elle affirme d'autre part que les trois *mappae mundi* détaillées (celles d'Albi 29, du Vat. lat. 6018, et de Beatus) ne sont pas des copies d'un modèle hypothétique classique, mais des créations médiévales issus de la "mise en carte" des textes d'Orose, Isidore de Séville et saint Jérôme: "The oldest world maps: classical sources of three VIIIth century *mappaemundi*," *The Ancient World* 24 (1993), 169-83. Mais le fait qu'elles ne soient pas des copies d'un seul modèle "classique", ce qui découle déjà de leur structure très différente, n'empêche nullement qu'elles remontent à des cartes élaborées avant le VIIIe siècle. C. R. Whittaker n'exclut pas l'origine "classique" des prototypes des cartes TO et des cartes des zones terrestres tout en affirmant qu'il n'y a pas de preuves qu'aucune des illustrations connues dérivent d'originaux romains: "Mental maps: seeing like a Roman," dans P. McKechnie (éd.), *Thinking like a lawyer. Essays on legal history and general history for John Crook* (Leyde-Boston-Cologne: Brill, 2002), 83. Dans le cas des cartes du *Commentaire* de Macrobe, la preuve existe dans le texte lui-même annonçant et décrivant le contenu de la figure.

[57] P. D. A. Harvey, *Mappa mundi. The Hereford world map* (Londres: The British Library, 1996), 22-24.

mundi furent fondés sur des mesures issues des techniques de l'arpentage romain, ce que rien ne permet de penser. Contrairement à ce qui est encore souvent avancé par suite de la reprise des affirmations des *Quellenforscher*, la carte d'Agrippa dont on ne connaît pas l'aspect mais qui n'était certainement pas bâtie sur un canevas ptoléméen, n'est pas à la source des *mappae mundi*, comme un exemple de cartographie encore "réaliste" qui aurait été perturbé par les copies et le mysticisme médiéval; plus exactement, il n'y a aucun moyen de savoir si des données de cette carte ont passé dans des cartes subséquentes, ce qui n'est évidemment pas exclu. Le récit de la *Cosmographia* de Julius Honorius sur l'envoi de géomètres dans les quatre parties du monde pour les mesurer est une fiction et, même si l'entreprise était avérée, il ne s'ensuit pas que le résultat ait pris une forme cartographique: le texte de la *Cosmographia* n'évoque pas de carte à ce sujet; aucun lien avec la carte d'Agrippa ne peut être établi par la documentation disponible.[58]

Ma conclusion pourrait être battue en brèche en partant de la théorie qui oppose comme exclusives l'une de l'autre, à propos de la géographie antique, une perception de l'espace, linéaire, qui serait caractéristique de la civilisation romaine et ne pourrait produire que des itinéraires, et une perception de l'étendue, qui seule permettrait la réalisation de véritables cartes, absentes dans l'Antiquité.[59] Mais cette hypothèse ne peut se concilier avec l'existence de témoins de *mappae mundi* d'origine antique attestées au VIIe (Colomban) et au VIIIe siècle (Vat. lat. 6018), qui ont pour ambition de représenter véritablement l'espace de l'*orbis terrarum* comme une étendue pour l'action ou pour la connaissance. Le fait que la carte du Vatican montre un grand nombre de lettres "c" pour ciuitas (seulement en Italie) ne prouve nullement que l'espace y est conçu comme une succession de points et non comme une continuité.[60]

On a observé qu'il y aurait eu, au IVe siècle, un changement dans le "Roman mental mapping," marqué par le rassemblement de

---

[58] C. Nicolet, "Les quatre sages de Jules César et la 'mesure du monde' selon Julius Honorius: réalité antique et tradition médiévale. I. La réalité antique," *Journal des savants* (1987), 157-83.

[59] Voir les remarques à ce sujet de C. Nicolet, *L'inventaire du monde. Géographie et politique aux origines de l'Empire romain* (Paris: Fayard, 1988), 89-90 et 252-53.

[60] C. Whittaker, "Mental maps," 103 ; mêmes idées dans "Itinerari romani e spazio in età antica," dans R. Greci (éd.), *Itinerari medievali e identità europea, Atti del Congresso internazionale Parma, 27-28 febbraio 1998* (Bologne: CLUEB, 1999), 46.

courts itinéraires en vue de composer des itinéraires plus complets s'étendant sur de grandes parties du monde romain, comme l'*Itinéraire de Bordeaux à Jérusalem* ou comme l'oeuvre d'Égérie.[61] Ce mouvement de tourisme religieux et de multiplication des voyages lointains accompagne, à la même époque, l'émergence des lieux saints et l'apparition d'une véritable topographie chrétienne ancrant dans le monde romain les réalités bibliques, établissant donc une continuité temporelle inscrite dans l'espace.[62] La création des *mappae mundi* tardo-antiques pourrait être comprise comme relevant en partie du même phénomène—à condition d'abandonner l'opposition entre ligne et étendue, entre itinéraire et carte, entre perception de l'espace à une et à deux dimensions.

La seule voie qui permettrait d'avancer davantage dans la connaissance précise de ces témoins de la cartographie de l'Antiquité tardive réside dans de minutieuses études de type philologique, analogues à celles initiées par K. Miller, mais contrôlées par une critique rigoureuse écartant toutes considérations et jugements a priori.

L'apport médiéval

En affirmant que les prototypes des mappae mundi comportant des détails de formes et de contenu toponymiques datent de l'Antiquité tardive, nous n'apprenons rien sur ce qui est spécifiquement "médiéval" dans les cartes que nous avons conservées. Examinons maintenant si le Moyen Age n'a été qu'un héritier passif, ou si le legs de l'Antiquité tardive a été infléchi, et dans quel sens. Ce problème conduit à aborder, à propos de cartographie, la "christianisation" des traditions antiques païennes. Cette question ayant été récemment traitée dans un ouvrage d'ensemble sur les transformations des savoirs antiques dans l'Antiquité chrétienne entre le Ier et le VIIe siècle, je m'y étendrai peu, sauf pour apporter quelques précisions.[63]

---

[61] C. Whittaker, "Mental maps," 96-97; "Itinerari romani e spazio in età antica," 41.

[62] R. A. Markus, "How on earth could places become holy ? Origin of the Christian idea of holy places," *Journal of Early Christian Studies* 2-3 (1994), 268.

[63] H. Inglebert, *Interpretatio christiana. Les mutations des savoirs (cosmographie, géographie, ethnographie, histoire) dans l'Antiquité chrétienne 30-630 après J.-C.* (Paris : Inst. d'Études Augustiniennes, 2001), 99-104. Les dates et les attributions de textes médiévaux que l'on rencontre dans les passages de cet ouvrage relatifs à la géographie et la

Mais ce qui est proprement "médiéval" dans les mappae mundi, ce ne sont pas des éléments factuels correspondant aux données de la Bible, mais plutôt des constructions idéologiques unifiantes qui justifient l'existence et l'efficacité intellectuelle et sociale des mappae mundi; c'est ce point, à mes yeux essentiel, que j'aborderai en second lieu.

Il faut d'abord récuser l'idée d'une "géographie des Pères de l'Eglise" qui, s'opposant à la science des païens, aurait été caractérisée par une régression. Il est clair aujourd'hui que cette idée, élaborée au XIXe siècle dans les milieux intellectuels sacrifiant au scientisme et à un certain anticléricalisme, est alimentée par la fiction des "dark ages" et de l' "obscurantisme" médiéval dont l'Église serait responsable. Le "positivisme" de ces jugements est encore bien vivant.[64] Les chrétiens des premiers siècles ont repris pour l'essentiel la géographie païenne, tant dans ses fondements et ses méthodes que dans ses résultats, en se bornant à apporter parfois de minimes corrections pour christianiser tel ou tel aspect mineur. Les caractéristiques de la géographie et de la cartographie chrétiennes sont fondamentalement les mêmes que celles adoptées par les païens, ce qui signifie que les données qui en provenaient ont été très rarement soumises à une critique formulée à partir des enseignements de la Bible.

Cela posé, quels détails spécifiquement chrétiens peut-on—parfois—trouver sur les *mappae mundi* ? Si l'on écarte la tripartition de l'oecumène, antérieure à l'ère chrétienne, et qui n'a que très exceptionnellement porté, au Moyen Age, le moindre symbolisme en rapport avec la Croix,[65] ils sont au nombre de quatre. La centralité de Jérusalem, tout d'abord. C'est devenu un lieu commun de la littérature sur les cartes: étant chrétiennes, elles sont nécessairement toutes

---

cartographie sont en général erronées. Entre autres: l'idée selon laquelle les "cartes hiéronymiennes" (c'est-à-dire les cartes présentes dans le ms. du XIIe siècle British Library, Add. 10049) seraient des fragments de mappemonde du IIe ou du IIIe siècle est controuvée (ibid., 100, n. 337); la carte d'Albi n'illustre pas l'œuvre de Julius Honorius (ibid., 103).

[64] Par exemple, H. Inglebert affirme que certains aspects de la géographie et de la cartographie des chrétiens "expliquent en partie le bien fondé de la critique positiviste" (ibid., 107).

[65] Cf. P. Gautier Dalché, "Principes et modes de la représentation de l'espace cartographique durant le haut Moyen Age," in *Uomo e spazio nell'alto medioevo 4-8 aprile 2002*, Settimane di studio del Centro italiano di studi sull'alto medioevo, 50 (Spolète: 2003), t. I, 133-34.

centrées sur Jérusalem. La réalité est tout différente. Cette centralité censée caractéristique de l'ensemble du Moyen Age chrétien est déjà présente dans les textes depuis le commentaire sur Ezéchiel de saint Jérôme. Elle suscita des réticences, comme chez Bède le Vénérable ou tel auteur d'époque carolingienne—pour se limiter au haut Moyen Age—, ce qui prouve qu'elle n'était pas universellement acceptée.[66] Elle n'apparaît qu'au XIIe siècle sur certaines *mappae mundi*, probablement à la suite de la Ire croisade à l'appel de laquelle le pape Urbain II, en suivant saint Jérôme, avait évoqué Jerusalem "in orbe medio posita."[67] Mais elle ne fut jamais systématique, à quelque époque que ce soit: comme l'avait déjà noté R. Uhden, même sur un groupe de cartes accompagnant l'oeuvre d'un même auteur, comme le chroniqueur Ranulf Higden (XIVe siècle), la situation de la ville peut changer.[68] Jérusalem n'est donc pas "a priori" au centre des schémas TO, comme l'affirmait hardiment Konrad Kretschmer:[69] bel exemple de construction de l'esprit puisque, quand la cité n'est ni nommée ni représentée, ce qui est fréquent, on ne saurait préjuger de la situation que le dessinateur lui conférait.

Il y a ensuite l'orientation à l'est, souvent présentée comme un trait spécifique du Moyen Age chrétien. En réalité, cette orientation privilégiée existe aussi bien, dans les textes, chez les auteurs païens que chez les chrétiens. En ce qui concerne les cartes, on doit constater que les premières conservées ne sont pas systématiquement orientées à l'est: c'est le cas de la carte d'Albi, mais celle du. Vat. Lat. 6018 a le sud placé dans la partie supérieure.[70] De plus, les diagrammes des zones de la sphère qui comportent souvent des contours géographiques dans la zone tempérée septentrionale, notamment dans le commentaire de Macrobe sur le *Songe de Scipion* et dans les

---

[66] "*In medio autem Hierusalem ... columna celsa stat, quae aestiuo solstitio umbram non facit. Vnde putant ibi mediam esse terram...*" Bède le Vénérable, *De locis sanctis*, II, 6, éd. I. Fraipont, Corpus Christianorum, Series Latina 175 (Turnhout: Brepols, 1965), 258; "Hierusalem ubi medius fertur mundus" (Eugenius Vulgaris, *Sylloga*, MGH, Poetae, t. IV, p. 434).

[67] Robert le Moine, *Historiens des Croisades*, t. III *Historiens occidentaux* (Paris, 1866), 717.

[68] R. Uhden, "Zur Herkunft und Systematik der mittelalterlichen Weltkarten," *Geographische Zeitschrift* 37 (1931), 332-33.

[69] K. Kretschmer, "Die mittelalterliche Weltkarte nach Anlage und Herkunft," *Petermanns Mitteilungen*, Erg. Heft 209 (Gotha, 1930), 62.

[70] Ce n'est pas un accident dû à la reliure, comme l'a montré L. Chekin, "Easter tables and the pseudo-Isidorean Vatican map," *Imago Mundi* 51 (1999), 13 et n. 4.

oeuvres du néoplatonicien Guillaume de Conches (XIIe siècle), ont le nord ou le sud placé en haut. Il n'est pas rare que, dans le même manuscrit, une carte de l'oecumène orientée à l'est voisine avec un schéma des zones à contenu cartographique orienté au nord. Même s'il arrive parfois que des cartes à zones présentent l'est en haut, la règle de leur orientation septentrionale est respectée dans la grande majorité des cas.

Le troisième élément est le peuplement des trois parties de l'*orbis terrarum* par les descendants des fils de Noé, selon la table des peuples de la Genèse.[71] Dans les textes exégétiques nombreux sur ce point, la notion n'est ni simple ni univoque. Il n'y a pas une association constante et nette entre Europe et Japhet, Sem et Asie, Cham et Afrique. Bien des descendants de Japhet sont associés à des régions de l'Asie, et l'identité de Cham comme ancêtre des peuples noirs est une création moderne. Cette notion se révèle donc à l'examen fort différente des simplifications produites par les historiens modernes.[72] Les cartes TO en revanche associent beaucoup plus étroitement *partes* et peuplement: sur certaines d'entre elles, à côté du nom de chaque partie de l'*orbis*, apparaît le nom du fils de Noé censé l'avoir peuplée. Mais ces diagrammes doivent être replacés dans le contexte général de l'ambiguïté de la géographie noachide, textuelle et figurée, dont ils ne constituent qu'une variante parmi d'autres.

Le problème est avant tout de situer le moment et le lieu où les Noachides furent introduits dans la représentation diagrammatique de l'*orbis terrarum*. On attribue parfois cette innovation à Isidore de Séville († 636). Il est vrai que, dans le livre XIV de ses *Étymologies*, le début du chapitre concernant l'*orbis terrarum* est illustré d'un schéma TO auquel est fréquemment joint un schéma quadrangulaire où les noms des fils le Noé sont inscrits et à côté duquel on lit: "Ecce sic diuiserunt terram filii Noe post diluuium." Mais le peuplement de la terre est abordé dans un livre différent de l'encyclopédie (le IXe) et Isidore n'associe jamais explicitement, *dans son texte*, les fils de Noé

[71] Étude fouillée des sources et de l'évolution doctrinale sur ce point d'une grande importance, au-delà des aspects géographiques et cartographiques, par H. Inglebert, *Interpretatio christiana*, 109-92.

[72] Voir les explications lumineuses de B. Braude, "The sons of Noah and the construction of ethnic and geographical identities in the medieval and early modern periods," *The William and Mary Quarterly* 3e série, 54 (1997), 103-42, sp. 108-15.

et les parties de l'*orbis terrarum*, pas plus que ne l'avaient fait, avant lui, tous ceux qui avaient traité ce thème, et notamment saint Augustin. Il y avait à ce refus une raison fondamentale: un schéma borné risquait de présenter comme clos une fois pour toutes l'extension de la parole du Christ.[73] On discernera donc deux étapes. D'abord eut lieu l'introduction dans les manuscrits des *Étymologies* du double schéma circulaire et quadrangulaire qui, on le remarquera, ne postule pas une identité entre chaque *pars* et un des fils de Noé; ensuite, l'intégration des noms des Noachides dans le schéma circulaire triparti. Cette opération ne fut qu'une généralisation et une simplification à but mnémotechnique n'excluant nullement les développements textuels plus détaillés et plus complexes que l'on pouvait trouver dans les textes répandus sur ce sujet, qui présentent rarement une telle simplification.

Le paradis terrestre est le dernier élément important spécifiquement chrétien. Dans les premiers siècles, l'interprétation du passage de la *Genèse* concernant l'Eden fut exclusivement allégorique. C'est seulement lorsque le paradis fut considéré avant tout comme un lieu réel qu'il put entrer dans les *mappae mundi*. La première description géographique apparaît dans l'encyclopédie isidorienne, et il figure déjà sous forme géométrique dans la carte du ms. Vat. Lat. 6018. Avec les images d'Adam, d'Eve et du serpent, il est un élément important de l'économie discursive des cartes de Beatus. Mais il est loin d'apparaître sur toutes les *mappae mundi* dessinées dans les manuscrits. En fait, le paradis terrestre est nécessairement à la fois un lieu situé dans l'*orbis terrarum*, et un lieu inatteignable, ce qui entraîne un *double bind* qui en rend la figuration délicate.[74]

Au total, ces éléments de christianisation de l'*imago mundi* d'origine antique ne sont pas d'emploi systématique: c'est qu'ils ne sont pas le plus important. D'autre part, les *mappae mundi* sont plus variées de contenu et de formes que les oeuvres de Miller et de Destombes ne le donnent à croire, et il est réducteur de caractériser tout un genre par un petit nombre de faits qui n'ont aucun caractère de généralité. L'empreinte chrétienne sur les représentations cartographiques ne

[73] A. Borst, *Der Turmbau von Babel. Geschichte der Meinungen über Ursprung und Vielfalt der Sprachen und Völker*, t. I, 1 (Stuttgart: Hiersemann, 1957), 401; P. Gautier Dalché, "De la glose à la contemplation....," 712-13.

[74] A. Scafi, *Mapping Paradise. A history of heaven and earth* (Londres: The British Library, 2006).

se révèle pas essentiellement par l'introduction de ces éléments de détails, mais par un remaniement idéologique qui donne un sens nouveau, valable dans le contexte chrétien, à des notions qui sont à la base de l'idée même de représentation cartographique de l'oecumène chrétienne.

Bien que les *mappae mundi* soient issues de prototypes de l'Antiquité tardive chrétiens ou non, elles n'ont plus rien d' "antique" du point de vue de la culture dans laquelle elles ont été produites. Contrairement aux présupposés de l'histoire de la cartographie à l'ancienne, qui se bornait à repérer les évolutions internes à la cartographie, il n'est pas possible de comprendre ces objets en les considérant en eux-mêmes et pour eux-mêmes. Qu'est-ce qui en fait des objets cartographiques spécifiquement médiévaux ? Il faut examiner pour cela le cadre de compréhension et d'interprétation dont disposaient ceux qui les dessinaient et ceux qui les contemplaient. J'examinerai les trois fondements qui rendent possible la représentation cartographique. Ce sont: la vision cosmique; la spatialisation de l'histoire du salut; la souveraineté comme condition de la connaissance. Ils sont étroitement liés et ne sont ici distingués que pour la clarté de l'exposé.

i. *Vision cosmique et contemplation monastique*[75]

Qu'est-ce qu'une *mappa mundi* (qu'elle figure l'oecumène ou bien la sphère), sinon ce qui est vu par un regard situé très haut dans l'espace ? Ce qui semble immense au regard humain, apparaît comme minuscule dans cette situation et de ce point de vue. Idée en réalité fort banale dans l'Antiquité. Le géographe grec, tel Denys le Périègète, accède par son art à une vision analogue à celle des dieux, "aérienne et synoptique."[76] La notion trouve des applications dans bien d'autres domaines de la pensée antique. Ainsi, pour les platoniciens et les stoïciens, le corps et la terre sont une prison qui retient l'âme à l'étroit, et le philosophe conscient de cette prison

[75] Assez nombreuse bibliographie; il suffira de renvoyer à l'article fondateur de P. Courcelle, "La vision cosmique de saint Benoît," *Revue des études augustiniennes* 13 (1967), 97-117.

[76] C. Jacob, *La description de la terre habitée de Denys d'Alexandrie* (Paris: Michel, 1990), 26.

souhaite s'envoler pour s'en échapper.[77] Reçue par les auteurs chrétiens, cette idée subit quelques transformations: c'est Dieu qui, placé dans son observatoire en haut de la voûte céleste, est le *speculator* qui peut embrasser d'un seul coup d'œil le monde tout entier, dans ses moindres détails. Or elle trouva un vaste champ d'application dans le monachisme bénédictin, élément structurant de la société et de l'idéologie chrétienne médiévales. Dans sa *Vie de saint Benoît*, Grégoire le Grand décrit pour la première fois cette faculté en l'attribuant à celui qui deviendra le modèle du monachisme occidental. Saint Benoît se tenant à une fenêtre en haut d'une tour voit une lumière surnaturelle qui éclaire le monde entier comme rassemblé sous un seul rayon de soleil: "*omnis etiam mundus sub uno solis radio collectus, ante oculos eius adductus est.*"[78] Le saint éprouve à cette vision un profond sentiment d'humilité. En somme, il a vu une *mappa mundi* et en a retiré un bénéfice spirituel. Ce texte n'est pas isolé: de nombreuses vies de saints reproduisent ce motif jusque bien avant dans le Moyen Age.[79] On a là le récit d'un événement de la vie du père du monachisme occidental révélateur d'une pratique spirituelle: la contemplation, dans l'intention de s'unir à Dieu ou d'éprouver la vanité des intérêts purement humains. Dans ce cas précis, qui n'épuise certainement pas les modes variés de la pratique contemplative, c'est la perception de la nature ambiguë de l'espace du monde, à la fois immense pour l'homme et minuscule pour Dieu, qui est le support de la méditation. La vision de Colomban, première manifestation de l'existence de *mappae mundi*, est une illustration de cette pratique, transformée par l'hagiographe en anecdote relative à l'espace où doit s'exercer la mission évangélisatrice.

Or ces textes hagiographiques et la pratique spirituelle qu'ils révèlent sont liés à des textes ayant une signification cartographique. En effet, Grégoire composa cet épisode de la vie de saint Benoît en ayant

---

[77] Voir par exemple R. M. Jones, "Posidonius and the flight of the mind through the universe," *Classical Philology* 21 (1926), 97-113.

[78] *Dial.* II, 35.

[79] Cf. P. Gautier Dalché, "De la glose à la contemplation...," 753-54; P. Dinzelbacher, "Voli celesti e contemplazione del mondo nelle letteratura estatico-vizionaria del medioevo," in *Cieli e terre nei secoli XI-XII. Orizzonti, percezioni, rapporti*, Atti della tredicesima settimana internazionale di studio, Mendola, 22-26 agosto 1995 (Milan, 1998), 215-33.

à l'esprit le *Commentaire sur le Songe de Scipion* de Macrobe,[80] écrit un siècle auparavant et qui comporte plusieurs diagrammes cartographiques de la sphère et de ses parties habitées. Le tableau géographique d'Orose qui fut utilisé par les cartographes montre la même thématique. L'historien présente ce tableau comme élaboré à partir d'un observatoire élevé, ce que désigne le terme *specula*, permettant d'embrasser d'un seul coup d'œil toutes les turpitudes du genre humain: la phrase d'introduction associe clairement la *specula*, les actions honteuses des hommes et la division de l'*orbis terrarum* en trois parties.[81] L'élaboration géographique est donc conditionnée par un but spirituel. Une problématique voisine insistant sur la contemplation se retrouve au début du XIIe siècle dans l'*Imago mundi* du moine Honorius Augustodunensis. Pour lui, l'*oculus corporis* qui se restaure par la vision de la carte (textuelle) est associé au *visus cordis*, le regard intérieur de contemplation et d'union avec Dieu qui se repose dans la contemplation de la *machina universitatis*. Or, pour sa partie géographique, l'*Imago mundi* est la copie d'une *mappa mundi*.[82] Un passage du *De vanitate* de Hugues de Saint-Victor, auteur de la *Descriptio mappe mundi*, établit nettement ce lien entre recueillement intérieur, regard sur le monde *e specula* et contemplation de l'action divine continue: "*Constitue igitur te quasi in quadam mentis specula, et eius aciem in arcam huius mundi circumquaque lustrandam dirige, ut totus contemplanti coram positus sit mundus, et inde tibi uniuersa demonstrabo.*"[83]

Dans cette relation entre deux fondements de la civilisation médiévale—et de la nôtre—, à savoir l'augustinisme historique d'Orose et les pratiques contemplatives du monachisme, il ne saurait s'agir de coïncidences. On est en présence, sous forme mystique (Grégoire) ou sous forme morale (Orose) d'un même système de justification et

---

[80] Voir Th. Delforge, "Songe de Scipion et vision de saint Benoît," *Revue bénédictine* 69 (1959), 351-54.

[81] "*Conflictationes generis humani et ueluti per diuersas partes ardentem malis mundum face cupiditatis incensum e specula ostentaturus necessarium reor, ut primum ipsum terrarum orbem quem inhabitat humanum genus, sicut est a maioribus trifariam distributus deinde regionibus prouinciisque determinatus, expediam.*" *Historiae adv. paganos*, I, 1, 15-16.

[82] "*...totius orbis tibi depingi formulam in qua sic oculum corporis ualeas reficere sicut uisum cordis soles in machina universitatis depascere.*" *Prologus*, *Imago Mundi*, V.I.J. Flint (éd.), *Archives d'histoire littéraire et doctrinale du Moyen Age* 49 (1983), 48. Cf. P. Gautier Dalché, "Maps in words: the descriptive logic of medieval geography, from the eighth to the twelfth century," dans P.D.A. Harvey (éd.), *The Hereford World Map: Medieval World Maps and their Context* (Londres: The British Library, 2006), 223-42.

[83] *Patrologia Latina*, t. 176, 704.

de légitimation chrétienne de l'existence des *mappae mundi*. La multiplication extraordinaire de ces objets au Moyen Age, qu'ils soient schématiques ou non, s'explique par la force agissante de ce système.

ii. *Représentation spatiale de l'histoire du salut*

La perspective du maître victorin est tout entière orientée par son herméneutique augustinienne fondée sur l'attention à la signification littérale de l'Écriture d'une part, et sur une conception complexe de l'histoire du salut dont la carte résume spatialement toute l'étendue, du paradis à la passion du Christ et au Jugement, d'autre part.[84]

Les *mappae mundi* contiennent, on l'a souvent remarqué, des toponymes de différentes époques, juxtaposant ainsi les noms antiques. Ainsi, dans la *Descriptio mappe mundi* de Hugues de Saint-Victor, on trouve Tiflis et Maroc (Marakech) aussi bien que Babylone et Persépolis. Cette caractéristique si contraire à notre notion de la carte donne encore lieu à deux types de réaction: soit l'on condamne l'incapacité des cartographes médiévaux à représenter leur propre espace; soit l'on mesure leur faible intérêt à moderniser leurs cartes face au poids de ce qu' il est convenu d'appeler la "tradition," et leur manque d'usage de sources d'information provenant de ce qu'il est convenu d'appeler l' "expérience." Cette opposition moderne entre "tradition" et "expérience," si fréquente dans les travaux d'histoire de la cartographie médiévale, est sans valeur heuristique, car elle se borne à retrouver les précurseurs, ou déplorer l'absence du référent implicite déjà présent constitué par nos concepts de "tradition" et d' "expérience," qui n'ont que fort peu en commun avec leurs équivalents médiévaux. L'une et l'autre approche sont en réalité identiques dans leur anachronisme puisqu'elles utilisent, pour juger leurs productions, des critères absolument étrangers à leurs auteurs: nos propres critères établis sur la domination absolue, et donc jugée naturelle, de la science et de la technique.

Pour comprendre de l'intérieur les raisons intellectuelles, culturelles ou idéologiques de cette surprenante caractéristique, il est utile de se référer au *Commentaire sur l'Apocalypse* du moine Beatus de Lié-

[84] P. Sicard, *Diagrammes médiévaux et exégèse visuelle. Le "Libellus de formatione arche" de Hugues de Saint-Victor* (Paris-Turnhout: Brepols, 1993), Bibliotheca Victorina, 4; P. Gautier Dalché, *La 'Descriptio mappe mundi' de Hugues de Saint-Victor.*

bana et à la *mappa mundi* qui l'accompagne, première représentation cartographique de l'évangélisation.[85] Cette carte joue un rôle particulier; elle est la seule figure à être annoncée explicitement et elle introduit directement au commentaire lui-même.[86] Elle met donc directement sous les yeux du lecteur la scène où se développe l'histoire orientée du salut, depuis le jardin d'Eden jusqu'à la Parousie, en passant par l'action des douze apôtres, chacun actif dans une région donnée. Dans un cadre ainsi défini, il est normal et aisément compréhensible que, sur certaines cartes de Beatus, soit dessinée une quatrième *pars terre* australe, présentée comme inconnue et inhabitable—mais peuplée de monstres tels que les antipodes ou les sciapodes. La contradiction n'est qu'apparente: selon la conclusion exprimée par saint Augustin dans la *Cité de Dieu*, ces êtres monstrueux sont des hommes pourvu qu'ils descendent d'Adam et qu'ils soient des animaux raisonnables mortels, donc susceptibles d'être évangélisés.[87] La mappemonde, par souci de complétude, figure donc un espace qui ne pourrait être exploré que dans l'avenir.

La présence de cette *pars australis* sur différentes cartes, en même temps que les réflexions cosmographiques sur la sphère ont été, durant tout le Moyen Age, un stimulant de l'exploration de l'*orbis terrarum* découlant de la nature même des *mappae mundi*: actualisation de l'histoire du salut par la pratique de la contemplation. Pour Beatus, la carte a une fonction spirituelle de support de la contemplation monastique, dans un contrepoint constant avec le texte. Le monde où les apôtres ont semé, après que les prophètes eurent labouré, est toujours *présent* aux yeux de celui qui regarde la carte: c'est une image spatialisée de l'Église éternelle, où la force agissante de la Parole est constamment active, et où ce qui est exposé dans l'Apocalypse comme relevant du passé est conçu par le lecteur comme se réalisant "nunc in ecclesia." Autrement dit, la *mappa mundi* contient toute l'histoire du salut tel qu'il s'étend progressivement sur l'ensemble du

[85] P. Gautier Dalché, "De la glose à la contemplation...," 693-764.

[86] "...*et quo facilius haec seminis grana per agrum huius mundi, quem prophetae laborauerunt et hi metent, subiectae formulae pictura demonstrat.*" Beatus, *Sancti Beati a Liebana Commentarius in Apocalypsin*, éd. E. Romero-Pose (Rome: Typis Officinae Polygraphicae, 1985), II, *Prol.* 3, 21, 193. Dans P. Gautier Dalché, "De la glose à la contemplation...," 751 et 752), j'ai commis une erreur d'écolier en interprétant les lignes qui précèdent ce texte; les apôtres ne "labourent" pas, ils sèment dans la terre que les prophètes ont labourée; cela n'affecte pas la démonstration.

[87] *Civ.*, XVI, 8.

monde terrestre; elle offre au regard intérieur cette histoire ramassée en un seul objet, qui l'actualise et la rend toujours présente au fidèle et, en particulier, au moine s'exerçant à la contemplation. La distinction des temps et des lieux selon nos critères n'a qu'une importance secondaire, d'où le mélange de toponymes de différentes époques.

iii. *Légitimation politico-religieuse de la représentation cartographique*

Le dernier fondement de la cartographie médiévale, qui concerne à la fois la souveraineté et la religion, est particulièrement mis en évidence sur la carte de Hereford. Dans le dessin qui figure dans l'angle inférieur gauche, l'empereur romain enjoint à trois "géomètres" d'aller par le monde et de le mesurer et d'en faire rapport au Sénat. Ce fait provient de la *Cosmographia* du Pseudo-Aethicus, la présence de trois "géomètres" au lieu des quatre de la *Cosmographia* n'étant qu'un effet mécanique de la tradition manuscrite particulière de ce texte dont certains manuscrits présentaient une lacune sur ce point dès le IXe siècle, et il est inutile de lui chercher des raisons particulières.[88] Au-dessus est transcrit un verset de Luc (2, 1) évoquant un *census* décidé par l'empereur. Cette constellation n'est pas propre à la carte de Hereford, on la retrouve en particulier au début du XIIe siècle dans le *Liber floridus* de Lambert de Saint-Omer qui contient de nombreuses *mappae mundi*.[89] L'idée selon laquelle la géographie antique, connue par les textes, est nécessairement parfaite puisque garantie par l'autorité politique, est au fondement de la reprise des textes de l'Antiquité par les auteurs médiévaux.

Que signifie pour un observateur médiéval la présence sur la carte de cette image où un recensement considéré comme pouvant aboutir à une représentation figurée selon l'usage gromatique émane de l'autorité politique ? Une *mappa mundi*, au Moyen Age, est toujours considérée sous deux aspects étroitement liés: c'est le résultat manifeste de la supériorité romaine et en même temps l'expression d'une vision chrétienne de l'oecumène en tant que lieu de l'histoire du salut promis par la naissance du Christ. On voit dans cette association tout d'abord le souvenir amplifié de l'agrimensure romaine, cette technique d'arpentage, de mesure et de représentation des terres

[88] Déjà signalé par K. Pertz, *De Cosmographia Aethici libri tres* (Berlin: 1853), 51 sqq.
[89] P. Gautier Dalché, "Les 'quatre sages' de Jules César...," 203-204.

exercée par de véritables "géomètres." Grâce à la conquête, l'autorité impériale s'est étendue à l'ensemble de l'oecumène, ce qui a rendu possible sa connaissance complète par les moyens gromatiques. D'autre part, l'étape finale de l'histoire du salut est marquée aussi, au moment de la naissance du Christ, par le recensement fait sur l'ordre d'Auguste. On n'a pas assez remarqué, jusqu'à présent, que le geste supposé de l'empereur, interprété au Moyen Age comme créateur de cartographie, est concomitant de l'Incarnation. La carte de Hereford signale de la façon la plus claire la réalité de cette association. L'empereur porte en effet une tiare papale munie d'une crosse, évidente représentation christianisée de César Auguste qui manifeste cette double réalité.[90]

La possession, la maîtrise intellectuelle de l'espace et la souveraineté ont toujours été étroitement en relation depuis l'Antiquité, que cela s'exprime par le moyen de textes ou de figures. Je n'en citerai que deux exemples. En 435, l'empereur Théodose II fit réaliser une carte par deux serviteurs suivant des modèles textuels ou figurés plus anciens. L'entreprise était rappelée par des vers transcrits dans deux manuscrits à la fin d'un texte schématique de l'Antiquité tardive intitulé *Divisio orbis terrarum*, peut-être héritier des *commentarii* dont Agrippa avait accompagné sa carte, et présentant des listes de provinces avec leurs dimensions.[91] Sous l'empereur Louis le Pieux, en 825, un savant proche de la Cour nommé Dicuil, dans un traité dont le titre *De mensura orbis terrae* ainsi que le contenu évoquent dans l'esprit du lecteur les techniques gromatiques, interpréta faussement ces vers comme la preuve d'une tentative de recensement général de

[90] S. Westrem, *The Hereford Map*, 8.

[91] P. Schnabel, "Die Weltkarte des Agrippa als wissenschaftliches Mittelglied zwischen Hipparch und Ptolemaeus," *Philologus* 90 (1935), 439-40. Les tentatives pour déterminer l'aspect et le sens de la réalisation sont purement conjecturales, donc sans intérêt. Pour W. Wolska, il s'agirait d'une mosaïque ou d'une peinture exposée dans un portique de l' "université" de Constantinople fondée par Théodose II et accompagnée d'un texte (la *Divisio orbis terrarum*) mis à la disposition des "élèves" pour faciliter la lecture des légendes: "Deux contributions à l'histoire de la géographie II. La carte de Théodose II: sa destination," *Centre de recherche d'histoire et de civilisation de Byzance. Travaux et mémoires* 5 (1973), 276-78. Pour E. Weber, il est "très vraisemblable" que le poème transmis par Dicuil se trouvait au début de la partie perdue de la *Tabula Peutingeriana*, dont la carte de Théodose serait l'ancêtre direct (*Tabula Peutingeriana,* 22). Il ne donne aucune preuve de cette vraisemblance; il suffit de rappeler le vers 11: "*...ac totum breuiter comprehendimus orbem*" qu'il paraît légèrement difficile d'appliquer à la *Tabula.*

l'oecumène. Le texte de la *Divisio orbis terrae*, très proche d'une carte, était pour lui l'aboutissement de cette entreprise effectuée sous l'impulsion de l'empereur. Dans un même contexte impérial, la tripartition de l'*orbis terrarum* est définie dans un texte de la fin du Xe ou du début du XIe décrivant le manteau de l'empereur: la ceinture doit se terminer par une *rota* portant la légende "Roma caput mundi tenet orbis frena rotundi;" au centre est inscrit le "tripharia thema orbis: Asia, Africa, Europa :" c'est une *mappa mundi* schématique.[92] Interpréter cet ornement cartographique du manteau impérial comme l'indice d'une prétention à la souveraineté universelle serait exagéré, comme l'a montré P. E. Schramm; plus simplement, il s'agit de l'expression d'une idéologie qui lie la souveraineté à la carte.

Ce n'est d'autre part sans doute pas un hasard si l'expression "mappa mundi" apparaît pour la première fois dans l'empire carolingien, en 821-822, dans le catalogue de la bibliothèque de la grande abbaye de Reichenau, sur le lac de Constance. On conjecture que ces mots désignent une copie de la *Tabula Peutingeriana*.[93] Par son aspect extérieur de carte itinéraire, elle pouvait passer pour un exemple parfait de l'art gromatique. Mais là n'est pas le plus important. La signification de *mappa (mundi)*, le choix de cette expression destinée à une fortune remarquable, renvoie nécessairement au souvenir de l'art gromatique conditionné, dans l'esprit des contemporains de Louis le Pieux, par la domination impériale sur l'ensemble des terres. En un mot, la puissance et la sagesse de Rome s'exprimaient de façon parfaite dans la carte de son empire. Etant donné la reviviscence de l'idéologie impériale dans le cadre de la *renovatio* voulue par Charlemagne, tous ces éléments associés, avec l'effet puissant de la culture monastique de la contemplation d'une part et de l'appel à l'évangélisation pour accomplir l'histoire du salut d'autre part, une constel-

---

[92] *Graphia aureae urbis Romae*, R. Valentini, G. Zuchetti (éds), *Codice topografico della città di Roma*, 3 (Rome: Tipografia del Senato, 1946), 102; voir P. E. Schramm, *Sphaira, Globus, Reichsapfel* (Suttgart: Hiersemann, 1958), 53. Nul besoin de rappeler ici les globes impériaux ('Reichsapfel') des XIIe et XIIIe siècles portant tripartition de l'*orbis terrarum*.

[93] P. Gautier Dalché, "Du nouveau sur la transmission et la découverte de la *Tabula Peutingeriana*," *Geographia Antiqua* 13 (2004), 79.

lation idéologique complexe permettant de comprendre l'existence même de la *mappa mundi* chrétienne avait été élaborée.[94]

Conclusion

L'examen des *mappae mundi* révèle qu'il y a eu, entre le IIIe et le VIIe siècle, en différents lieux et en différents moments, plusieurs opérations de "mise en carte" de textes géographiques qui, par leur structure même, pouvaient se prêter aisément à cela. Ces opérations se comprennent dans le cadre des pratiques scolaires d'une part, et selon les critères de la culture aristocratique vouée à l'*eruditio* d'autre part. Des comparaisons minutieuses des formes apparaissant sur les *mappae mundi* permettraient d'approfondir et d'enrichir cette affirmation. Ces représentations tirent leur prégnance mentale et leur efficacité, plutôt que d'une christianisation des contenus qui n'a jamais été ni systématique ni générale, de fondations idéologiques élaborées durant le haut Moyen Age, le processus trouvant son aboutissement dans la *renovatio* carolingienne. A partir de la "mise en carte" des textes et de la cristallisation de systèmes idéologiques entre Antiquité tardive et temps carolingiens, l'entreprise cartographique chrétienne a été progressivement légitimée. La force de cette justification explique sans doute l'efficacité *pratique* séculaire des *mappae mundi*—selon les critères de l'efficacité pratique que partageaient leurs créateurs et leurs utilisateurs. Ainsi, l'héritage antique est constamment présent dans la cartographie des *mappae mundi*, qui sont en même temps des créations originales.

---

[94] Sur les usages des textes géographiques antiques dans le cadre de l'idéologie impériale carolingienne, voir N. Lozovsky, "Roman geography and ethnography in the Carolingian Empire," *Speculum* 81 (2006), 325-64.

PROCESS AND TRANSFORMATION ON THE SEVERAN MARBLE PLAN OF ROME

Jennifer Trimble

> [A map is] a social construction of the world expressed through the medium of cartography. Far from holding up a simple mirror of nature that is true or false, maps redescribe the world—like any other document—in terms of relations of power and of cultural practices, preferences, and priorities. What we read on a map is as much related to an invisible social world and to ideology as it is to phenomena seen and measured in the landscape.[1]

The Severan Marble Plan is an extraordinary map. Carved between 203 and 211 CE, it represented the imperial city of Rome in plan view, depicting every temple and warehouse, every street and alley, every tenement building and luxury dwelling, every ground-floor room, doorway and internal staircase (Figs 1, 9). Centered on the Capitoline and carved at a scale of 1:240, it mapped an area of more than 13.5km^2. Its accuracy and detail make it a crucial resource for understanding the imperial city. Unfortunately, only about 12% of

---

[1] John Brian Harley, "Text and Contexts in the Interpretation of Early Maps," in David Buisseret (ed.), *From Sea Charts to Satellite Images: Interpreting North American History through Maps* (Chicago: University of Chicago Press, 1990), 4. Frequently cited works are abbreviated below as follows:

Stanford Project	Stanford Digital Forma Urbis Romae Project http://formaurbis.stanford.edu
FUR 2006	R. Meneghini and R. Santangeli Valenziani (eds.), *Formae Urbis Romae: Nuovi Frammenti di Piante Marmoree dallo Scavo dei Fori Imperiali. Bullettino della Commissione Archeologica Comunale di Roma Suppl.* 15 (Rome: "L'Erma" di Bretschneider, 2006).
FUA 2002	Emilio Rodríguez-Almeida, *Formae Urbis Antiquae: Le Mappe Marmoree di Roma tra la Repubblica e Settimio Severo.* Collection de l'Ecole Française de Rome 305 (2002).
AG 1980	Emilio Rodríguez-Almeida, *Forma Urbis Marmorea. Aggiornamento Generale 1980* (Rome: Quasar, 1981).
PM 1960	Gianfilippo Carettoni et al. (eds), *La Pianta Marmorea di Roma Antica. Forma Urbis Romae* (Rome: Comune di Roma, 1960).
Reynolds	David West Reynolds, "*Forma Urbis Romae:* The Severan Marble Plan and the Urban Form of Ancient Rome" (diss., University of Michigan, 1996).

the original survives, in 1,194 fragments.[2] This fragmentation makes the map monumentally difficult to work with, and research has necessarily focused on identifying and piecing together individual fragments. Moreover, scholars have found this map significant primarily for its depiction of monuments attested in ancient texts, and for its help in reconstructing the topography of Rome.[3] Productive as this work has been, it has too often meant treating the map as an objective source of information, an unproblematic reflection of the physical record. Little attention has been paid to this monument as a map, a complex visual image in its own right, and one that raises important questions about the control of information and the conceptual representation of space.

In this article, I draw on the approach of J.B. Harley, cited above, to explore the Severan Marble Plan as a constructed image of the city as much as an indicator of what stood where.[4] This was no

---

[2] The Stanford Project, in collaboration with the Sovraintendenza Comunale di Roma, has placed photographs and 3D models of 1,186 fragments, many with scholarly commentary, on a public website (http://formaurbis.stanford.edu). Thirty-one newly-discovered fragments from the recent excavations in the Imperial Fora have now been published in FUR 2006, 13-26 and 37-39 (23 of these can be seen on the Stanford Project's website; all have the inventory number prefix "fn"). The figure of 12% was generated by David Koller of the Stanford Project. For a description of the Project's work, see David Koller, Jennifer Trimble, Tina Najbjerg, Natasha Gelfand and Marc Levoy, "Fragments of the city: Stanford's Digital Forma Urbis Romae Project," in Lothar Haselberger and John Humphrey (eds.), *Imaging Ancient Rome: Documentation—Visualization— Imagination, Journal of Roman Archaeology Suppl. 61* (2006), 237-52. On the late antique and medieval spoliation of the map, AG 1980, 39-43; see now Riccardo Santangeli Valenziani's discussion in light of the new excavations, in "Distruzione e Dispersione della *Forma Urbis* Severiana alla Luce dei Dati Archeologici," in FUR 2006, 53-59.

[3] Recent decades' scholarship is reviewed in Tina Najbjerg and Jennifer Trimble, "The Severan Marble Plan since 1960," in FUR 2006, 75-101. For the Stanford Project's new fragment matches and identifications, see David Koller and Marc Levoy, "Computer-aided Reconstruction and New Matches in the Forma Urbis Romae," in FUR 2006, 103-25. Working with more traditional methods, Pier Luigi Tucci has recently contributed two new sets of identifications: "Eight fragments of the Marble Plan of Rome shedding new light on the Transtiberim," *Papers of the British School at Rome* 72 (2004), 185-202, and "L'Arx Capitolina: tra mito e realtà," in *Imaging Ancient Rome: Documentation—Visualization—Imagination*, 63-73. Finally, several new fragment locations are proposed, although not explained, in FUA 2002, pl. XII.

[4] A precedent for this approach is the work of Richard Talbert on the Peutinger map. See elsewhere in this volume, as well as Richard Talbert, "Rome's Marble Plan and Peutinger's Map: Continuity in Cartographic Design," in F. Beutler and W. Hameter (eds.), '*Eine ganz normale Inschrift*'...*und Ähnliches zum Geburtstag von Ekkehard*

ordinary map in the modern sense, easily consulted and read primarily for the quality of its information—excellent as that information was. It was carved onto 150 marble slabs mounted in eleven rows on a wall in a large room inside the Templum Pacis in the political heart of Rome.[5] Measuring about 18 x 13m, the map was set high on the wall, beginning over 4m above the floor and extending upward to the height of a modern four-storey building. This vast size and imposing installation are easily forgotten because of the map's poor survival and the detailed presentation of its individual fragments in modern publications. However, they imposed major constraints on visibility, and made the map impossible to consult in any straightforward way. In Harley's terms, "relations of power, cultural practices, preferences, and priorities" are visibly and vividly in play.

Specifically, I focus here on the Severan Marble Plan's relationship to earlier Roman urban maps. Several such fragments are known, and they attest to an established cartographic tradition with its own conventions and particular emphases. Studies of these predecessors have shown that the Severan map belongs firmly within this cartographic tradition, but also deviates from it in certain respects. I argue here that the scope of those changes has not been fully taken into account, or their implications explored. The Severan map shows selective adoptions and adaptations of that cartographic tradition, making it unique in the surviving evidence. These adaptations are visible in the map's use of line, depiction of features, treat-

Weber (Vienna: Eigenverlag der Österreichische Gesellschaft für Archäologie, 2005), 627-34; "Cartography and Taste in Peutinger's Roman Map," in R. Talbert and K. Brodersen (eds.), *Space in the Roman World: Its Perception and Presentation* (Münster: LIT, 2004), 113-41.

[5] The evidence is the pattern of clamp holes still visible on the wall, which was repaired with Severan masonry after a fire in 192 CE (the fundamental study is still Lucos Cozza's, published in PM 1960, 177-95, with a detailed drawing of the wall at pls. LXIa and b; see the updates in FUA 2002, pl. VIII. This is now the outside back wall of the Church of SS. Cosma and Damiano, visible from the Via dei Fori Imperiali; the majority of the surviving fragments of the Severan Marble Plan were found in 1562, fallen at the base of this wall. Once mounted, the edges of some slabs were smoothed to align with adjacent slabs, and the incisions were carved over this smoothing, proving that the map was carved *in situ* (PM 1960, 201, with examples at n. 25). On the Templum Pacis itself as a showcase of spectacular monuments, see Filippo Coarelli, "Pax, Templum," in Eva Margareta Steinby (ed.), *Lexicon Topographicum Urbis Romae* IV (Rome: Quasar, 1999), 67-70.

ment of inscriptions, and interactions of image and text. Tracing them illuminates both process and transformation.

First, examining the relationship between its predecessors and the Severan map points to a range of processes and levels of decision-making at work on the latter, with a marked difference between overarching directives and localized executions. We normally cannot see the processes by which Roman monuments took visual and conceptual form; this map is extraordinary partly because those processes are exactly what it puts on display. Second, this map's transformation of existing urban cartographic traditions resulted in a particular representation of the city of Rome. Public space was mapped in relational terms of time and space; by comparison with its predecessors, this map flattened the architectural and social hierarchy. The treatment of major monuments, and the way in which inscriptions were combined with visual emphases, required a lower degree of map-reading literacy and drew on conventions of public spectatorship to create a particular kind of collective viewership. Analyzing the Severan Marble Plan's cartographic relationship to its predecessors thus offers a fuller understanding of how this particular map was made and how it may have worked within its original social, historical and visual context.

The first step is to situate the Severan Marble Plan within Roman cartography. Roman maps included representations of the world, cadastral land maps, and architectural maps of urban segments and single properties, as well as more pictorial or symbolic forms of the representation of space.[6] Of particular interest in evaluating change and continuity on the Severan Marble Plan are the surviving urban map fragments. At least five are known; all are dated to the first or

[6] Still important is O.A.W. Dilke, *Greek and Roman Maps* (Ithaca, NY: Cornell University Press, 1985). Emilio Rodríguez-Almeida's study of architectural maps includes an important discussion of two map fragments depicting aqueduct sections and spelling out individual property owners' rights to draw water from them (FUA 2002, 23-36). On the differences between land surveyors and architectural surveyors, see the comments in Reynolds, 39-42. A recently discovered marble fragment, seemingly an architect's sketch, depicts a section of the Forum of Augustus in plan: see now Roberto Meneghini, "La Nuova *Forma* del Foro di Augusto: Tratto e Immagine," in FUR 2006, 157-171, and Elisabetta Carnabuci, "La Nuova *Forma* del Foro di Augusto: Considerazioni sulle Destinazioni d'Uso degli Emicicli," in FUR 2006, 173-95.

Fig. 1. Fragment 10g of the Severan Marble Plan, showing streets lined by one-room shops, apartment buildings, open courtyard spaces, and internal doorways and staircases (marked by V-shaped symbols) in the Subura neighborhood on the Oppian hill. Where the two larger streets join at center left, the acute angles of the architecture are probably due to the difficulty of matching up different survey areas. This fragment's maximum length is 0.71m. (Stanford Digital Forma Urbis Romae Project and Sovraintendenza Comunale di Roma).

Fig. 2. Urban map fragment found in the Via della Polveriera in Rome in 1890, depicting parts of several buildings with the individual property owners' names inscribed. A street is just visible at top, inscribed with a length measurement in Roman feet. (Sovraintendenza Comunale di Roma).

second century CE.[7] All are marble, or are modern drawings of marble fragments now missing. Additional maps in other, more perishable media must also have existed but do not survive.

A fragment of a marble slab was discovered in 1890 in the Via Polveriera on the Oppian hill; it depicts the rooms within several buildings and part of a street (Fig. 2).[8] Another marble map fragment found in Amelia in Umbria now survives only as a drawing in an early seventeenth century manuscript; it depicts rooms belonging to several buildings, a passageway, and a colonnaded sidewalk and street (Fig. 3).[9] In 1983, in the Via Anicia in Trastevere, a large portion of a marble slab came to light; it depicts commercial and monumental buildings along the right bank of the Tiber (Fig. 4).[10] These features include the Temple of Castor and Pollux in the Circus Flaminius; usefully, the area depicted overlaps with fragments 32ghi of the Severan Marble Plan. Another marble map fragment was discovered in the Isola Sacra necropolis of ancient Portus at the mouth of the Tiber; it depicts two streets, a row of tabernae, additional rooms and apartments, and perhaps a colonnaded sidewalk.[11] Finally, a marble map fragment was discovered in 1995 packed into the original paving of the Forum Transitorium in Rome; it includes

[7] Emilio Rodríguez-Almeida has now made the most complete study: FUA 2002. Single-property architectural maps are not treated in this analysis because they differ from the urban fragments in important respects. On these, see PM 1960, 207-10; FUA 2002, 37-41; and FUR 2006, 26-37.

[8] Now in the Capitoline Museums in Rome. PM 1960, 207 and pl. Q, fig. 47; FUA 2002, 41-43; FUR 2006, 27-28.

[9] The drawing is in the Biblioteca Ambrosiana, Milan: Codex H. 180 inf., folio 48v - 49r. It is reproduced and discussed in PM 1960, 208-209 and pl. Q, fig. 49; in FUA 2002, 51-56; and in FUR 2006, 35-36.

[10] Now in the Epigraphy Museum in Rome. FUA 2002, 43-49; FUR 2006, 26-27. Key publications include Marisa Conticello De' Spagnolis, *Il Tempio dei Dioscuri nel Circo Flaminio* (Rome: De Luca, 1985); Ferdinando Castagnoli, "Un nuovo documento per la topografia di Roma antica," *Studi Romani* 33 (1985), 205-11; and Emilio Rodríguez-Almeida, "Un frammento di una nuova pianta marmorea di Roma," *Journal of Roman Archaeology* 1 (1988), 120-31. Pier Luigi Tucci has proposed that the building at bottom right (also visible on fr. 32i of the Severan map) was the museum built under Augustus to house the ship of Aeneas: "Dov' erano il tempio di Nettuno e la nave di Enea," *Bullettino della Commissione Archeologica Comunale di Roma* 98 (1997), 15-42.

[11] First published in PM 1960, 208 and p. Q, fig. 48. See also FUA 2002, 56-59, and FUR 2006, 34-35.

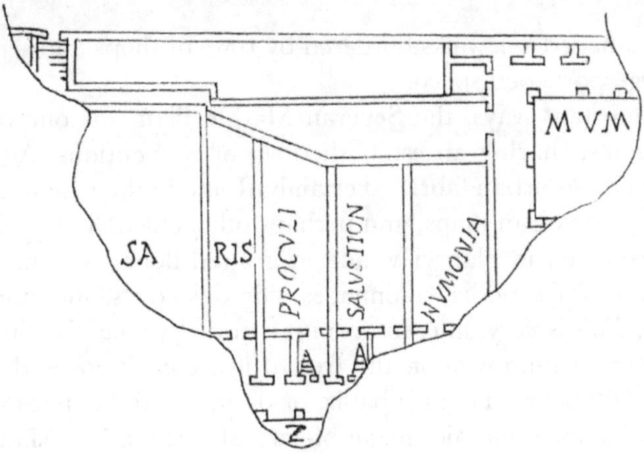

Fig. 3. Early 17th c. drawing of an urban map fragment found in Amelia in Umbria, now lost, depicting parts of several buildings with owners' names inscribed, a passageway and external staircase at upper left, and a street with (probably) a length measurement in Roman feet inscribed along the edge of the sidewalk. (FUA 2002, fig. 17, reproduced by permission).

Fig. 4. Urban map fragment found in the Via Anicia in Rome in 1983, depicting the Temple of Castor and Pollux in the Circus Flaminius and several privately-owned buildings along the Tiber, with the owners' names inscribed on each. Length measurements in Roman feet are inscribed along the riverfront colonnade. This fragment overlaps with fragments 32ghi of the Severan Marble Plan. (Soprintendenza Archeologica di Roma).

two colonnaded sidewalks bordered by rows of shops and separated by a large porticoed space.[12]

In important ways, the Severan Marble Plan was one of these urban maps, sharing an established set of conventions. All depict segments of an urban fabric—certainly Rome in the case of the Via Anicia and Severan maps, and perhaps others as well. The features are represented in plan view and at ground floor level: individual rooms, doorways, podia, colonnades, sidewalk edges, and other built features. This is very different from the bird's eye view of cities seen on Trajan's Column or in the fresco of a city discovered on the Oppian Hill below Trajan's baths, or on late antique maps such as the sixth century mosaic image of Jerusalem found at Madaba in Jordan.[13] It is also very different from the way in which the late Roman Peutinger map was conceived (see elsewhere in this volume). These urban maps are also totalizing images, showing an interest in mapping every segment of built space down to individual doorways and internal staircases, and implying the marshaling of resources of knowledge, skill and will.

On all these urban maps, including the Severan, a scale of 1:240 seems to be a consistent convention.[14] It is employed across the entire surface, which is treated as a unified cartographic space. Only built features are depicted; there is no explicit indication of hilly terrain, natural features such as the river Tiber (note the empty, river-shaped

[12] FUA 2002, 61-66; FUR 2006, 28-29; see also Monica Ceci, "Una *Forma* Privata dal Foro di Nerva," in FUR 2006, 197-200.

[13] On the fresco from the Oppian hill, G. Caruso and Rita Volpe, "Preesistenze e persistenze delle terme di Traiano," and Eugenio La Rocca, "L'affresco con veduto di città del colle Oppio," both in Elizabeth Fentress (ed.), *Romanization and the City: Creation, Transformations, and Failures*, *Journal of Roman Archaeology Suppl.* 38 (2000), 42-56 and 57-71.

[14] A scale of 1:240 is now accepted by most scholars, including PM 1960, 75-76 and 78, and Emilio Rodríguez-Almeida in various of his writings. *Contra*, arguing for a scale of 1:250, is P.M. Lugli, "Considerazioni urbanistiche sulla Pianta Marmorea del Foro della Pace," *Bollettino di Archeologia* 16-18 (1992), 19-31. A shared scale of 1:240 can be confirmed only on the Via Anicia plan, which overlaps frs. 32gh and i of the Severan map. The other surviving maps look comparable (FUA 2002, *passim*). Other kinds of Roman feature-mapping employ very different scales; much smaller ones for land cadasters and much larger ones for single-property maps. Moreover, on at least one of the latter kind, now in Perugia, a funerary monument and two floors of the custodian's house are depicted in plan view on the surface of a marble plaque; each of these employs a consistent scale internally, but none of the three scales is the same: see FUA 2002, 37-41.

swathe defined by riverfront architecture in Fig. 7), or garden plantings.[15] Streets, plazas, courtyards and doorways are articulated by the adjacent walls, sidewalk edges and colonnades. Dots or small squares normally represent columns. Staircases take two forms: a V shape usually denoting an internal staircase and a rectangular shape with horizontal lines, usually depicting external stairs. On several of the urban maps, including the Severan, the V-shaped symbol sometimes has horizontal lines between the bars. These may be purely conventional, although one scholar has suggested that they indicate the number of upper stories belonging to that building.[16]

It is not clear whether any of these earlier map fragments were part of a spectacular monumental presentation like that of the Severan Marble Plan. The other known urban maps survive as fragments of marble slabs, presumably for display on a wall. However, we do not know whether these maps depicted the entire city or only certain sectors, or whether they were mounted in a more accessible and functional way than the Severan Marble Plan. Differences in the carving and in the use of inscriptions suggest that the other urban maps were intended for consultation in a way that the Severan was not (see below), but this does not exclude monumental display. The evidence is simply too scanty to know.[17] Certainly there was a long

---

[15] The Via Anicia map (Fig. 4) perhaps depicts the Tiber bank at bottom; yet given the conventions seen everywhere else on the fragment, it is more likely that this line indicated a built feature. On the Severan map, certain fragments like those depicting the Adonaea (frs. 46a-e and 68ab) include dense arrays of dots that do not make sense as columns and accordingly have been interpreted as trees or a form of trellis; again, this would mark so strong an exception to the rule that more positive evidence is required. The new excavations in the Templum Pacis courtyard, for example, have shown that the linear features depicted on the Severan map are in fact built statue bases (frs. 15abc). Eugenio La Rocca, "La Nuova Immagine dei Fori Imperiali," *Mitteilungen des Deutschen Archäologischen Instituts, Römische Abteilung* 108 (2001), 171-213.

[16] Luigi Pedroni, "Per una lettura verticale della Forma Urbis Marmorea," *Ostraka* 1.2 (1992), 223-30. This is an intriguing suggestion, and may support the cadastral function of the earlier urban maps, as explained below; however, it is difficult to substantiate for the Severan map. The majority of V-shaped symbols on the map have no bars, even when a building's floor plan looks like that of a multistory apartment building. Conversely, the bars occasionally occur when there cannot plausibly be several upper floors, as within an open courtyard inside a commercial building on fr. 28a in Transtiberim (Fig. 8). This may be a convention from earlier maps that was retained on the Severan map, but not consistently so.

[17] Rodríguez-Almeida argues for a monumental Augustan map of the city on the basis of the later urban map fragments (FUA 2002), but this remains speculative.

Roman tradition of the public display of monumentalized information, including displays of the Fasti, or religious calendar, as well as totalizing representations like Agrippa's map of the world.[18] And certainly cadastral maps must also have existed in more easily used and updated media than marble.

In other ways, the Severan Marble Plan is unique among the known urban maps. Not least, it depicts information belonging to different time periods. The Septizodium appears; this monument was dedicated by the emperor Septimius Severus in 203 CE and provides a *terminus post quem* for the map's date (Fig. 5).[19] By contrast, other structures are represented in distinctly pre-Severan forms. The monumental entrance of the Porticus Octaviae is shown in its Augustan form of approximately two centuries earlier (Fig. 6).[20] The area of the Crypta Balbi depicts a pre-Severan situation (frs. 30a-f), as does the Porticus Aemilia (Fig. 7).[21] The Lacus Iuturnae seems to be shown in a phase belonging to the end of the first century CE rather than the start of the 3rd, and fr. 18a has accordingly been argued to be the only known fragment of an earlier, Vespasianic marble map.[22] A different explanation is more likely. The clear implication is that

---

Tina Najbjerg and Jennifer Trimble, "Ancient maps and mapping in and around Rome," *Journal of Roman Archaeology* 17 (2004), 577-83. There is an ongoing discussion about whether the Severan Marble Plan replaced an earlier map of the city mounted in the same location in the Templum Pacis, but there is no direct evidence for this either, and no particular need to think it did. On the problem of fr. 18a, see below, n. 25.

[18] Claude Nicolet, *Space, Geography, and Politics in the Early Roman Empire* (Ann Arbor: University of Michigan Press, 1991), a translation of *L'Inventaire du Monde: Géographie et Politique aux Origines de l'Empire Romain* (Paris: Fayard, 1988).

[19] No ancient text mentions this map, so dating relies on the masonry of the wall, shown to be Severan, and on internal evidence. Besides the Septizodium, an inscription naming the emperors Septimius Severus and Caracalla on frs. 5Ab-e suggests a *terminus ante quem* of Severus' death in 211: PM 1960, 213-18.

[20] Paola Ciancio Rossetto, "Rinvenimenti e restauri al portico d'Ottavia e in piazza delle Cinque Scole," *Bullettino della Commissione Archeologica Comunale di Roma* 97 (1996), 267-79.

[21] On the Crypta Balbi phasing, Daniele Manacorda, *Archeologia Urbana a Roma: Il Progetto della Crypta Balbi*, 1 (Florence: All'Insegna del Giglio, 1982), 19-20. On the Porticus Aemilia and others, G. Cressedi, "Contributo per la datazione della Forma Urbis," *Bullettino della Commissione Archeologica Comunale di Roma* 73 (1949-50), 91-95.

[22] Eva Margareta Steinby, "Il frammento 18a della *Forma Urbis Romae*," in *Lacus Iuturnae* I (Rome: De Luca, 1989), 24-33. However, the fragment includes features found only on the Severan Marble Plan, including the use of single rather than double lines to depict walls, and the use of inscriptions only for public buildings; Emilio Rodríguez-Almeida, "Euristica materiale e Forma marmorea. Alcuni falsi

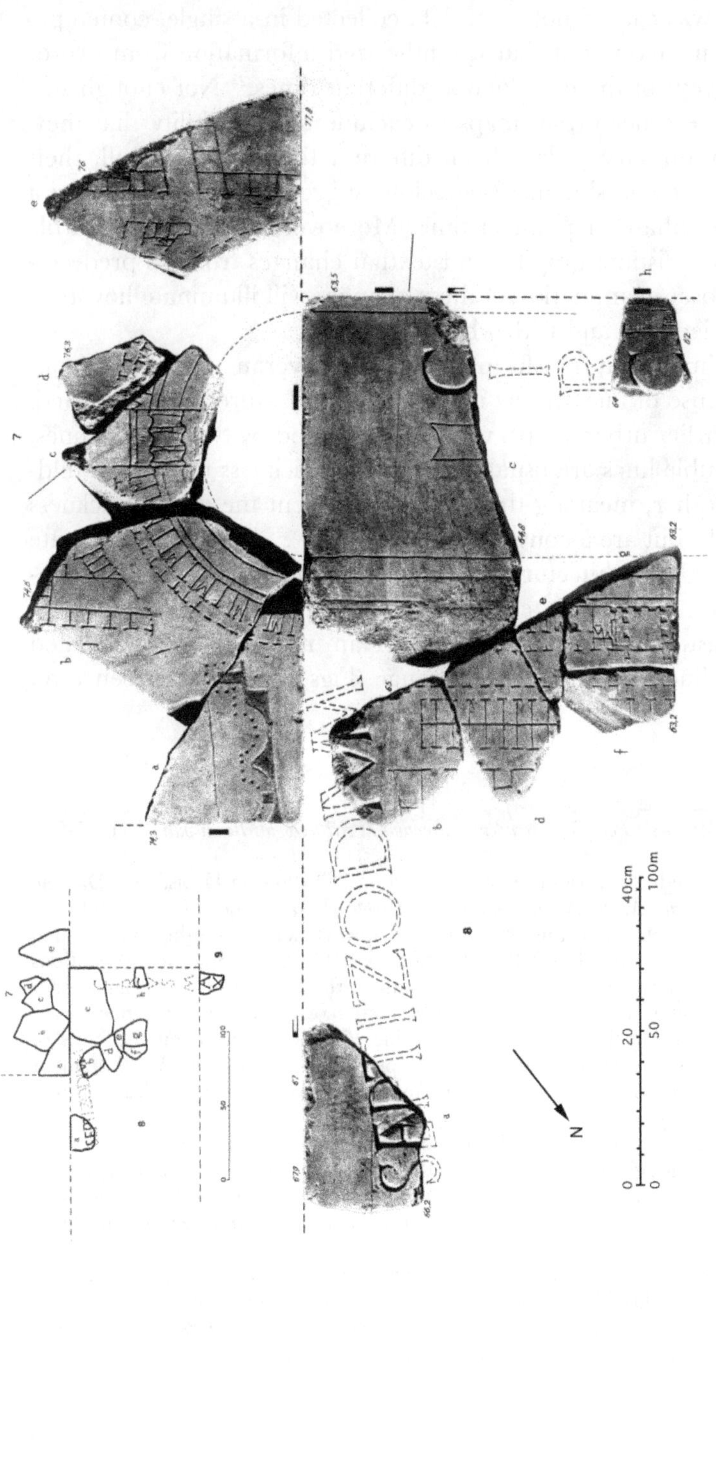

Fig. 5. Area of the Septizodium and Circus Maximus as depicted on the Severan Marble Plan, showing the multiple forms of visual emphasis given to the Septizodium, a dynastic monument dedicated by the emperor Septimius Severus in 203 CE: recessed double lines for the triple-bayed fountain and a large inscription extending almost twice as far as the length of the building itself, in addition to the plaza in front that marked the terminus of the Via Appia running downward from the top of the map. (Detail of PM 1960, pl. 17, reproduced by permission).

this map was carved not from data collected in a single, contemporary urban survey, but that it synthesized information from two or more surveys of the city done at different times.[23] Not enough survives of the other urban maps to exclude the possibility that they, too, drew on survey data from different time periods. Still, their inclusion of owners' names (see below) suggests a far greater interest in a single phase or point in time. Moreover, the Severan Marble Plan shows distinct graphic and textual changes from its predecessors. A closer examination of these changes will illuminate how they were carried out, and with what results.

A first important transformation on the Severan Marble Plan concerns the use of line and its impact on how features were depicted. On the earlier urban maps, walls are indicated by two parallel lines. These double lines are usually of uniform thickness from one building to another, meaning they do not represent the actual thickness of the walls but are a convention (Figs 2, 3, 4).[24] Single lines indicate other kinds of architectural features, including colonnade and sidewalk edges, individual steps, podium edges, and some thresholds.[25] By contrast, on the Severan Marble Plan, most walls are indicated by single lines, a fundamental change (Figs 1, 8). The presence on

problemi," *Atti della Pontificia Accademia Romana di Archeologia: Rendiconti* 68 (1995-96), 3-20.

[23] G. Cressedi "Contributo per la datazione della Forma Urbis," 95; Daniele Manacorda *et al.*, *Museo Nazionale Romano: Crypta Balbi* (Rome: Electa, 2000), 15. Rodríguez-Almeida thinks that the Severan map was made by synthesizing existing sector maps of the city: FUA 2002, 74. The new IRC-Oxford-Stanford excavations in the area depicted on fr. 18a, directly south of the Lacus Iuturnae, suggest that different phases were depicted on different parts of the fragment; hence this is additional evidence that the Severan Marble Plan was created on the basis of multiple urban surveys from different time periods. Jennifer Trimble, "Fragment 18a in light of the new excavations," in Andrew Wilson et al. (eds.), *Excavations in the Roman Forum post Aedem Castoris* (forthcoming).

[24] When these double-line walls intersect, either one set of lines continues across the other, as though the latter abutted the first, or the outlines connect, leaving uninterrupted empty space between, as though the walls joined. The evidence is too fragmentary to know whether this difference had architectural or legal meaning.

[25] Reynolds interprets some of these single lines at the edge of a colonnade as roof edges (Reynolds 33 and 70-71). However, this would mark an exceptional use of line and hence requires more secure evidence; Reynolds' cases of "roof lines" can more plausibly be understood as changes of level on the ground.

some fragments of faintly incised, double guidelines prove that this decision was proper to the making of the Severan map.[26]

One effect is of simplification and loss of information. Walls can no longer be distinguished at a glance from other features; thresholds now look like walls and must be eliminated if the doorway is to be seen; any meaning attached to different visual treatments of intersecting walls is lost. Similarly, doorway widths are simplified on the Severan map, in contrast to their different widths on, for example, the Via Anicia map (Fig. 4). In fact, the great Severan map appears less carefully carried out than its predecessors in many respects. Carvers of different parts of the map worked with varying degrees of attention and ability; for example, there is a marked contrast between the wavering lines and relatively careless execution of the Vicus Stablarius area in the southwestern Campus Martius (frs. 37f-i and 40a-h) and the clean lines and sharp details of frs. 11e-I on the slopes of the Viminal. Individual carvers also employed different versions of a given convention, especially visible in staircases and columns, suggesting that strict graphic consistency was not considered crucial.[27] Errors range from tiny and highly localized to large and visible. [28]

Simplification is not a sufficient explanation; this change also allowed new visual hierarchies to take cartographic form. Double lines exist on the Severan map, but are employed only for public buildings, most often the *cella* walls of temples (e.g. Fig. 6), but also some baths, spectacle buildings, and commercial structures. These double lines do not indicate a wall's actual thickness, but are a means of creating visual emphasis.[29] The surface within the double lines was

[26] PM 1960, 200 and pl. R, fig. 53.

[27] V-shaped staircases could have horizontal bars or none, as in the contrast between the empty Vs of fr. 10g in the Subura (Fig. 1) to the barred V's on frs. 37Aa-l in Transtiberim just below the Janiculum (Fig. 8). Columns rendered as squares sometimes have dots inside and sometimes not.

[28] On errors and corrections, PM 1960, 200 and pl. R; AG 1980, 44-53. Line segments are occasionally erased to create needed doorways; on the Temple of Minerva on the Aventine, the steps leading up to the podium have not been carved (fr. 22b); features that ought to be rectangular are occasionally rendered with non-right angles, probably where different survey segments were brought together (Fig. 1). The most serious error is the 21-degree misorientation of the Templum Divi Claudi.

[29] At a scale of 1:240, a 15mm-thick line would, implausibly, indicate a wall 3.60m thick (PM 1960, 201-202). More importantly, many large and thick-walled

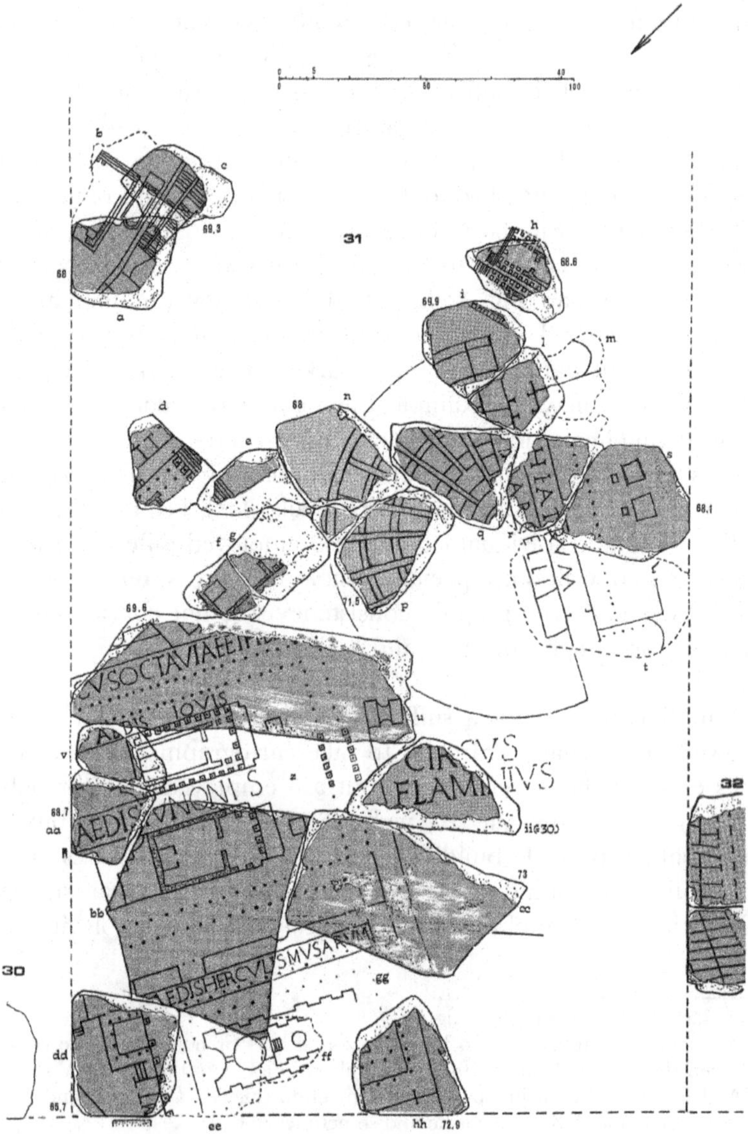

Fig. 6. The well-preserved slab V-12 of the Severan Marble Plan of Rome, showing different forms of added emphasis in an intensely monumental zone: the depiction of outlined seating sections on spectacle buildings rather than the standard ground-floor view, the frequent use of recessed double lines for temple *cella* walls and squares for their external columns, and the inscription of labels on many public buildings. Visible on this slab are two temples at the edge of the Capitoline hill at upper left, two temples in the Forum Holitorium at upper right, the Theater of Marcellus at middle right, the Temples of Bellona and Apollo Sosianus at middle left, and below them the Porticus Octaviae, including the Temples of Jove, Juno, and Hercules and the Muses. (AG 1980, pl. 23, reproduced by permission).

sometimes chiseled down, creating thickened, recessed bands that will have caught the light with stronger shadows and a more marked visual effect (Figs 6, 7).[30] However, not all public buildings were given double lines, and not all doubled lines were recessed in this way. For example, although all three surviving imperial bathing complexes are rendered at least partly with double lines, only the Baths of Trajan and Titus, which are adjacent to one another on the Oppian, show recessed areas (frs. 10wxy, 10z, 565 and frs. 110, 113). The Baths of Agrippa, some distance away in the Campus Martius, do not (fr. 38). Here again, the details of carving attest to a difference between generalized principles and their local execution, in this case, between the practice of highlighting of certain public buildings and the discretion of individual carvers in how that was actually done.

This selective use of double lines distinguishes the highlighted buildings from the architectural backdrop of the city, but finding a rationale behind this distinction is not straightforward. The few buildings highlighted in this way do not closely correspond to any known program of urban interventions carried out under the ruling emperor, Septimius Severus. Not all public buildings are treated this way, or even all examples of a given type. Moreover, different kinds of public buildings are graphically emphasized in different ways.

Some part of this visual emphasis follows precedent and may be conventional. Only one of the earlier map fragments includes a public building in sufficient detail: on the Via Anicia map, the Temple of Castor and Pollux in the Circus Flaminius is depicted quite differently from the surrounding buildings (Fig. 4). The temple's walls

---

buildings on the Severan map are rendered with a single line. Reynolds is surely right in concluding that these occasional double lines are normally employed to highlight certain buildings visually, and not to record the walls' actual thickness: Reynolds, 71-72, 77.

[30] If these doubled and recessed bands were originally filled in with red paint, as was long thought the case for all lines on the map, they will have stood out even more. However, analysis by the Capitoline Museums has shown that the surviving red paint within the incisions was added in modern times; Laura Ferrea of the Sovraintendenza Comunale, pers. comm. Color was visible on the fragments discovered in 1955-56 (PM 1960, 207, without stating exactly where). Red paint was applied in ancient times in at least one case: the surface of the monumental street along the north side of the Circus Maximus. FUR 2006, 15-16 and Fig. 2 (inv. 36395); for its placement, see David Koller and Marc Levoy, "Computer-aided Reconstruction and New Matches in the Forma Urbis Romae," in FUR 2006, 108-110 and figs. 8-11.

are significantly thicker than the fragment's other walls, and they include two rectangles with an X inside, incised on the *cella* walls to either side of the porch, perhaps windows. The columns of the porch are also indicated in unusual detail, as circles within squares. In this case, the temple is treated very differently from the surrounding, privately-owned architecture. On the Severan map, temples do not receive this level of detail, although their columns are often indicated by squares rather than dots, in addition to the double lines of the *cella* walls.[31] However, their visual emphasis vis-à-vis the surrounding architecture is similar to that seen on the Via Anicia map (compare Fig. 6 to Figs 1, 7 and 8). At the very least, the carving of the Severan Marble Plan included the decision to maintain this kind of distinction, even as the most fundamental way of depicting architectural features was changed.

Similarly conventional may be the treatment of spectacle buildings. These are also carved with double lines, but to very different effect. The Theater of Pompey, the Theater of Marcellus (Fig. 6), the Flavian Amphitheater (Colosseum) and part of the Circus Maximus are shown not at ground-floor level but by an abstracted overhead view of their tiers of seating.[32] In the theaters, polygons indicate the different seating sections, or *cavea*; in the Circus and Amphitheater, concentric rounded lines indicate the tiers of seating. No other Roman map of a spectacle building survives, so the relationship to existing graphic conventions is unknowable. However, these examples again suggest both an overall decision (whether new or conventional) to render these buildings in an overhead view, and individual carvers' decisions, visible in the slightly different ways each one is carried out.[33]

Public buildings other than temples and spectacle buildings are given double lines on the Severan map, but not with any perceptible consistency. The walls of at least one imperial display monument,

---

[31] Reynolds, 77, counts 28 temples treated in this manner in contrast to eight that are not; he points out that the latter are all exceptional in some other way as well.

[32] PM 1960, 201.

[33] Reynolds, 82-92, offers a detailed discussion of these differences; he includes the Ludus Magnus and the Ludus Dacicus. He speculates that the depiction of seating differs partly according to optimal audience viewpoints in theaters as opposed to other spectacle buildings, and partly because of an emphasis on relative rank among spectators.

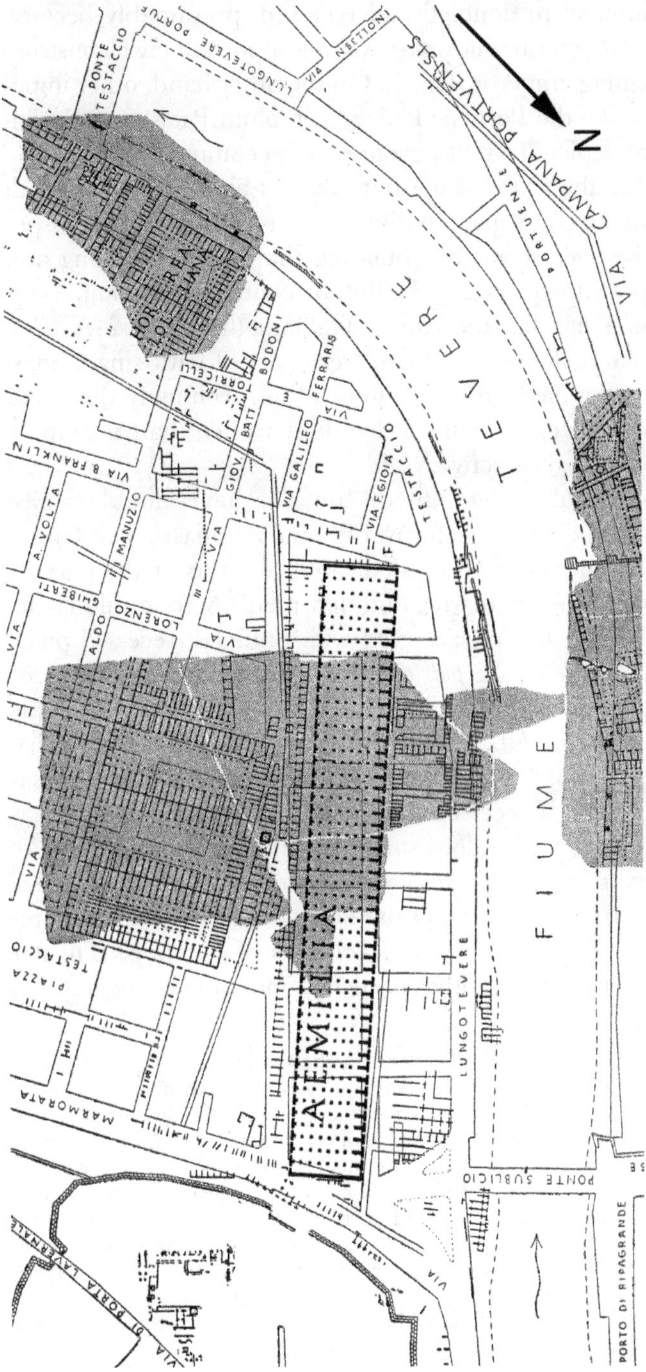

Fig. 7. The Emporium zone along the Tiber south of the Aventine. Against a white and black background showing the modern city streets, the shaded areas depict the surviving fragments and incisions of the Severan Marble Plan in this area. The exact function of the long, narrow, rectangular Porticus Aemilia, built in the early 2nd c. BCE, is not known, but it must have played a role in processing the huge quantity of goods and foodstuffs shipped to Rome. On the Severan map, it was given outlined and recessed piers and an inscription, and so will have stood out visually in an area of mostly single lines and no inscriptions. (Detail of PM 1960, 95, reproduced by permission).

the Septizodium, were doubled and recessed, presumably because this monumental dynastic showcase was an important civic construction of the reigning emperor (Fig. 5). On the other hand, other imperial complexes like the Porticus Liviae, Templum Pacis, and Forum of Trajan were depicted with single lines. One commercial structure, the Porticus Aemilia in the Emporium district along the river south of the Aventine hill, was given outlined and recessed piers, an especially distinctive treatment in a building that was 487m long and dense with supporting piers (Fig. 7). But, by contrast, other important warehouse complexes like the Horrea Galbana (frs. 24ac, 24A, 24B) and the Horrea Lolliana (frs. 25ab) were carved with single lines. The depiction of aqueducts varies most of all, seemingly the result of a variety of carvers' localized decisions rather than a graphic convention or unified directive.[34]

Indeed, most public monuments do not have double walls, although some received other forms of visual emphasis. The Circus Maximus included a triple arch dedicated to Titus at the curved, eastern end, here depicted with outlined piers. A room along the northern side of the Circus is carved with outlined, recessed piers; this probably represents the *pulvinar,* which held statues of the gods and was sometimes used as an imperial box (Fig. 5).[35] The Temple of the Deified Claudius (4b, 5a-h) and at least two other, unidentified buildings (11fg, 219b, 232abc) have what look like built garden features, depicted with a recessed double line. Columns, depicted as squares rather than simple dots, sometimes monumentalize parts of larger complexes, including the gateway to the Porticus Octaviae (Fig. 6), the central colonnades of the Porticus of Pompey's Theater (frs. 38a, b, d), and the *scaenae frons* columns of Pompey's Theater (38e). Such nodal locations within particular monuments suggest that

---

[34] A portion of the Aqua Claudia between the Caelian and Palatine hills is represented with squares for the piers with pairs of concave lines between to represent the arches (fr. 8i). The Arcus Neroniani, an offshoot of the Aqua Claudia that brought water to the Templum Divi Claudi on the Caelian hill, is rendered in a mix of plan and elevation views, including squares for the piers alternating with concave lines, but also dashed lines along the line of the aqueduct—the standard depiction of an arcade on the map—and short, perpendicular lines of different lengths extending off to the right (fr. 4a). The Aqua Alsietina, by contrast, seems to be shown entirely in elevation (frs. 37Babcdefg, 37Cabc). PM 1960, 201; AG 1980, 144-45.

[35] Paola Ciancio Rossetto, "Circus Maximus," in E.M. Steinby (ed.), *Lexicon Topographicum Urbis Romae* I (Rome: Quasar, 1993), 272-77.

these are deliberate points of emphasis, not random variations in the use of outlined details.[36]

To sum up: the Severan Marble Plan is unique among the known Roman urban maps in its use of single lines to depict walls, and its only occasional use of double lines. This practice highlights certain public buildings while subsuming everything else, including almost all private property, into the city's background mass of architecture. There is no single explanation encompassing the buildings so treated: some are ideologically important (the Septizodium), some seem at least partly motivated by convention (the depiction of temples), others are emphasized at the discretion of their particular carvers. Process is thus an important factor, and the map's overall effect is of a loose plan carried out in many different ways. Scholarly interpretation must therefore be cautious; any single element on the map, although belonging to its overall conception and effects, may be the result of a very localized carving decision and not of a high-level plan or systematic agenda.

A second unique aspect of the Severan Marble Plan is its changed use of inscriptions. The other urban maps are characterized by two kinds of inscriptions: owners' names, and length measurements. Except on the Isola Sacra map, individual buildings and properties are labeled with names in the genitive (Figs 2, 3, 4). Interestingly, the only full example of a public building is inscribed in the same way as the individually-owned properties surrounding it. On the Via Anicia map, Castor and Pollux are named within their temple in the genitive, like Cornelia and her associates across the row of *tabernae* directly below (Fig. 4). The gods' names, like everyone else's, are also subject to the constraints of space: POLLVCIS is written in much smaller letters than CASTORIS to fit into the small spaces on either side of the cult statue base along the back wall of the *cella*. On these maps, the owners' inscriptions were added after the features were incised, seemingly by different carvers, and were normally placed in

[36] Similarly, features at the corners of the courtyard of the Ludus Magnus are depicted in outline form (6bf). The squares indicating piers on one of the map's aqueducts were recessed (fr. 517d, now identified by Pier Luigi Tucci as part of the Aqua Marcia just north of Trajan's Forum: "L'Arx Capitolina: tra mito e realtà," 63-73). The edge of the Capitoline (frs. 31ac, Fig. 6) seems to include a retaining wall depicted with a double line. Outlined details were added to additional fragments as well.

the center or upper left of the available space.[37] A name could cross room boundaries to suggest the extent of that owner's property (Fig. 4). The names also vary considerably in size, seemingly depending on the space available, not the importance of the property or its owner. These names do not add to the legibility of the architectural features, but provide an additional layer of information about ownership.

Moreover, the urban maps—except the Severan—include numerals corresponding to lengths in Roman feet (Figs 2, 3, 4). These are inscribed along stretches of external walls or street frontages: on the Via Anicia plan, the measurements are found along different segments of the riverfront colonnade.[38] Like the names, these numbers are cartographically unnecessary but support an emphasis on the relationship of individual properties to one another within the urban fabric. The force of all these inscriptions is cadastral; they point to an administrative function for these maps related to the precise recording of property ownership. Commentators have further pointed out that this form of mapping is time-specific; any change in the size or shape of the properties (e.g. through sale, extension, or collapse), or in the identity of the owners would make such a map obsolete.[39]

In physical respects, the Severan map's inscriptions are similar to those on the earlier urban maps. They were carved after the Marble Plan's 150 marble slabs had been mounted on the wall and after the features had been incised, probably by different carvers.[40] They were oriented for legibility while following the lines of the labeled struc-

[37] These names were added after the cartographic incisions were done, as is clear from the placement of CORNELIAE ET SOC. on the Via Anicia plan and other similar instances. The inscriptions sometimes employ a deeper line, suggesting that they were incised by different carvers (FUA 2002, 43). On the Via Polveriera map, the letters slant so as to compensate for the angle of the depicted features, indicating that the inscriptions were carved after the slab had been mounted on a wall.

[38] They may have corresponded to those riverfront segments that the corresponding property owners were responsible for maintaining: FUA 2002, 43-44.

[39] Rodriguez-Almeida argues that the Isola Sacra map, because it has length measurements but no owners' names, is functionally removed from this sense of the here and now; FUA 2002, 58.

[40] PM 1960, 167. The slope of the inscribed letters tends to compensate for the angle of the inscription relative to the horizontal, showing that they were incised once the slabs were mounted on the wall; FUA 2002, 74.

ture; when this was vertical or nearly so, the letters could themselves be arranged vertically, as in the Circus Maximus (Fig. 5). They were often placed at center or upper left (e.g. on Fig. 7). The size of the letters varies from 20 - 85mm;[41] these different sizes mostly depend on the space available. For example, the Temple of Jove within the Porticus Octaviae was labeled AEDIS IOVIS in comparatively small letters that fit the narrow space (Fig. 6).

In their content, however, the inscriptions on the Severan map are fundamentally different. They include neither owners' names nor any measurements (Figs 1, 5, 6, 7, 8, 9).[42] Instead, they label some, but not all, public buildings: certain temples, entertainment structures, imperial display monuments, major bath buildings, aqueducts, roads, and large warehouses, as well as some more minor sites. An immediate implication is that the Severan Marble Plan was not cadastral. The very inscriptions that allow the other urban maps to be interpreted as administrative documents concerning property ownership are absent; moreover, the removal of specific owners' names in favor of unchanging public building names removes much of the city from a specific sense of time.[43] Beyond this key implication, however, and generalizations about the map's public emphasis, previous commentators have said little about how the Severan map's inscriptions actually work, and how they relate to this map's other transformations of urban mapping conventions.

Before exploring those questions, an important consideration emerges here again; as in the graphic highlighting of certain buildings, different levels of decision-making were visibly at work. There was clearly an overarching directive to eliminate private ownership inscriptions and measurements, and to label public buildings instead. There is also evidence of individual and localized decisions about

---

[41] PM 1960, 167.

[42] Personal names are sometimes used, but as identifiers of the buildings. Private owners are not inscribed, with a possible exception on fr. 3A, whose inscription has been reconstructed as [DOMVS C]ILONIS and might therefore label the house of Lucius Fabius Cilo, the urban prefect at the time (first proposed in PM 1960, 215-16). However, the fragment itself is lost, and survives as no more than a sixteenth century drawing which records only ILONIS and may not be entirely accurate, especially at the initial I at the very edge of the drawn fragment; this reconstruction is tenuous, therefore.

[43] Reynolds, 115-27 and Rodriguez-Almeida in FUA 2002, 4, 72-74 have made this point compellingly.

whether to label a particular structure, and how. All the largest public monuments seem to have been given inscriptions, but the labeling of smaller or more obscure ones shows no unified or systematic rationale. On slab XI-6 at the top of the map, for example, not only the broad Via Appia (this may be the Via Nova instead), but also the MVTATORIVM and AREA RADICARIA are labeled (frs. 1a-e). Although the evidence is fragmentary, other places on the map do not seem to have been labeled in as much detail. Moreover, the content of an inscription sometimes appears to have been determined separately from the depictions on the map. Both the Porticus Aemilia (Fig. 7) and the Templum Pacis (frs. 15abc, 16a) have ample space for long inscriptions but are inscribed at upper left, respectively, with the comparatively abbreviated [AEMI]LIA and PACIS. There are localized changes of plan too; the letters of the BAL(nevm) SVRAE inscription are 39mm tall, although the inscription's guidelines are only 32mm apart (frs. 21abcd).[44] All this means that within the general plan there was scope for individual details and localized decisions; not every instance was significant or meaningful in the same way or to the same degree.

A third way in which the Severan Marble Plan is unique concerns the interactions of image and text across the map's surface. Like the changed use of line, the changed use of inscriptions here entailed not only a simplification and loss of information, but also new visual emphases and effects. These inscriptions, with their relative freedom from topographic constraints, played an important visual role in shaping space. The map's cartographic content was largely determined by the conventions of the urban mapping tradition and by the source material; every wall and built feature was included, and this in itself did not leave much leeway for manipulations of visual effect. Some graphic highlighting and shaping of emphasis, even if not of content, was done through the selective use of double lines, recessing, and other features, as seen above. The inscriptions offered even more leeway in characterizing both specific places and larger

[44] PM 1960, 201, n. 21; there is a list of guidelines for inscriptions ibid., 167. Odd spellings may also fit here: the Temple of Minerva on the Aventine is inscribed MINERBAE (fr. 22b), and the branch of the Aqua Claudia bringing water to the Templum Divi Claudi is labeled AQVE/DVCTI/VM (fr. 4a). Rodríguez-Almeida has argued that some of these odd spellings may reflect contemporary pronunciation at Rome (AG 1980, 26).

areas in the city. This flexibility can be seen in individual decisions about how to label a building, in combinations of inscriptions with other forms of highlighting in a particular area, and in the characterization of entire areas of the city through the use of inscriptions.

One way in which the inscriptions affected the cartographic content was by emphasizing the unity of a given structure or complex. For example, the words HORREA LOLLIANA were incised within the bounds of a large, double warehouse structure along the Tiber (frs. 25ab). The words could have occupied one courtyard, extended evenly across the structure, or have been written one in each courtyard; in fact, each word was divided between both courtyards, with HOR/LOL in one and REA/LIANA in the other. Interacting in a different way were the two inscriptions labeling the Theater and Porticus of Pompey in the Campus Martius; HECATOSTYLVM runs downward to the right along the lower edge of the Porticus (frs. 39a), leading the eye to the horizontal THEATRVM [POMP]EI (frs. 39ef) and around the outside of the theater, unifying the entire complex. These highly localized decisions may, or may not, have been deliberately executed with this goal in mind, but the visual effects are similar.

Inscriptions could also be combined with other features to create focal areas. The Septizodium, a monumental, three-storey fountain dedicated by Septimius Severus in 203 CE as a dynastic showcase, stood below the southeast corner of the Palatine as the Roman terminus of the Via Appia (Fig. 5). On the Severan map, the monument's three bays were depicted with a recessed double line (frs. 8ab and 7ab). The associated inscription was one of the largest on the map, and—uncharacteristically for its inscriptions—was almost twice as long as the monument it labeled, with letters about as high as the building itself was deep. Just above the Septizodium was an inscription naming Septimius Severus and Caracalla.[45] An array of associated features, whether coincidental or deliberate, added visual emphasis. The Circus Maximus lay directly adjacent, with an even

[45] SEVERIETAN / TONINIAV[GG] / NN, "of Severus and Antoninus [i.e. Caracalla], our emperors" or "in the name of our emperors," perhaps labeling an adjacent monument whose depiction does not survive. Depending on the translation, this may be an imperial exception to the lack of personal ownership information on this map. With the Septizodium, it offers modern scholars the second internal piece of dating evidence, because Septimius Severus died in 211.

larger inscription running vertically down its arena. Directly above, the Via Appia ran downward from the top of the map to a large plaza directly in front of the Septizodium, drawing the eye to this key point.[46] Finally, a roughly horizontal band of important inscriptions crossed the map at this level (the third and fourth rows of slabs from the top), and may also have helped draw the eye to the Septizodium. The largest of these inscriptions, in combination with the visually most distinctive spaces, are the Porticus Liviae, Thermae Traianae, Amphitheater, Ludus Magnus, Templum Divi Claudi to the left of the Septizodium and Circus; there are smaller inscriptions along the same band as well. To the right (unfortunately, most of the map is missing here), was the inscription labeling the very large complex of the Praedia et Horrea Galbana.

A very different effect was created in areas without inscriptions or visual highlighting. The largely commercial and residential buildings in Transtiberim, the right bank of the Tiber, contain no known inscriptions, double lines, or other visual emphases (Fig. 8). In this context, the label associated with the Aqua Alsietina will have stood out partly because of its isolation. Across the river, the Porticus Aemilia was visually emphasized not only by its great size, inscription, and outlined and recessed piers, but also because the riverside warehouses and commercial installations immediately around it were not highlighted or inscribed (Fig. 7).

These visual effects extended to entire areas of the city. Of course, given the architectural mapping conventions adopted by the Severan map, different zones were already distinguished simply by different building sizes and combinations of space. Depictions of hillside areas with narrow, winding streets (e.g. Fig. 1 at right) looked different from those that mapped relatively open, rectangular blocks on flat land (Fig. 8, top). Areas containing warehouses or open industrial spaces (Fig. 7) were visually different from packed apartment buildings (Fig. 8, bottom left). Nonetheless, this map's use of public inscriptions, in conjunction with other forms of visual emphasis, added a strong characterization to certain parts of the city over and above their cartographic differences.

---

[46] Jennifer Trimble, "Rome as Souvenir: the Septizodium and the Severan Marble Plan," in C. Mattusch et al. (eds.), *Common Ground: Archaeology, Art, Science and Humanities. Proceedings of the XVI International Congress of Classical Archaeology, Boston, August 23-26, 2003* (Oxford: Oxbow Books, 2006), 106-109.

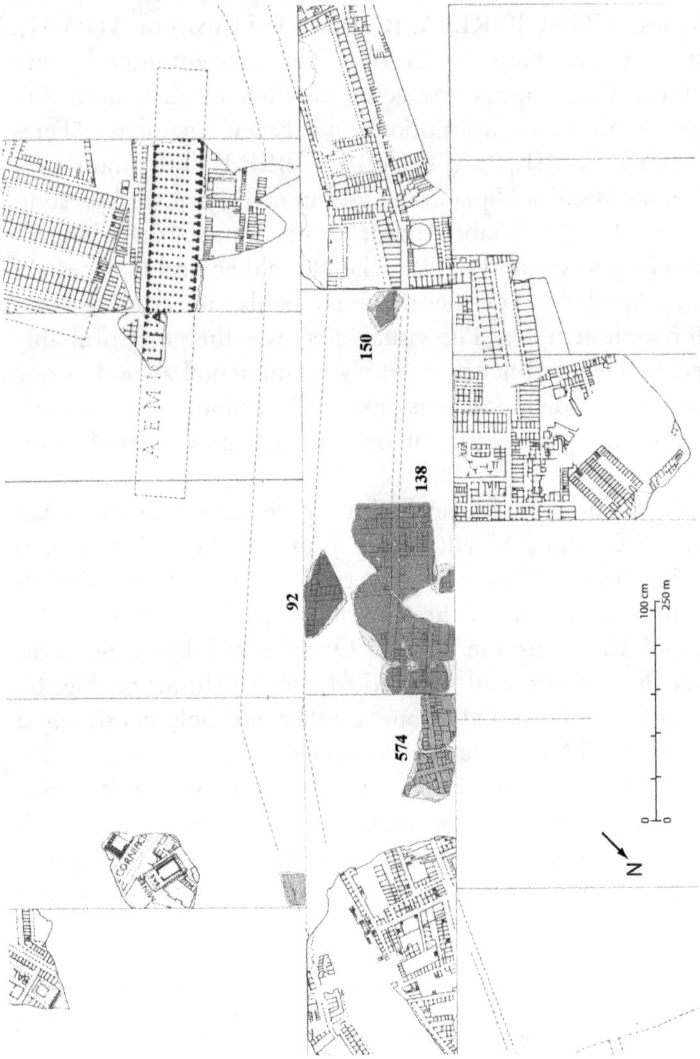

Fig. 8. Identified parts of the Severan Marble Plan in Transtiberim near the river, including recent fragment locations (see Tucci 2004 on frs. 138a-f and 574b, and Koller and Levoy 2006 on frs. 150 and 92). On the right bank of the Tiber, opposite the Porticus Aemilia, the southern part of the city included warehouses and other commercial structures flanking the Via Campana-Portuensis, which ran parallel to the river. (Image by David Koller, after PM 1960 and AG 1980).

For example, little survives of the Roman Forum area on the Severan map (Fig. 9).[47] Even so, on the surviving fragments, there are no fewer than eight inscriptions: [B]ASILIC[A AEMILIA], [R]EGIA, (Aedes) [C]ASTORIS, B[ASILICA J]VLIA, (Aedes) [SAT]VRNI, (Aedes) [CONC]ORDIA, [G]RECOST[ASIS or ADIVM], IANVS (frs. 16e, 17, 18a-g, 19, 212b).[48] The adjacent imperial fora are even more fragmentary respective to their original area, but again there are numerous inscriptions on very few fragments: (Templum) [PA]CIS, BASIL[ICA] VLPIA, LIBERTATIS (naming a hemicycle of the Basilica Ulpia in the Forum of Trajan), and presumably more, now lost (frs. 15abc, 16a-d, 29a-h). None of these inscriptions is especially large, and the associated architectural features are not otherwise highlighted besides the use of double lines and squared columns for some temples. The visual effect was therefore probably of a densely labeled and hence intensely monumental zone, but one with no particular individual focus or overall emphasis—a venerable and monumental space, but not one of this map's visual "hot-spots."

A similarly monumental aspect with a different visual effect can be seen in the Campus Martius (Figs 6 and 9). This area, too, is dense with inscriptions. The well-preserved slab V-12 alone, depicting the southernmost part of the Campus, contains inscriptions for the Theater of Marcellus, the Porticus Octaviae et Filippi, the Aedis Iovis, the Aedis Iunonis, and the Aedis Herculis Musarum (Fig. 6). The latter four are stacked atop one another and only gently tilted off the horizontal within available colonnades and passageways inside the Porticus Octaviae. Not all nine temples visible on the slab have inscriptions, but most are given double and recessed walls, as well as squares for the external columns. The Theater of Marcellus is depicted with outlined seating sections, adding to this area's dense

[47] A great hole was punched through the wall in this area, perhaps seriously damaging the map already in the fifth century CE (AG 1980, 39-41). However, it now seems more likely that the wall was already denuded at the time this hole was made, perhaps in the eighth or ninth centuries. Riccardo Santangeli Valenziani, "Distruzione e Dispersione della *Forma Urbis* Severiana alla Luce dei Dati Archeologici," in FUR 2006, 53-59, esp. 56-57.

[48] Lawrence Richardson has reconstructed [ ]ORDIA on fr. 19, which survives only as a Renaissance drawing, as [VENVS VERTIC]ORDIA instead, which would mean repositioning the fragment. Lawrence Richardson, "The Approach to the Temple of Saturn," *American Journal of Archaeology* 84 (1980), 51-62.

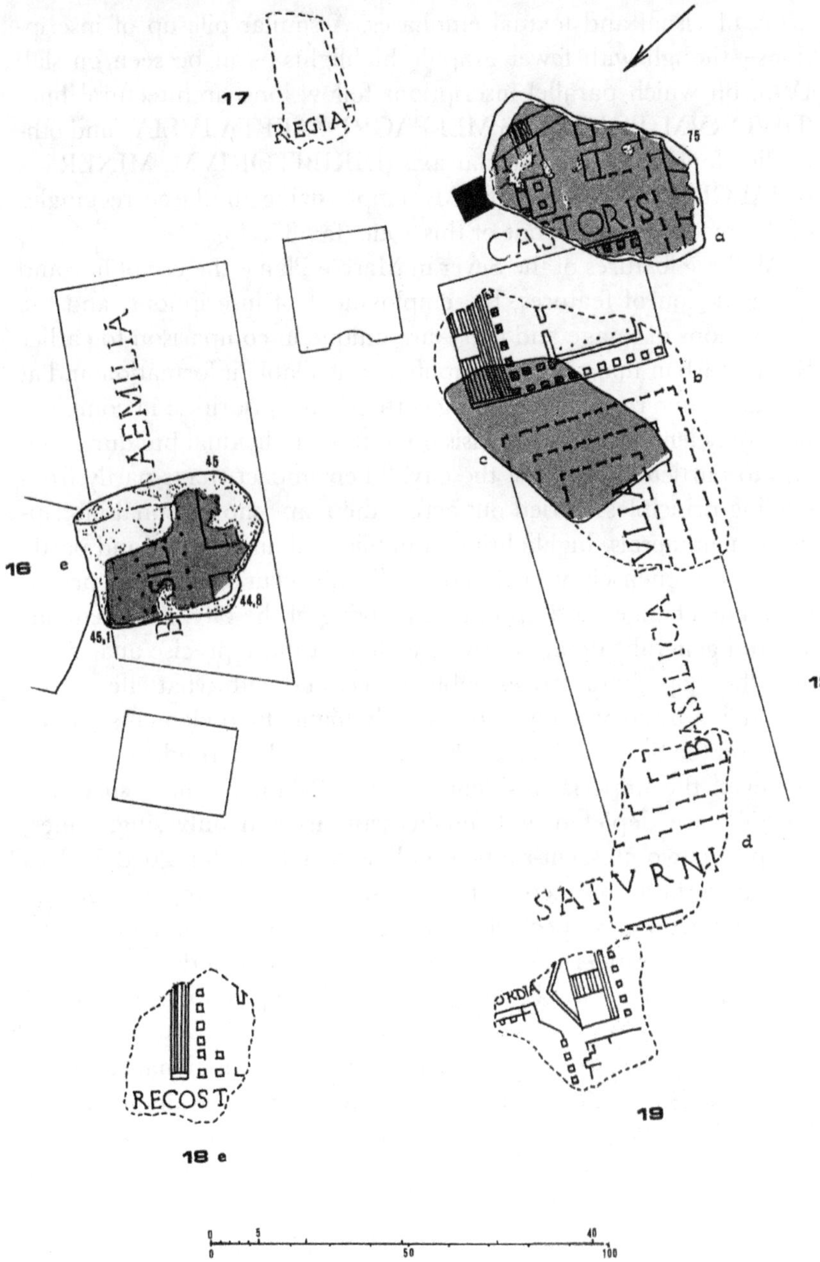

Fig. 9. The area of the Roman Forum. This part of the map was heavily damaged in antiquity, but the density of the inscriptions is clear. (AG 1980, pl. 13, reproduced by permission).

array of visual and textual emphases. A similar pile-up of inscriptions—though with fewer graphic highlights—can be seen on slab IV-5, on which parallel inscriptions follow long architectural lines (DIVORVM, PORTICVS MELEAGRI, SAEPTA IVLIA), and others lie along the perpendicular axis (DIRIBITORIVM, MINERVA CHALCIDICA, SERAPAEVM), emphasizing the large rectangles of the monumental layout of this zone (frs. 35, 36).

All these features of the Severan Marble Plan—the use of line and the depiction of features, the employment of inscriptions, and the interactions of image and text—are unique in comparison to earlier Roman urban maps. They simplify the available information and at the same time rearrange existing cartographic priorities; in combination, they lend special emphasis not only to individual buildings, but also to particular zones of the city. Their impact stems partly from guiding principles carried out across the map—no personal inscriptions or numerals, highlighting of public buildings only—and partly from the immensely varied and localized execution of those principles. The changes add up to a re-imaging of the city. This remains a bland generalization, however, without a more precise analysis of how the map characterizes public space, and with what effects.

Judging by its visually and textually highlighted elements, public space on the Severan Marble Plan does not include residential space, not even the imperial residence on the Palatine, whose surviving portions are depicted with no inscriptions and only single lines, except for two crescent-shaped outlines visible on frs. 20ad. It does include certain commercial buildings, as well as temples, political and legal structures, spectacle buildings, baths, streets and aqueducts, not to mention imperial monuments. Nor does it include every example of these categories. Public space on this map is not a typological category.

Instead, the map constructs public space as relational. Emphasized structures are characterized in terms of the surrounding areas; these buildings do not exist in a cartographic vacuum, but are highlighted in terms of Rome's architectural fabric as a whole. The effect is realized in different ways. Some buildings are more "public" than others in that they are given larger inscriptions, stronger graphic emphasis, or added visual connections to nearby structures. Graphically and textually, the Porticus Aemilia and the Septizodium are similarly highlighted, but with very different contextual effects. The first is visually isolated within the warehouse district south of the Aventine, while the second is embedded in a monumental zone filled with different kinds of visual emphases (Fig. 7).

Public space on this map is also relational in time, articulated here in a flattened social hierarchy. Private properties, with neither inscriptions nor visual highlighting, are both anonymous and synchronic. Unlike their depiction on the earlier cadastral maps, they are not identified by specific owners or located in any particular time, and no distinction is made in this regard between tenements and luxurious houses, or between rich and poor residents (Figs 1, 7, 8). Likewise, many small public buildings and neighborhood facilities are neither named nor visually emphasized, although some are. Some major temples and entertainment buildings are visually distinguished against this background, but with their unchanging names are still relatively synchronic (Figs 6, 7). Very different are the brand-new Septizodium and the inscription naming the current emperors (Fig. 5). These are neither anonymous nor unchanging, with their visual and epigraphic emphases, and their explicit link to the current emperors and a particular time. In short, public space on this map is characterized not in terms of isolated monuments or types of space, but rather within the city's entire urban fabric, in terms of a social hierarchy of space and time.

One effect of this relational mapping emerges from consideration of its viewers' textual and cartographic literacy. In contrast to the familiarity of map reading in modern industrial societies, cartographic literacy was surely rare in Rome. It is impossible to know exactly how many people in early third century Rome could read at all. William Harris, in the fullest existing study, estimates that perhaps ten per cent of adult males in the Empire could read and write.[49] Other scholars, while generally agreeing with this low number, have emphasized the many levels and practices of literacy other than full fluency in reading and writing.[50] These included the ability to read simple block letters but not cursive writing, as well as recognition of the role and nature of a piece of writing even if its detailed content was not decipherable (an election slogan, for example, or honorific

[49] William Harris, *Ancient Literacy* (Cambridge, MA: Harvard University Press, 1989).

[50] Several responses to Harris' study are collected in J.H. Humphrey (ed.), *Literacy in the Roman World*, *Journal of Roman Archaeology Supp.* 3 (1991).

inscription). People had access to written content through the help of a relative or associate, or within a crowd of onlookers in which one person read aloud.

The Severan Marble Plan attests to a comparatively broadened readership. All its predecessors (except the Isola Sacra map) include an owner's name on every property (Figs 2, 3, 4), creating a mass of unfamiliar written detail that will have required some degree of reading ability to decipher. Indeed, these cadastral maps were relatively restricted and specialized in function, and must have been primarily used by—and legible to—those most concerned with property administration: owners and bureaucrats. On the Severan map, the use of inscriptions fostered a broadened cartographic as well as textual literacy. These inscriptions were far fewer; they were often larger because of the increased space available within public buildings; they were mostly short, and written primarily in the nominative, not the genitive (Figs 5, 6, 7, 9). Moreover, the largest and most familiar labels were associated with distinctively-shaped spaces—the enormously long upside-down U of the Circus Maximus, the large oval of the Flavian Amphitheater (Colosseum), or the semi-circles of the Theaters of Pompey and Marcellus.

In this way, the city's best-known public complexes were made the most recognizable to the greatest number of viewers. Their recognition did not require extensive familiarity with cartographic conventions. These easiest-to-read combinations then showed viewers how to decipher other sites on the map, at least to some degree. Any viewing of the map in groups meant that decipherment of the different inscriptions could be shared. Moreover, once viewers grasped that the inscriptions labeled public structures, they did not need to decipher every single one to understand the general import of the map's inscriptions. In these ways, the Severan map's emphasis on public buildings lowered the literacy requirements of the earlier urban maps, broadening viewers' access to the highlighted information. This broadening, through every new viewing, reading aloud of an inscription, or identification of a structure, also confirmed the map's particular presentation of the city.

All this strengthened the Severan map's emphasis on collective experience and spectatorship. The inhabitants of imperial Rome surely maintained personal mental mappings of the city—their daily routes, good spots in the neighborhood to obtain work or foodstuffs, certain places marked dangerous and to be avoided—built from an array of lived experiences and spatial knowledge. The Severan Marble Plan instead emphasized a single, shared experience of space. Its highlighted places were known through spectacles in the theaters or Circus, processions and group rituals focused on the temples, the movement of many people along major roads, the public fountains and water spectacles supplied by the aqueducts, the large number of goods and foodstuffs that passed through the Porticus Aemilia and other warehouse structures.

In these ways, the map's viewers were constructed as part of a public, collective audience with a shared experience of the city, not as individuals or small groups with diverse and personal perceptions. This was by no means a democratic vision of the city. Those viewers were not constructed as collective owners of Rome, or as equals within it. This map lowered the required levels of literacy in comparison to other urban maps, but it did so to involve the viewer in a fundamentally hierarchical characterization of Rome. Viewers did not participate in naming, shaping or changing the city's space; rather, Rome was delivered to them in a particular and spectacular form, and they were visually posited in turn as collective, admiring, and reactive. In an ancient forerunner of modern fascism, this map exalted Rome in a way both populist and authoritarian.

Thus an examination of the Severan Marble Plan's transformations of existing mapping conventions clarifies how it worked as a map, and with what implications. The dominant scholarly approach has been to focus entirely on this map's information content, treating it as a transparent and objective topographic record of the city's buildings. This approach is possible—and remains fruitful—precisely because of the high quality of the information employed in creating the map. However, the Severan Marble Plan also represents the monumentalization of information. It allows us to glimpse the processes and decisions at work in that process—among them, the interactions of existing cartographic conventions, directives to take key steps differently, and localized decision-making by individual carvers. Altogether, these resulted in a vision of the city that was comparatively simplified, but that also reconfigured the older maps' visual hierarchies of space and persons. In turn, the map mapped its viewers and their responses, flattening out social hierarchies below the imperial level, emphasizing collective experience of the city's public structures, and reaffirming authoritarian control over the city's inhabitants as well as its spaces.

CONSTRUCTING A DIGITAL EDITION FOR THE PEUTINGER MAP[1]

Tom Elliott

The preparation of scholarly editions of primary sources has long been a fundamental activity of classical scholarship. For literary and documentary works, an edition comprises the systematic presentation of a text, carefully constituted from available exemplary manuscripts and variously augmented by scientific apparatus, textual and historical commentary, translation and imagery. Without proper editions for seminal works, historians and philologists are hobbled in their inquiries, whether the goal is synthetic historical analysis, thematic literary study or linguistic reference. A surprising number of works from classical antiquity still lack satisfactory modern editions.

An outstanding example of a work too long ill-served is the large-format, small-scale world map known commonly as the Peutinger map (see Fig. 1).[2] A seamless cartographic composition, it originally spanned the limits of Rome's claims (or dreams) of control and influence from the Atlantic Ocean to Sri Lanka. This precious artifact, held since 1738 in the Austrian National Library in Vienna, constitutes the only extant evidence for Roman mapping at this combination of scale and geographic extent.[3] The surviving copy was probably produced around AD 1200. No statement concerning the identity of

[1] The author wishes to express his deep appreciation to the members of the Peutinger Map Project team at the Ancient World Mapping Center at the University of North Carolina in Chapel Hill: Nora Harris, Andy Hull, David O'Brien, Graham Shepherd and Gannon Hubbard. Without their hard work, imagination and dedication, there would have been no paper to write. Moreover, the advice of Richard Talbert, who leads the project, has been essential at every stage. Where I correctly characterize the Peutinger Map and its significance, I owe my understanding to him.

[2] The Latin moniker "Tabula Peutingeriana" is a modern invention. Whatever title and summary components might have originally accompanied the map were lost before its modern appearance.

[3] For an overview of the map's history to c. 1600, see P. Gautier Dalché, "La trasmissione medievale e rinascimentale della *Tabula Peutingeriana*," in F. Prontera (ed.), *Tabula Peutingeriana: Le Antiche Vie del Mondo* (Florence: L. S. Olschki, 2003), 43-52.

Fig. 1. A small extract from the Peutinger Map, from the 1888 photographs by Carl Angerer and Alexander Göschl. *Peutingeriana tabula itineraria in Bibliotheca Palatina Vindobonensi asservata nunc primum arte photographica expressa* (Vienna, 1888).

the copyist or the original cartographer is incorporated; nor is any clue offered to the location or date of its initial composition. The original map is commonly held to have been a Roman production, originating perhaps in the fourth or fifth century. A serious recent opponent of this view, however, has argued that its composition may date to the Carolingian age or even later.[4] At least there can be no dispute that its maker drew on a vast store of now-lost historical material dating as far back as the first century AD.

Given the extent of terrain that the map aims to represent, its most striking aspect is the elongated frame, which originally measured no more than 13 inches (34 cm) in height by nearly 28 feet (850 cm) in width. The rolled map was already missing its left-hand end when it came into the possession of Konrad Celtes sometime around 1500. He later bequeathed it to his friend Konrad Peutinger, whence its name. In 1863, the roll was separated into its eleven parchment segments. These divisions hold no significance for the map's content, however; the size of a parchment sheet was dictated by the skins available and by the working needs of the copyists.

The selection of such an oddly unbalanced frame necessitated a sophisticated and flexible approach to the selection and layout of the map's elements. The difficulty of this task was further increased by evident design dictates which place Rome at the center of the composition and incorporate an extensive collection of routes. Altogether these link nearly 2,700 places from the Roman empire and beyond to the east, although the geographic distribution is very uneven. Through the presentation of physical landscape and the placement of principal settlements, the mapmaker demonstrates a geographical awareness that must derive from the Hellenistic tradition inspired by Eratosthenes. Nonetheless, the challenges presented by the map's unique form require a fluid approach to such "concrete" issues as mathematical projection, consistency of scale, orientation of the compass points, and preservation of shape or area. In particular, most bodies of water are compressed to mere ribbons of green, and land forms with extensive north-south axes are rotated and elongated as necessary to accommodate both the map frame and the desired cul-

[4] E. Albu, "Imperial Geography and the Medieval Peutinger Map," *Imago Mundi* 57 (2005), 136-48, with dissent by B. Salway, "The nature and genesis of the Peutinger map," *Imago Mundi* 57 (2005), 119-35.

tural and physical features. Thus the Italian peninsula appears to tip progressively more on its side as it unfolds from geographic north to south. Moreover, Italy consumes far more space on the map than the geographically more extensive region of Persia to the east.

Any historical map constitutes a challenge for the scholar contemplating an edition. As graphic depictions of geographic information, maps are generally large-format artifacts incorporating significant small details. Accordingly, photographic and facsimile editions are regularly employed in communicating important historical maps to a wider audience, and these often need to be produced at a scale of 1:1. The Peutinger map has received such treatment, beginning with a series of widely reproduced engravings, essentially hand copies.[5] In 1888, a high quality photographic edition was produced by Angerer and Göschl in Vienna.[6] From the outset, this was a rare, expensive specialty item, and in recent decades it has proved virtually inaccessible, since only a small number of libraries worldwide hold a copy and these holdings are largely unrepresented in online catalogs. It was not until 1976 that a color replacement appeared, accompanied by a modest commentary and gazetteer.[7]

Although a map is a form of graphic communication, it is nonetheless usually replete with textual material as well. Consequently, extensive commentary is necessary to identify and contextualize the features and text presented there. A traditional printed edition, with its apparatus and commentary, typically performs this task well for textual sources, but leaves much to be desired with respect to maps. The chief difficulties arise from the print format itself. An edition of a text can readily follow the ordered, one-dimensional organization of the work. By contrast, textual treatments of maps must impose this structure on a work that is spatially ordered in at least two dimensions. Linear and regional features (like rivers and territories) create their own, separate orderings in space. Networked features, too, like the route network in the Peutinger map, cannot be reduced

[5] Reproductions of the map since 1753 are treated in an Appendix (pp. 132-41) to R. Talbert, "Cartography and taste in Peutinger's Roman map," in R. Talbert and K. Brodersen (eds.), *Space in the Roman World: its Perception and Presentation* (Münster: LIT: 2004), 113-41.

[6] *Peutingeriana tabula itineraria in Bibliotheca Palatina Vindobonensi asservata nunc primum arte photographica expressa* (Vienna: Angerer and Göschl, 1888).

[7] Ekkehard Weber, *Tabula Peutingeriana: Codex Vindobonensis 324* (Graz: Akademische Druck- und Verlagsanstalt, 1976).

to one-dimensional presentation without exposing the user to significant limitations in comprehension and utility. Although indices and cross-references can facilitate navigation from entry to entry in a book, a map generally includes so many of these cross-structural linkages as to present serious obstacles for the user of a printed edition.

The inevitable inadequacy of text in print constitutes one of the major shortcomings of the sole large-scale commentary on the Peutinger map, Konrad Miller's massive *Itineraria Romana* (Stuttgart, 1916; hereafter ItMiller). The print medium effectively forced Miller to divide up the map's near-seamless route network, substituting in his commentary a series of itineraries with start- and end-points determined by himself. Granted, he had little other practical choice, but still his approach undermines the fundamental cohesiveness of the route network as presented on the map, and imposes upon it subdivisions and other characteristics that are not present in the original.

Impressive though Miller's work remains for its time, there are further reasons why a replacement for it is long overdue. His predilection for roads and route networks has served to overshadow many of the map's other features, and has had the harmful effect of diverting attention away from the representation of the physical landscape and its implications for the history of Roman cartography. Miller, of course, also had to contend with the lack of any comprehensive reference work for Roman geography, a deficiency progressively worsened by a century of scientific archaeology and scholarly publication. No satisfactory synthesis was forthcoming until 2000 with the publication of the *Barrington Atlas of the Greek and Roman World*.[8] Its accompanying *Directory* now provides a comprehensive catalog of locations and attested names against which the testimony of the Peutinger map can be compared. Moreover, the Atlas furnishes accurate geographic bases on which the map's features can be accurately mapped.

The time is right, therefore, for a new edition and commentary for Peutinger's Roman map. Today moreover, digital technology, in the form of hypertext document collections and interactive graphics

[8] R. Talbert, ed. (Princeton and Oxford: Princeton University Press, 2000).

applications, provides new media adequate to the challenges presented by the map. A hypertext document collection permits the presentation, discussion and analysis of each discrete feature depicted. The thematic and topological relationships between features (such as connectivity via route 'stretches') can also be modeled without subjecting these to arbitrary groupings or order. Via hyperlinks, users can move from one feature to the next, following without impediment whatever connection or relationship interests them. Because a hypertext document *is* also text, the longer sections of narrative commentary can be supported as they would be in a printed book, with the added value of easily-traversed linkages between discussion and presentation of individual features. In addition, both static and manipulable images are readily incorporated anywhere at low cost, equipping users with a complete visual record of the map's content that is easy to use.

Recently, with my own collaboration and that of an interdisciplinary team of students at Chapel Hill's Ancient World Mapping Center, Richard Talbert has been developing just such an edition of the Peutinger map. He has assembled a technology framework to support the research, and an entire draft database is now nearing completion. While the final delivery mechanisms have yet to be determined (CD-ROM, for example, or via the web), the structure and presentation modes of the edition are sufficiently well-defined to be outlined here. Although this is not the place to explain technological underpinnings in detail, it should be noted that the edition is designed for users to interact with it via standard web browser software.[9] For research needs, users will also be given full access to the underlying data.[10] Four classes of material make up the edition: an annotated database of features and names; an extended thematic commentary;

[9] The prototype has been tested with Mozilla Firefox and Microsoft Internet Explorer.

[10] Textual materials, including the Database of Map Features and Names, are encoded in accordance with the Text Encoding Initiative's (TEI) *Guidelines for Electronic Text Encoding and Interchange* in the eXtensible Markup Language (XML): http://www.tei-c.org/Guidelines2/. Raster graphics are provided as Tagged Image File Format (TIFF), Portable Network Graphics (PNG), or Joint Photographic Experts Group (JPEG) files as appropriate for the application. Vector graphics are provided as Scalable Vector Graphics (SVG) files. Geographic data are to be provided as ESRI Shapefiles.

high resolution images of the map; and a collection of geographic data related to the routes.

The database of features and names provides a record for each salient component of the map. Each record, delivered to the user's browser as an HTML page, treats its subject completely (see Fig. 2 for an example).[11] A color image is always displayed. These images are centered on the feature in question, and are bounded so as to include relevant context such as route linkages and connected features. Where an 1888 photograph is especially useful for discerning detail or reading text, a similar image is extracted from it in addition. Any label text associated with the feature is transcribed too. Where appropriate, notes on readings or oddities of presentation are inserted. The following additional analytical components are also included:

- type of feature, according to a newly devised classification system
- type of symbol, according to a classification system which has been the standard since its development in the 1960s[12]
- identification with modern geographic location (where possible) by reference to the *Barrington Atlas*
- identification with a place named in a Roman itinerary (in particular, the Antonine Itinerary)

The location, context and topology of each feature on the map are addressed. Where a feature is connected to others by way of route stretches, each of these others is indicated together with distance figures marked, river crossings and other relevant components. Notes concerning features to which stretches link are readily accessible via hyperlinks, as are those notes treating rivers crossed and other components discussed. In order to provide a reference system for citation, a grid has been imposed across the entire map. Further, in the case

[11] Hypertext Markup Language (HTML) is the "language of the web", used to encode the information formatted and displayed by web browsing software. Talbert's Peutinger Map edition will employ XHTML (Extensible Hypertext Markup Language), conformant with the provisions of Steven Pemberton *et al.*, *XHTML™ 1.0 The Extensible HyperText Markup Language (Second Edition): A Reformulation of HTML 4 in XML 1.0*, W3C Recommendation 26 January 2000, revised 1 August 2002, http://www.w3.org/TR/xhtml1.

[12] A. Levi and M. Levi, *Itineraria picta. Contributo allo studio della Tabula Peutingeriana* (Rome: Bretschneider, 1967).

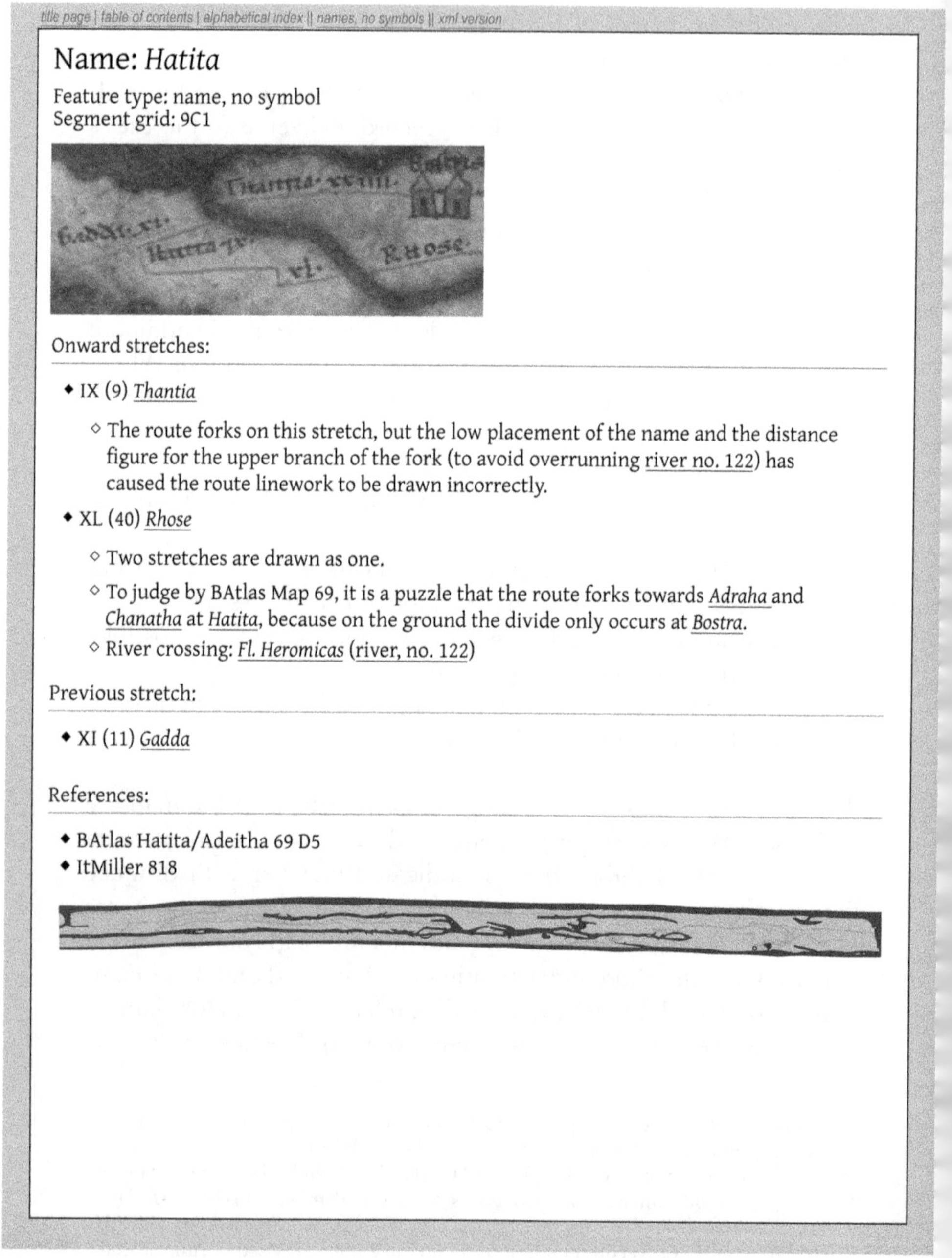

title page | table of contents | alphabetical index || names, no symbols || xml version

Name: *Hatita*

Feature type: name, no symbol
Segment grid: 9C1

Onward stretches:

- IX (9) *Thantia*
  - The route forks on this stretch, but the low placement of the name and the distance figure for the upper branch of the fork (to avoid overrunning river no. 122) has caused the route linework to be drawn incorrectly.
- XL (40) *Rhose*
  - Two stretches are drawn as one.
  - To judge by BAtlas Map 69, it is a puzzle that the route forks towards *Adraha* and *Chanatha* at *Hatita*, because on the ground the divide only occurs at *Bostra*.
  - River crossing: *Fl. Heromicas* (river, no. 122)

Previous stretch:

- XI (11) *Gadda*

References:

- BAtlas Hatita/Adeitha 69 D5
- ItMiller 818

Fig. 2. Example record from the Database of Features and Names.

Contents:

- Introduction
- Map features and notices
  - Networked names, no symbols
  - Networked symbols, named
  - Networked symbols, unnamed
  - Unnamed or illegible features
  - Isolated names
  - Isolated symbols, named
  - Isolated symbols, unnamed
  - Islands
  - Mountains
  - Peoples
  - Regions
  - Rivers
  - Special features and notices
  - Water, other than rivers
- Appendix: Symbol Classification
- Alphabetical index of all named features on the Peutinger map
- Concordance of References
- Works Cited

Fig. 3. List of "Map Features and Notices" from the Database of the Peutinger Map Features and Names.

of unlabeled features, a system of reference numbers has been developed for use throughout. The user is kept oriented to the position of any current feature in two ways. First, the entry for each feature incorporates an outline sketch of the entire map, on which is superimposed a small rectangle corresponding to the footprint of the context image of the feature. Second, the contextual color image is rendered clickable by means of a hyperlink; this mechanism provides a fluid transition to the seamless map image browser (see below).

The database of features and names is augmented by a series of indices that provide ready access to referenced records by way of hyperlinks. The database "table of contents" constitutes one such index, organizing access to records according to feature type such as names, symbols, rivers (see Fig. 3). An alphabetical index of all named features provides an alternative means of exploring the database, as does an appendix of symbols reflecting the classification system mentioned above. A Concordance of References makes it possible for users to access records in the database by way of a *Barrington Atlas* placename, a citation from a Roman itinerary, or a column number in ItMiller (see Fig. 4). The edition includes four extensive chapters by Talbert, variously treating the history of the surviving copy of the map and prior scholarship on it; the cartography of the map, its design, thematic content and method of production; the copy's likely fidelity to the lost Roman original; and finally the sources, date, authorship and purpose of that original.

Alongside the contextual images provided for every feature and name in the database, the edition aims to offer a high-resolution, color image of the entire map. A browser-based graphical application provides access to this digital scan in seamless fashion, permitting zooming and panning across the map's entire extent irrespective of the boundaries of the parchment segments. This presentation is enhanced by a series of thematic overlays that highlight such individual cartographic elements as rivers, routes and mountain ranges. Each of these overlays can be turned on and off at will by the user. Similar control can be exercised over a grid layer that corresponds to the citation system used in the database and elsewhere, as well as over layers bearing the numeric labels for rivers, unlabeled features and the like.

The edition is reinforced by an assemblage of further digital files. A complete set of high-resolution scans of the 1888 photographs is included; these, however, are not integrated into the seamless browser application. Geographic data files, together with an appropriate browsing utility, mark the 'real-world' locations of features (wherever possible), and trace their connections to each other along the map's route networks. A similar set of files for the Antonine Itinerary is included for comparison. These geospatial data components all relate the features and routes to a seamless digital map base derived from the *Barrington Atlas*.

Scholarly editions and commentaries preserve classical sources and make it possible to assess, interpret and compare them in informed ways. Within the past few years the creative application of emerging technologies has enhanced this process, at long last offering

Concordance: Barrington Atlas to Peutinger Map (H)

- Hadia 89 unlocated = *Hadia* 10C3 (name, no symbol)
- Hadra 20 unlocated ?= *Hadre* 4A4 (name, no symbol)
- *Hadriani 40 C3 = *Radriani* 3B5 (name, no symbol)
- Hadrianopolis 38 B1 = *Hadrianopol(is)* 7C4 (symbol, named)
- Hadrianopolis 54 B2 = *Hadrianopoli* 6B3 (name, no symbol)
- Hadrianoupolis 51 H1 = *Hadrianopoli* 7B3 (symbol, named)
- Hadrianoutherai 56 F2 = *Hadrianvteba* 8B3 (name, no symbol)
- (H)Adriaticum/Superum Mare 1 F2 = *HADRIATICVM PELAGVS* 6B2-7B5 (water)
- Hadrumetum/Iustinianopolis 33 G1 = *Hadrito* 5C2 (symbol, named)
- Halaesa 47 E3 = *Halesa* 6B1 (name, no symbol)
- (H)Alata 49 B1 = *Nalata* 6A1 (name, no symbol)
- Halex? fl. 46 C5 ?= (unnamed river, no. 69) 6B2 (river)
- Hali 84 unlocated ?= *Hale* 8A2 (isolated name)
- Halia 67 unlocated = *Halia* 10B3 (name, no symbol)
- Haliakmon fl. 50 A4 ?= (unnamed river, no. 66) 6B5 (river)
- Halys fl. 87 A3 ?= (unnamed river, no. 107) 9A1-9B1 (river)
- Hammeum 21 D6 = *Hammeo* 6A3 (name, no symbol)
- (H)Apsos fl. 49 B3 = *Hapsvm Fl.* 6A2 (name, no symbol)
- Harenatium 10 C5 = *Arenatio* 1A4 (name, no symbol)
- *Hares/*Haribus/*Har/*Phoar? 35 unlocated = *Haribvs* 6C1 (isolated name)
- Haris 89 unlocated = *Haris* 10B1 (name, no symbol)

Fig. 4. Extract from the Concordance of *Barrington Atlas* references to Peutinger Map features.

unprecedented potential for the presentation and analysis of works whose character defies the traditional print edition. Understanding of the Peutinger map can derive exceptional benefit from a judicious combination of the old approaches and the new. Its first comprehensive re-examination in almost a century is now well on the way to completion. It differs radically from Miller's great work of 1916, and amply reflects the extraordinary advances of all kinds that have been made since, during recent decades in particular.

RETHINKING THE PEUTINGER MAP

Emily Albu

The Peutinger map is a map of the inhabited world from Britain to Sri Lanka, drawn c. 1200 C.E. onto a parchment roll nearly seven meters long and 32 to 34 centimeters high (Plate I).[1] Because it was created from Roman itinerary lists and features some 70,000 Roman miles of Roman roads—with hundreds of Roman sites identified by their Roman names and with mileage between sites marked, mostly in Roman miles—historians of cartography have long assumed it to be a copy of a Roman map. Recently I have suggested, however, that our medieval map had a Carolingian prototype, clearly meant for display and not intended as a road map for ancient travelers.[2] Carolingian rulers had ample motivation for commissioning a map to display their Roman imperial ambitions, while ninth-century scribes had the expertise and resources necessary for creating an antiquarian work based on Roman itinerary lists. This paper extends that argument by highlighting the probable site for the production of both the Carolingian original and its extant "copy," our Peutinger map. These not-very-Roman identifications further situate the map in its medieval context.

Let us begin with a few words about Roman and early medieval world maps. Specifically, why do we see an explosion of world maps in the early Middle Ages after what appears to be a long dry spell?

---

[1] Codex Vindobonensis 324, now separated into its eleven individual leaves, is in the Austrian National Library, Vienna. It is perhaps best known through Konrad Miller's color lithograph, in *Die Weltkarte des Castorius genannt die Peutingersche Tafel* (Ravensburg: Maier, 1887/1888), reproduced in Francesco Prontera (ed.), *Tabula Peutingeriana: Le antiche vie del mondo* (Florence: Leo S. Olschki, 2003). Richard Talbert (University of North Carolina and the Ancient World Mapping Center) is producing an electronic edition. For more information on this project, see www.unc.edu/awmc/rttpeut.html. In collaboration with Professor Talbert, Benet Salway and I are preparing text and commentary on the map for a concise print edition.

[2] "Imperial Geography and the Medieval Peutinger Map," *Imago Mundi* 57 (2005),136-48.

The answer will also help us understand the map's medieval identity.

No world maps survive from the Roman era. Daring critics, with Kai Brodersen in the lead, have even challenged long-held assumptions that Romans made frequent use of world maps.[3] The Romans, in fact, did not even have a distinct word for "map," so we are left to debate whether their *itineraria picta* were painted lists or maps proper. Similar problems surround the Roman terms *descriptio mundi* and *tabula*. The famous example is "Agrippa's map," displayed in the Porticus Vipsania, in what is now the Via del Corso area of Rome.[4] What form did this "map" take? Kai Brodersen has assembled the widely diverging judgments of scholars:

> . . . depending on which opinion one follows, it was a globe or a 'large scale map,' executed as a mosaic, painted in colour, engraved in bronze, or hewn into marble; it was circular, oval, or rectangular, and it measured 6-10 m in height, 9 m in width and 18 m in height, or 24 m in width and 12 m in height (on a pediment of 5 m), or 75 m wide, but only 4.5 m high. At the top was east, south, or north, and it resembled the *Tabula Peutingeriana*, a schematic medieval *mappamundi*, or an early modern portolan chart—*quot homines tot sententiae.*[5]

Was it even what we would call a map, or might it have taken some other form, like an itinerary list? On this, too, we really cannot be certain.

Whatever world maps the Roman Empire produced, like "Agrippa's map" these existed for display, not as an aid for travelers. Such world maps were, at best, a rarity in the Roman world and fraught with weighty baggage. They could be exhibited only in service to the Roman *imperium*. A private person who possessed a world map did so at his own peril, as he was clearly plotting rebellion. So when Mettius Pompusianus committed the "cartographic crime" of painting a depiction of the world on his bedroom wall, the emperor Domi-

[3] Kai Brodersen, *Terra Cognita: Studien zur römischen Raumerfassung* (Hildesheim: Georg Olms, 1995; ed. 2, 2003).

[4] Kai Brodersen, "Mapping the Ancient World," *Ad Familiares: Journal of the Friends of Classics* 17 (Autumn, 1999).

[5] Kai Brodersen, "Mapping (in) the Ancient World," *Journal of Roman Studies* 94 (2004), 185. For his evidence, see his *Terra Cognita*, 269-70.

tian (81-96 C.E.) ordered his execution on the assumption that he was harboring imperial ambitions.[6]

Our best evidence for any lost Roman world maps comes from late antique rhetoricians. From these sources, Benet Salway has presented the two lost maps for which we have the most persuasive testimony.[7] The first, erected in a portico at Augustodunum (Autun) in the waning days of the third century, is described in a contemporary oration before a provincial governor as a very publicly displayed teaching aid and advertisement of Roman dominion. The orator Eumenius concluded his brief remarks on that display map with this assessment: "For now, now at last it is a delight to see a picture of the world, since we see nothing in it that is not ours."[8] The later map, or perhaps collection of local maps, was commissioned in 435 by the Emperor Theodosius II—"whom the whole world scarcely contains," as our fifth-century source announces.[9] World maps like these were a kind of imperial propaganda, a demonstration of Roman

---

[6] Pascal Arnaud, "L'affaire Mettius Pompusianus, ou le crime de cartographie," *Mélanges de l'École française de Rome: Antiquité* 95 (1983), 677-99. The details derive from Cassius Dio and Zonaras. Suetonius (*Domitian* 10.3) has Pompusianus carrying the parchment map on his person.

[7] Benet Salway, "The Nature and Genesis of the Peutinger Map," *Imago Mundi* 57 (2005),128-29.

[8] "Nunc enim, nunc demum iuuat orbem spectare depictum, cum in illo nihil uidemus alienum": Eumenius, *Oratio pro instaurandis scholis*, 9(4), in Roger A.B. Mynors (ed.), *XII Panegyrici Latini* (Oxford: Clarendon Press, 1964), 20.2-21.3. C. Edward V. Nixon and Barbara S. Rodgers (trans.), *In Praise of Later Roman Emperors: The Panegyrici Latini* (Berkeley: University of California Press, 1995), 176-77.

[9] "Hoc opus egregium, quo mundi summa tenetur,
aequora quo, montes, fluuii, portus, freta et urbes
signantur, cunctis ut sit cognoscere promptum
quidquid ubique latet, clemens genus, inclita proles,
ac per saecla pius, totus quem uix capit orbis,
Theodosius princeps uenerando iussit ab ore
confici, ter quinis aperit cum fascibus annum."

("This outstanding work—in which the whole world is included, in which seas, mountains, rivers, harbours, straits and towns, are indicated, so that all might know where any feature lies—the kind natured, nobly born, and forever pious emperor Theodosius (whom the whole world scarcely contains) from his reverend mouth ordered to be made, when he opened the year with his fifteenth consulship.")

For this poem by Probus, see the *Diuisio orbis terrarum* in Alexander Riese (ed.), *Geographi Latini Minores* (Heilbronn: Henninger Brüder, 1878), 19-20; and Dicuil, *Liber de mensura orbis terrae*, ed. J. J. Tierney and Ludwig Bieler (Dublin: Dublin Institute for Advanced Studies, 1967), 5.4. Translation by Benet Salway, "The Nature and Genesis of the Peutinger Map," 128.

power closely tied to claims of Roman *imperium*, and they could be created only for sanctioned use.

Christianity changed all that. The creation story in *Genesis*, universally disseminated through popular sermons and learned commentaries, taught that God gave the earth to mankind for the use of human beings.[10] Jews and Christians understood that the inhabitable world belonged to humankind. Christianity presented a further, distinctively non-Roman concept—namely, if God entered the physical world through the incarnation, and dispatched his apostles throughout the world with the exhortation to spread the Gospel among all peoples, then surely every Christian had the right to see or even display the world that those holy men traversed and converted to Christianity. This right is precisely the point of the often copied Beatus maps, fifteen of which survive in tenth- to fifteenth-century manuscripts, showing evangelism to the far corners of the earth.[11] No longer could secular rulers demand the unique privilege of displaying the realm over which they exerted unique authority. The Church began to exhibit the earth as its inheritance, as for instance when Pope Zacharias (741-752) commissioned a world map for the dining room at the Lateran Palace.[12] That map, or others implying Christian claims, may have inspired Charlemagne to present his silver maps, one of them a map of the world, as a cartographic counterclaim.[13] By the ninth century, when the term *mappa mundi* is first attested, we see a proliferation of world maps. Such maps still held their formidable power to demonstrate ownership and majesty. But legitimate claims extended far beyond the political realm. Dueling authorities of Church and secular court produced the intense activity of map production in the early Middle Ages.

If, as I have argued, the prototype of our Peutinger map was one result of that Carolingian activity, precisely where was it made? In a 1974 article, Hans Lieb identified a likely candidate for the Caro-

[10] *Genesis* 1:26-30.

[11] John Williams (ed.), *The Illustrated Beatus: A Corpus of the Illustrations of the Commentary on the Apocalypse*, 5 vols. (London: Harvey Miller Publishers, 1994-2003).

[12] On the Lateran map and papal claims to universal power, see Marcia Kupfer, "Medieval World Maps: Embedded Images, Interpretive Frames," *Word and Image* 10 (1994), 262-88.

[13] On Charlemagne's maps and his geographical curiosity, see Emily Albu, "Imperial Geography and the Medieval Peutinger Map," 139-40.

lingian prototype, a *mappamundi* listed in the early ninth-century catalogue of the library at Reichenau abbey.[14] Elsewhere I have summarized his neat detective work in uncovering evidence for the map's presence in the Reichenau library in the mid-eleventh century, when Hermann Contractus found on the map the old Roman name for the nearby Black Forest (*silua Marciana*).[15] The depiction of this forest and only one besides, the *silua Vosagus* just across the Rhine, is another of the clues pointing to the map's creation in this region, as also is the distinctive formation of regional names along this stretch of the Rhine.[16] Here the river broadens to form Lake Constance (Plate II). Accumulating evidence suggests that the Peutinger prototype was created at Reichenau, on an island in that lake, and stayed there at least until our version of c. 1200 was drawn, also in the Reichenau scriptorium. The new map remained there or at least close to its home, to become the model for another version made nearby in 1265, until Konrad Celtes spirited it away at the end of the fifteenth century and bequeathed it in 1508 to Konrad Peutinger, for whom it is named.

Reichenau, a Benedictine monastery in southern Germany, was one of the great centers of the Carolingian renaissance. Founded by Pirmin in 724, it soon attracted the support of Charles Martel and, in turn, his successors, who made it essentially an imperial foundation. The monastery acquired considerable properties from Charlemagne, Louis the Pious, and Charles the Fat (who was buried in the monastery church). An early abbot and founder of its library, Waldo (abbot 786-806), was Charlemagne's confessor and one of his chief counselors. The celebrated scholar and poet Walafrid Strabo received his early education as a child-oblate at Reichenau, until he was sent to the abbey of Fulda for advanced training under Hrabanus Maurus. There he attracted the attention of the emperor Louis the Pious and the empress Judith, who appointed him tutor of their son Charles. As a reward for his nine years as mentor to the young prince

---

[14] Hans Lieb, "Zur Herkunft der Tabula Peutingeriana," in Helmut Maurer (ed.), *Die Abtei Reichenau: Neue Beiträge zur Geschichte und Kultur des Inselklosters* (Sigmaringen: Thorbecke, 1974), 31-33.

[15] Emily Albu, "Imperial Geography and the Medieval Peutinger Map," 140-41.

[16] Benet Salway summarizes this evidence in "The Nature and Genesis of the Peutinger Map," 123-24.

at Aachen, Louis appointed him abbot of Reichenau in 838. So close were the bonds between monk and prince that Walafrid had to flee into exile when war erupted among the royal heirs at the death of Louis in 840. After two years, Louis the German finally allowed Walafrid to return to Reichenau, where he again served as abbot until his death in 849, as he was returning from a meeting with his former student, then King Charles the Bald. Such strong ties with the Carolingian rulers made Reichenau a likely site for a royal commission to create an archaizing Roman map that would suit secular and imperial propaganda. Reichenau was also, not incidentally, on the Roman road network itself, on a highway to Italy and thus visited by many travelers from faraway places including Greece, Ireland, Iceland—pilgrims who brought relics and books as gifts. Not surprisingly, in his survey of pre-Carolingian Latin manuscripts produced in what is now Switzerland, E. A. Lowe discovered that "most seem to have come from a rather narrow section of that country, namely from that part which follows the well-trodden pilgrim routes along which lay Rheinau (Reichenau), St. Gall, Pfävers, Chur, and Disentis."[17] It seems quite likely that among the manuscripts in the library at Reichenau were some containing the Roman itinerary lists essential for the creation of the Peutinger prototype.[18]

Reichenau's scriptorium certainly had the other resources required for such an ambitious undertaking. Nearby St. Gall is better known today, in large part because its magnificent library still survives,[19] while the Reichenau collection was dispersed early in the nineteenth century. The two monasteries were closely linked, sometimes even sharing the same abbot. Their scriptoria, too, had intimate ties to one another. Together these scriptoria developed the distinctively broad Alemannic script. Indeed, the hands of St. Gall and Reichenau

[17] E. A. Lowe (ed.), *Codices Latini Antiquiores: A Palaeographical Guide to Latin Manuscripts prior to the Ninth Century* (Oxford: Clarendon Press, 1956), v.

[18] I am now tracing the histories of these itinerary lists and identifying, wherever possible, the early medieval archives that held Roman itineraries. On the library of Reichenau, see Rosamond McKitterick, *The Carolingians and the Written Word* (Cambridge: Cambridge University Press, 1989), 179-82.

[19] For details on this library see Johannes Duft, *The Abbey Library of Saint Gall*, James C. King and Petrus W. Tax, trans. (St. Gallen: Verlag am Klosterhof, 1985). On Reichenau and St. Gall, see Walter Berschin, *Eremus und Insula: St. Gallen und die Reichenau im Mittelalter—Modell einer lateinischen Literaturlandschaft* (Wiesbaden: Reichert Verlag, 1987).

are so similar that it is sometimes impossible to determine which of the two monasteries produced a given manuscript.[20] Yet Reichenau had certain advantages over its brother monastery. Sheltered on an island in Lake Constance, Reichenau offered better protection in times of distress, as in the tenth century when Hungarian assaults threatened St. Gall and much of its library was temporarily moved to the island monastery for safekeeping. Reichenau thus sometimes found itself the dominant partner and the more prolific creator of manuscripts.

The Reichenau scriptorium nurtured the Benedictine passion for manuscript production as a meditative activity that cared for the soul even as it produced valuable and often beautiful texts. Not only did monks copy both Christian and secular works, but some composed their own poems, saints lives, and chronicles. There Walafrid Strabo wrote his hexameter *Vision of Wettin* as well as verses on the Irish saint, Blaithmaic. Reichenau also came to be celebrated as the foremost school for Ottonian book illumination.[21] As early as the 830s or even a decade earlier, however, the Reichenau scriptorium produced the Plan of Saint Gall (Plate III), a unique artifact described as "the only major architectural drawing to survive from the period between the fall of Rome and the thirteenth century."[22] Reichenau monks sewed together five pieces of parchment to make a rectangular sheet measuring about 78 cm by 112 cm. On this was drawn an entire monastic compound—churches and dormitories, kitchens and stables, a brewery, and an infirmary with a separate structure for bloodletting—with ground plans in red ink and inscriptions in brown. Here is tantalizing confirmation that Reichenau monks in the early ninth century were working in a medium similar to that of the more ambitious Peutinger prototype.

---

[20] E. A. Lowe (ed.), *Codices Latini Antiquiores*, ix.

[21] C. R. Dodwell disputed that claim in "Reichenau Reconsidered: A Reassessment of the Place of Reichenau in Ottonian Art," *Warburg Institute Surveys, II.* The Warburg Institute, University of London, 1965; reprinted in id., *Aspects of Art in the Eleventh and Twelfth Centuries* (London: Pindar Press, 1996), 228-59.

[22] Edward A. Segal, "Monastery and Plan of St. Gall," *Dictionary of the Middle Ages* 10 (New York: Charles Scribner's Sons, 1988), 617. On the dating and intended use of the Plan, see Richard E. Sullivan, "What Was Carolingian Monasticism? The Plan of St Gall and the History of Monasticism," in Alexander Callander Murray (ed.), *After Rome's Fall: Narrators and Sources of Early Medieval History. Essays presented to Walter Goffart* (Toronto: University of Toronto Press, 1998), 251-87.

If the itinerary map was also made in the ninth-century Reichenau scriptorium, it remained there to be seen by Herman Contractus in the mid-eleventh century and to be "copied" (though not precisely, as I have argued elsewhere) in its scriptorium around 1200.[23] On paleographical evidence, Bernard Bischoff thought that the surviving Peutinger map came from the region of medieval Swabia / Alamannia, that is, from the general area of the imperial Reichenau monastery.[24] Evidence recently presented supports a growing consensus that affirms Bischoff's view. This evidence comes from the lost Padua map, acquired by Venetian ambassadors to the Council of Basel (1431-1449) and later displayed on a wall in the antechamber of the episcopal palace in Padua. Before this map went missing at the end of that century, one section of it was sketched by Pellegrino Prisciani (1435-1518), humanist polymath and librarian for d'Este princes.[25] This section, as Patrick Gautier Dalché has shown, almost exactly duplicates the corresponding segment of the Peutinger map.[26] Slight deviations, however, would seem to rule out the possibility that the Padua and Peutinger maps were one and the same. Prisciani's drawing includes Forum Alieni, here near Ravenna, but absent in the Peutinger map. Prisciani also claimed to have seen Greek letters on the Padua map. These, too, are nowhere on the Peutinger map. These differences have raised the possibility that the Padua map was "a now-lost 'sister' of the Peutinger map," derived from a common ancestor, and thus evidence that the Peutinger map was part of a larger cartographic tradition going back to late antiquity.[27]

This Padua map, first attested in Basel, may well be the world map that Patrick Gautier Dalché has identified as produced there

---

[23] Emily Albu, "Imperial Geography and the Medieval Peutinger Map," 141-42.

[24] Ekkehard Weber reported Bischoff's conclusion in *Tabula Peutingeriana: Codex Vindobonensis 324, Kommentar* (Graz: Akademische Druck- und Verlagsanstalt, 1976), 11.

[25] Pellegrino Prisciani, *Annales Ferrarienses* (Modena: Archivio di Stato, MS 129), fol. 44v.

[26] Patrick Gautier Dalché, "La trasmissione medievale e rinascimentale della *Tabula Peutingeriana*," in *Tabula Peutingeriana: Le antiche vie del mondo*, 48. See also Benet Salway, "The Nature and Genesis of the Peutinger Map," 123-24.

[27] For the suggestion, see Benet Salway, "The Nature and Genesis of the Peutinger Map," 122, 127.

on twelve parchment sheets in 1265.[28] Was its model then the Peutinger map of c. 1200, made at Reichenau, just 115 kilometers up the Rhine?[29] It seems quite likely that it was "copied" from the Peutinger map, with the sorts of alterations that Ingrid Baumgärtner has shown occurred in the "copying" of Beatus maps.[30] She has demonstrated that medieval mapmakers felt no compunction whatsoever to reproduce their prototypes exactly. So the addition of new sites and Greek lettering on the Padua "copy" would be perfectly natural occurrences if they suited the interests of the Padua mapmakers—and let them show off their knowledge of Greek. This scenario makes the Padua map the daughter, rather than the sister, of the Peutinger map.

In sum, our Peutinger map may be quite thoroughly the product of the Reichenau scriptorium, its prototype the result of Carolingian imperial designs articulated by a Benedictine establishment closely connected to the ruling dynasty. We know it as the Peutinger map because of its bequest in 1508 from Konrad Celtes to his friend Konrad Peutinger. Celtes, the librarian to the Emperor Maximilian I, was a learned man educated by Greek and Italian humanists, and a large-scale book thief, who roamed monastic libraries and episcopal palaces removing books and manuscripts for his emperor (and for himself and his friends). Perhaps it is no coincidence that both he and Peutinger were in the vicinity of Padua when that map disappeared.[31] To the riches of the Reichenau library, Celtes had even readier access. We should not continue to honor this pilfering by calling the extant map "Peutinger." Why not call it the Reichenau map?

[28] Patrick Gautier Dalché, "La trasmissione medievale e rinascimentale della *Tabula Peutingeriana*," 47-48.

[29] Benet Salway, "The Nature and Genesis of the Peutinger Map," 127, notes the proximity of Basel to Reichenau.

[30] Ingrid Baumgärtner, "Die Welt im kartographischen Blick: Zur Veränderbarkeit mittelalterlicher Weltkarten am Beispiel der Beatustradition vom 10. bis 13. Jahrhundert," in Wilfried Entrecht et al. (eds.), *Der weite Blick des Historikers: Einsichten in Kultur-, Landes-, und Stadtgeschichte. Peter Johanek zum 65. Geburtstag* (Cologne, Weimar and Vienna: Böhlau Verlag, 2002), 527-49. For a brief discussion of her work, and its significance for study of the Peutinger map, see Emily Albu, "Imperial Geography and the Medieval Peutinger Map," 142.

[31] Patrick Gautier Dalché, "La trasmissione medievale e rinascimentale della *Tabula Peutingeriana*," 48-49.

THE *BOOK OF CURIOSITIES* AND A UNIQUE MAP OF THE WORLD

Yossef Rapoport and Emilie Savage-Smith

A rectangular map of the world unlike any other recorded ancient or medieval world map has been preserved in a recently discovered Arabic treatise dating from around 1200 and containing a total of seventeen maps and cartographic designs. Together, these maps present several different perspectives on the world considered at that time to be inhabited. The anonymous Arabic treatise containing these maps is a cosmography, in 48 folios (96 pages). The treatise consists of two books: Book I, on the heavens in ten chapters, and Book II, on the earth, in twenty-five chapters. Its title, *Kitāb Gharā'ib al-funūn wa-mulaḥ al-ʿuyūn* translates loosely as *The Book of Curiosities of the Sciences and Marvels for the Eyes*. The original treatise does not survive, but in June 2002 the Bodleian Library in Oxford acquired the only known copy.[1] Prior to being offered for sale at auction in London on October 10, 2000, this manuscript, and even the treatise it contained, was totally unknown to scholars.[2] It is now the subject of a joint research project of the Oriental Institute and the Bodleian Library; it is published in its entirety on a website hosted by the Bodleian (http://www.bodley.ox.ac.uk/bookofcuriosities), and in book form at a later date.[3]

[1] Oxford, Bodleian Library, MS arab. c. 90. The acquisition of the *Book of Curiosities* was made possible by generous donations from the Heritage Lottery Fund, the National Arts Collections Fund, the Friends of the Bodleian Library, ARAMCO (Saudi Arabia), several Oxford colleges (All Souls, Merton, New, Nuffield, St Antony's, St Cross, St John's, Wadham, Wolfson), and a number of private individuals. Continued work on the edition, translation and publication of the unique manuscript has been supported by the Heritage Lottery Fund and the Arts and Humanities Research Council (AHRC).

[2] At auction (Christie's, London, *Islamic Art & Manuscripts*, lot 41) the manuscript was purchased by Sam Fogg, a well-known London dealer in rare books and manuscripts. Not long thereafter he offered it to the Bodleian Library at a price well under the true market value.

[3] For earlier discussions of other portions of the treatise, see J. Johns and E. Savage-Smith, "*The Book of Curiosities*: A Newly Discovered Series of Islamic Maps," *Imago Mundi* 55 (2003), 7-24; Y. Rapoport and E. Savage-Smith, "Medieval Islamic

Though not giving his name, the author provides a considerable amount of information about himself. From references to various events, it is possible to place the composition between 1020 and 1050.[4] Furthermore, it is evident that the author lived in Egypt. He was literate, but not scholarly. It is possible that he was self-taught, for such a form of education was not uncommon at the time, the argumentative Cairene physician Ibn Riḍwān, who was his exact contemporary, being a prime example. The treatise is dedicated to an unnamed patron, presumably to gain some favour, position, or possibly even backing for a venture. In preparing the work, the author drew, in magpie fashion, upon a range of written sources, over twenty of which he names. This material he supplemented with his own personal information, as well as including accounts from traders and travellers along the East African coast. This mix of sources for his book reflects the diverse and complicated milieu in which he was operating.

Among the seventeen maps in this impressive *Book of Curiosities* is the rectangular world map (Plate IV and Fig. 1). It constitutes the entire second chapter of Book II, with no additional text. Its only title occurs at the bottom of the preceding page (folio 23a), where it reads simply "It [Chapter 1] is followed by Chapter 2 on the depiction of the earth (*fī ṣūrat al-arḍ*)". The map occupies a full opening (folios 23b-24a) and measures 32.4 x 49 cm (12.7 x 19.2 ins). The

---

View of the Cosmos: The Newly Discovered *Book of Curiosities*," *The Cartographic Journal* 41 (2004), 253-59; and E. Edson and E. Savage-Smith, *Medieval Views of the Cosmos* (Oxford: The Bodleian Library, 2004).

[4] There are four features that indicate its composition to have been after 1020 and before 1050: (1) Sicily is described as being under Muslim rule; thus the treatise was composed before the Norman invasion of Sicily in AD 1070; (2) the Banū Qurrah are mentioned as still inhabiting the lowlands near Alexandria; since Fāṭimid authorities waged several campaigns against them in 1050-1051, eventually banishing them from the region in 1051-1052, our author is writing before 1050 (442 H); (3) in the chapter on Tinnīs, it is stated that six large buildings for merchants were constructed in 1014-1015 (405 H), bringing the total number of merchant inns and covered markets in the city to 56; for this reason our author must have been working after 1015; and (4) al-Ḥākim bi-Amr Allāh, the Fāṭimid ruler of Egypt and Syria from 996 to 1021, is referred to as if he were no longer reigning, meaning that our author is writing after the year 1020, since al-Ḥākim died on February 13, 1021 (27 Shawwāl 411 H).

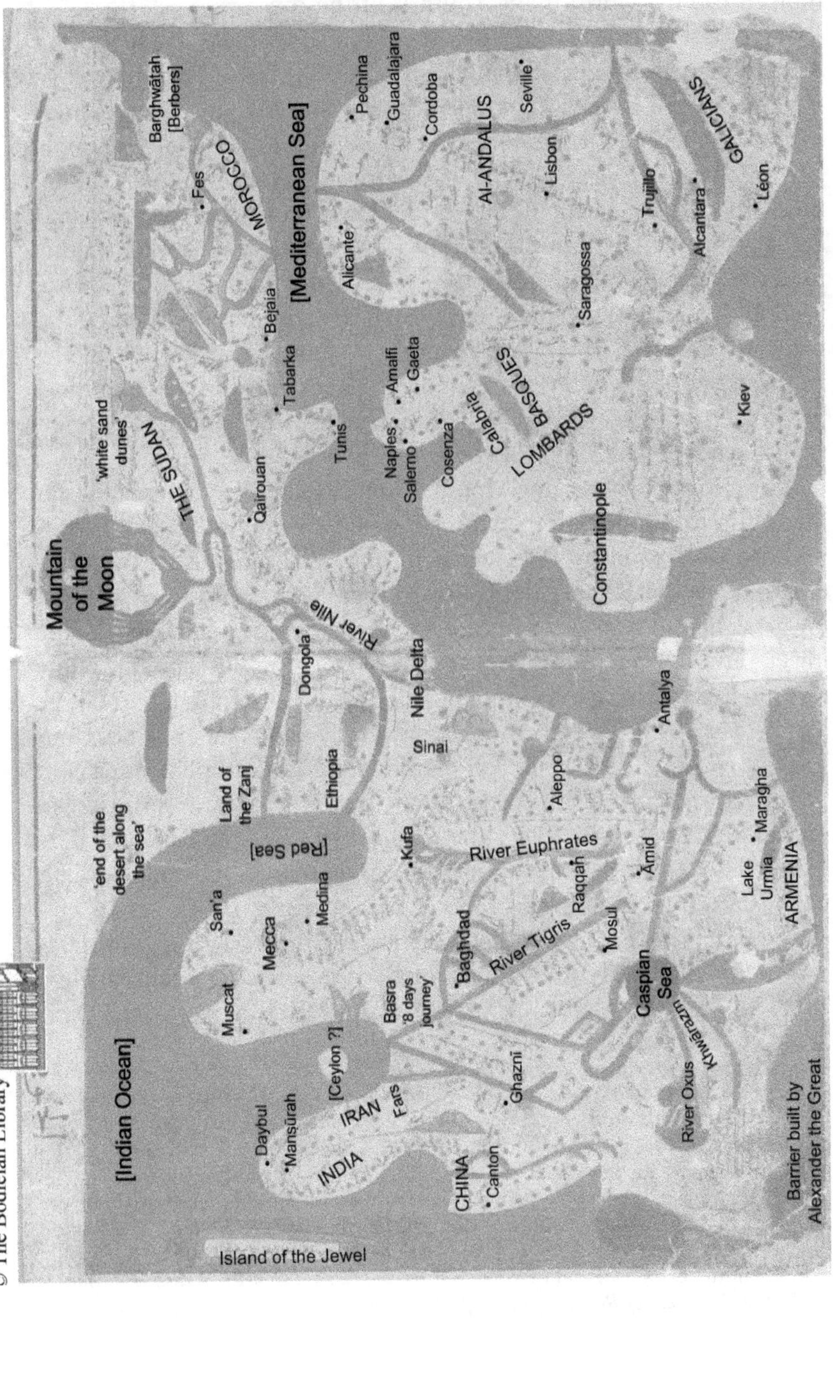

Fig. 1. Annotated version of the rectangular world map from the anonymous *Book of Curiosities* compiled about 1020-1050. Oxford, Bodleian Library, MS arab. c. 90, fols. 23b-24a. Undated, *c*1200 (?). Reproduced with permission from the Bodleian Library.

map is a bifolio—that is, one continuous sheet of paper.[5] As is common in Islamic world maps, south is at the top.

The most unique feature of the map is a carefully executed graphic scale. The "Mountain of the Moon" (*jabal al-qamar*), considered by medieval Arabic writers to be the source of the Nile, has been painted over the scale to the right of centre, and the green ocean at the upper left obscures the eastern portion.[6] Infra-red and ultra-violet lamps revealed that the graticule lies under the green paint of the ocean and the brown "Mountain of the Moon". A red ink similar to that used for the scale (and possibly the same drawing instrument) was used to delineate the general outlines of the land masses. Thereafter, black ink was used to retrace the larger outlines and delineate smaller features, before pigments were applied. The redrawing of the outlines by the person using the black ink is particularly evident in the upper-right corner of the map.

The "Mountain of the Moon" is represented by a semicircular mountain from which ten streams diverge, five on either side, pouring into two circular pools which in turn feed into one lake before emerging as the River Nile. Europe is depicted as a landmass in the lower right of the map, with a very large Iberian peninsula. In the lower right corner of the map there is a four-line inscription, badly damaged, describing the extent of al-Andalus. The Italian and Greek peninsulas are indicated, though only Italy bears labels, and Constantinople is marked at the left extremity of the European continent behind a brown masonry wall. Beneath the lengthy inscription to the right of Constantinople is the single city of Kiev (*al-Kūyābah*). The unlabelled "Gulf of Constantinople" appears to communicate with northern waters, making Europe into a large island.

No islands are found in the Mediterranean Sea, although a separate map of the Mediterranean in the manuscript details a huge

[5] There is a guard on the back that serves as a repair to strengthen the fold rather than to join two leaves together. On the front (inside) fold, some areas have repairs or guards, but others do not, and in the latter places it is clear that it is a bifolio. We wish to thank Sabina Pugh, of the conservation department at the Bodleian Library, for her careful analysis of the manuscript's structure.

[6] For a recent analysis by raman spectroscopy of the pigments used in the manuscripts, see Tracey D. Chaplin, Robin J. H. Clark, Alison McKay, and Sabina Pugh, "Raman spectroscopic analysis of selected astronomical and cartographic folios from the early 13th-century Islamic *Book of Curiosities of the Sciences and Marvels for the Eyes*," *Journal of Raman Spectroscopy* 37 (2006), 865-77.

number of islands, including Sicily and Cyprus, which are also shown individually on yet further additional maps. The Indian Ocean occupies the upper left quadrant of the map, and is also devoid of islands except for an unlabelled circular island, curiously outlined in a metallic glitter, that may have been intended as Ceylon (Sri Lanka).[7] Jutting into the Indian Ocean are two peninsulas, the larger representing Arabia, with Persia and India subsumed into the smaller one.

On the large Arabian peninsula, the city of Mecca is distinguished by a yellow horseshoe-shaped symbol. All other cities on the map are indicated by simple red dots of uniform size. A scattering of unlabelled red dots, along the coasts and elsewhere, presumably reflects either ignorance or negligence on the part of the copyist, or the corruption of his source.

The red line down the middle of the smaller peninsula may indicate the Indus River, depicted as rising from a large red mountain and ending near the coast at al-Manṣūrah, the capital of Muslim Sind. To the east (left) of the Indian/Persian peninsula, another river flows into the ocean; the curved coastline to the left of this river is China. The two highly stylised and complicated river systems between and below the two peninsulas represent the Euphrates and the Tigris.

At the lefthand margin of the map, the easternmost limit of the inhabitable world is represented by a brown landmass bearing an inscription, encircled in red, reading: "Island of the Jewel, and its mountain surrounds it like a basket". The map also depicts, in the lower left corner, a gate in the legendary wall constructed by Alexander the Great to imprison Gog and Magog.[8] To the right, the Caspian Sea is represented by a dark circle. A number of tributaries

---

[7] The periphery of the circular island is decorated in red and gold. The red pigment has been analysed as cinnabar (or vermilion), but the analysis of the reflective gold decoration was inconclusive. This is probably metallic gold, but its actual presence cannot be detected by raman spectroscopy. Similar reflective golden flakes were used on one of the comets in Book 1 (fol. 14a). See Chaplin, *et al.*, "Raman spectroscopic analysis," 866 and 870.

[8] It is labelled "Barrier (*sadd*) which was built by Alexander *Dhū al-Qarnayn* (the Possessor of Two Horns, a common epithet for Alexander the Great)." For early accounts of this rampart, see A. R. Anderson, *Alexander's Gate, Gog and Magog, and the Inclosed Nations* (Monographs of the Medieval Academy of America 5; Cambridge, Mass.: The Medieval Academy of America, 1932); and E. van Donzel and Claudia Ott, "Yad̲j̲ūd̲j̲ wa-mād̲j̲ūd̲j̲," in H.A.R. Gibbs et al. (eds.), *Encyclopaedia of Islam*, ed. 2 (Leiden: Brill, 1960-2005), 11: 231-35.

converge on it, one of which flows from Gog and Magog's mountain enclave. Unlike European mapmakers, the Islamic cartographers always showed the Caspian as an inland sea.

The presence of the scale raises the possibility that the present copy of the rectangular world map is derived from a map plotted on the basis of coordinates of longitude and latitude. Early Islamic scholars, following Ptolemy (2nd cent. AD), showed much interest in compiling longitude and latitude tables, and discussed methods of producing maps based on these coordinates. However, there is not one surviving example of a mathematically plotted map, world-map or other, from the early Islamic period.[9] In fact, the rectangular world map in the *Book of Curiosities* is the earliest surviving example of a map carrying a scale, not only in Islamic cartography but in European cartography as well. If the rectangular world map was indeed plotted, or at least was based on a mathematically plotted prototype, this means that it is also the earliest surviving example of its kind. If not, then what is the scale doing there, and what did the mapmaker intend it to signify?

In itself, the presence of the scale does not prove that the map was mathematically plotted. It is in fact obvious that the copyist of the manuscript discussed here, and perhaps also the author before him, did not understand the purpose of the scale. Only the cells on the right-hand folio are numbered, with *abjad* letter-numerals which increase cumulatively in units of five degrees.[10] The numbering begins on the right, and 135° is the last visible number before the

[9] A circular world map prepared by Ibn Faḍl Allāh al-'Umarī (d. 1349) carries a superfluous graticule added later over the map. It is preserved in a copy of his encyclopaedia said by Fuat Sezgin to date from about 1340, though others consider it a sixteenth century manuscript. This manuscript also includes a crudely drawn rectangular world-map with a vertical scale to the right for latitude and a horizontal scale. For illustrations of maps from Ibn Faḍl Allāh al-'Umarī's *Masālik al-abṣār fī mamālik al-amṣār*, see D.A. King, *World-Maps for Finding the Direction and Distance to Mecca: Innovation and Tradition in Islamic Science* (London: al-Furqān Foundation; Leiden: Brill, 1999), 35, Fig. 1.7.5, and 93, Fig. 2.8.3; G. R. Tibbetts, "Later Cartographic Developments," in *The History of Cartography* vol. 2 part 1, 153 Fig. 6.14; and F. Sezgin, *Mathematische Geographie und Kartographie im Islam und ihr Fortleben im Abendland. Historische Darstellung,* Teil I and Kartenband (Geschichte des arabischen Schrifttums, 10 and 12; Frankfurt am Main: Institut für Geschichte der Arabisch-Islamischen Wissenschaften an der Johann Wolfgang Goethe-Universität, 2000), 1: 20-22 and Kartenband, 23 no. 1a (rotated with North at top).

[10] *Abjad* letter-numerals are the letters of the Arabic alphabet given numerical values.

scale is over-painted with the "Mountain of the Moon". Had the numbering continued as presented here, with five degrees to a single large division and each subdivision equalling one degree, the last number written at the lefthand margin of the folio (that is, in the gutter of the manuscript), would have been 175, with the scale covering 180°. The scale on the facing folio remains unnumbered, presumably because the copyist realised his error—namely that he should have been counting two large divisions as equal to five degrees, and each of the smaller divisions as equal to half a degree—and so simply stopped. In other words, the scale should show 180° across the open bifolio. There is an additional anomaly in that the graticule on the lefthand page was laid out using a slightly larger proportional scale, so that it has only 28 larger divisions while the righthand side has 35. Other features, however, confirm that the two halves of the map, in the form it exists today, were constructed and drawn separately.

One must question the purpose of the scale, therefore. It could merely be a superfluous and decorative detail of no significance to the way the map was produced. Its inclusion could suggest that maps produced by such a grid were circulating. Crucially, the existence of this scale may, or may not, point to the existence of a mathematically plotted prototype.

In spite of the corruption of the scale as it exists on the present copy, it seems clear that the mapmaker of the *Book of Curiosities* did indeed use a mathematically plotted world map as a chief, though not exclusive, source in producing his rectangular world map. The plotted prototype was in all likelihood derived from a treatise of mathematical geography written in the ninth century by Muḥammad ibn Mūsá al-Khwārazmī, even if not actually copied directly from al-Khwārazmī's work. The argument rests, first of all, on evidence that the rectangular map itself provides. Secondly, other parts of the treatise also show reliance on early Islamic works of mathematical geography. Finally, we believe there is sufficient evidence to suggest that plotted maps were circulating in the eleventh-century Egyptian milieu in which the treatise was produced, and that one such map could have been available to the author of the *Book of Curiosities*.

The earliest extant maps from the Islamic world are found in a copy made in 1037 (428 H) of the *Book of the Depiction of the Earth* (*Kitāb ṣūrat al-arḍ*) by al-Khwārazmī (d. *c* 847), who was at the court of al-Ma'mūn in Baghdad. His treatise, as preserved in a single but incomplete copy, presents latitudes and longitudes for 545 cities, and is illustrated by four regional maps showing the Island of the Jewel, the Sea of Darkness (*al-baḥr al-muẓlim*), the Sea of Azov, and the River Nile.[11] Al-Khwārazmī's treatise consists of tables of geographic names grouped under five headings: towns, mountains, seas, islands, and springs or rivers. Under each heading, the names are arranged in terms of climes, beginning south of the Equator and working north. For each entry, the longitudes and latitudes are provided in degrees and minutes of arc. At least twice in his treatise of geographic tables, al-Khwārazmī states that he did not include a name for a locality because it was not given "on the map" (*fī al-ṣūrah*).[12] The tables provided by al-Khwārazmī strongly resemble those of Ptolemy, but there are substantive differences. Al-Khwārazmī adds some place-names, his latitudes vary considerably, and he positions the base meridian ten degrees to the east of Ptolemy's. The historian C. A. Nallino has explained these differences by suggesting that al-Khwārazmī was not working directly from Ptolemy's text, but rather from a world map based on Ptolemy.[13] If al-Khwārazmī did use such a map, then presumably it was a now lost Syriac or Arabic map and not a Greek one, and was not accompanied by tables; otherwise he would have used the tables themselves to compile his lists. In any case, it is evident that al-Khwārazmī had access to maps and thought in terms of maps. Even the title of his treatise could just as well be translated as *Book of the Map of the Earth*, since *ṣūrah* was a common term for "map" in addition to designating the general form or depiction of any object.

---

[11] Strasbourg, Bibliothèque Nationale et Universitaire, MS 4247. The four maps (on fols. 11b, 21a, 30b-31a, 47a) are reproduced in the edition of the text by Hans van Mžik, *Das Kitāb ṣūrat al-arḍ des Abū Ǧaʿfar Muḥammad ibn Mūsā al-Ḫuwārizmī* (Bibliothek Arabischer Historiker und Geographen, 3, Leipzig: Harrassowitz, 1926), Tafel 1-4. The maps are also reproduced in G.R. Tibbetts, "The Beginnings of a Cartographic Tradition," in *The History of Cartography* vol. 2 part 1, 105-106 and Plates 4-5.

[12] See al-Khwārazmī, *Das Kitāb ṣūrat al-arḍ*, 139, line 4, reading "a city that has no name on the map," and 77, line 9, "other (rivers ?) which are not named on the map."

[13] C.A. Nallino, "Al-Ḫuwārismī e il suo rifacimento della Geografia di Tolomeo," in M. Nallino (ed.), *Raccolta di scritti editi e inediti*, 6 vols. (Rome: Istituto per l'Oriente, 1939-48), 5:458-532, esp. 481-93; G.R. Tibbetts, "The Beginnings of a Cartographic Tradition," 100.

Lists of geographical coordinates, all influenced to a large degree by the Ptolemaic tables, acquired considerable importance in subsequent decades, and a large number of such tables exist in medieval Arabic literature.[14] In the following centuries, several scholars discussed methods of projection—that is, how to produce on a flat surface a map of the spherical earth, employing longitude and latitude tables. For example, Suhrāb (d. 930), also working in Baghdad, proposed and illustrated a rectangular grid for use in making a world map. These instructions were given in his *Marvels of the Seven Climes to the End of Habitation* (*ʿAjāʾib al-aqālīm al-sabʿah ilá nihāyat al-ʿimārah*), a catalogue of coordinates for 492 localities, heavily dependent upon the work of al-Khwārazmī. Suhrāb advocated placing a lateral scale of 180 degrees at the top and at the bottom of a map, and a vertical scale down each side, divided into 110 degrees with 90 degrees to the north and 20 degrees to the south of the Equator. Once the seven climes were marked on the map, the towns were to be plotted from their coordinates with the aid of a pair of weighted strings. The result of such a procedure would have been an orthogonal projection maintaining distances along the Equator and the meridians, but having a greater east-west stretch in the temperate zones than the projection proposed by Marinus of Tyre (fl. AD 100). No maps constructed in this manner have been preserved, but a diagram explaining Suhrāb's method is preserved in the unique manuscript copy of his treatise.[15] The scale at the top of the rectangular world map in *The Book of Curiosities* is strikingly suggestive of the method of mapmaking proposed by Suhrāb.

While no early Islamic map produced by any sort of projection is known to have survived, quite a few are reported in medieval sources. The most famous of them is the world map made for al-Maʾmūn (r.

---

[14] See E.S. Kennedy and M. H. Kennedy, *Geographical Coordinates of Localities from Islamic Sources*, Veröffentlichungen des Instituts für Geschichte der Arabisch-Islamischen Wissenschaften, Reihe A: Texte und Studien, 2 (Frankfurt am Main, Institut für Geschichte der Arabisch-Islamischen Wissenschaften an der Johann Wolfgang Goethe-Universität, 1987).

[15] Suhrāb, *Das Kitāb ʿaǧāʾib al-aḳālim as-sabʿa*, ed. Hans von Mžik (Bibliothek arabischer Historiker und Geographen 5; Leipzig: Harrossowitz, 1930). See also G.R. Tibbetts, "The Beginnings of a Cartographic Tradition," 104-105; King, *World-Maps*, 33 n. 61; and E. S. Kennedy, "Suhrāb and the world-map of Ma'mūn," in J. L. Berggren and B. R. Goldstein (eds.), *From Ancient Omens to Statistical Mechanics: Essays on the Exact Sciences Presented to Asger Aaboe* (Acta Historica Scientiarum Naturalium et Medicinalium 39: Copenhagen, University Library, 1987), 113-19.

813-833) by a group of scientists, including the above-mentioned al-Khwārazmī. The actual form of *al-ṣūrah al-Ma'mūnīyah* (the map of al-Ma'mūn) has been the subject of considerable speculation.[16] The encyclopaedist al-Mas'ūdī (d. 956), an authority frequently cited by the author of the *Book of Curiosities*, stated that the map made for the caliph al-Ma'mūn had climate boundaries. Few other details are available:

> I have seen these climates represented [*muṣawwarah*] in various colours in several books. The best that I have seen are in the book *Jughrāfiyā* [Geography] of Marinus and the commentary to *Jughrāfiyā* on the divisions of the earth. [I have seen them also] in *al-ṣurah al-Ma'mūnīyah* that al-Ma'mūn ordered to be constructed by a group of contemporary scholars to represent the world with its spheres, stars, lands, and seas, the inhabited and uninhabited regions, settlements of peoples, cities, etc. This was better than anything that preceded it, either the *Geography* of Ptolemy, the *Geography* of Marinus, or any other.[17]

Because al-Mas'ūdī compares al-Ma'mūn's map to the works of Ptolemy and Marinus, it seems plausible that it was based on tables of longitude and latitude. This suggestion is supported by the tables of longitude and latitude compiled by al-Khwārazmī, who was one of the scholars at al-Ma'mūn's court.

However, the only extant Islamic world maps that can be dated before AD 1100 come from the "Balkhī school". The rubric "Balkhī school" is used for four scholars of the tenth century, taking the name from Abū Zayd Aḥmad ibn Sahl al-Balkhī, who died in 934, having spent most of his working life in Iraq, particularly Baghdad.[18] His treatise on geography, titled *Illustration of the Climes* (*Ṣuwar al-aqālim*), contained one world map and twenty-one regional maps. No copies of his treatise are preserved today, but there are many copies of the illustrated treatises compiled by his three followers: al-Iṣṭakhrī (d. *c* 961), Ibn Ḥawqal (d. *c* 990), and al-Muqaddasī (d. *c* 1000).

---

[16] See G. R. Tibbetts, "The Beginnings of a Cartographic Tradition," 104-105; and F. Sezgin, *Mathematische Geographie*, 1: 73-140.

[17] al-Mas'ūdī, *Kitāb al-Tanbīh wa-al-ishrāf*, ed. M. J. de Goeje (Bibliotheca Geographorum Arabicorum 8; Leiden: Brill, 1894), 33 (compare 44); quotation (in slightly different translation) given by G.R. Tibbetts, "The Beginnings of a Cartographic Tradition," 96.

[18] G. R. Tibbetts, "The Balkhī School of Geographers," in *The History of Cartography* vol. 2 part 1, 108-36.

All of the treatises were illustrated with a world map and twenty-one regional maps. The focus of the "Balkhī school" treatises was different from that of the *Book of Curiosities*, because the authors were concerned with land routes, not sea routes, and in particular with land routes useful to trade, pilgrimage and the postal system (*barīd*) operative in the eastern lands, a system that depended upon relays of postal carriers over designated land routes. Moreover, the detailed regional maps covered only the world of Islam, with the world map acting as a loose outline or guide to how the Muslim world related to non-Muslim regions.

Plate V shows the inhabited world as configured in a typical world map produced by the "Balkhī School", in this case by the earliest of the group, al-Iṣṭakhrī.[19] It is a circular rather than rectangular map, with South also at the top. The outside ring enclosing the map represents the "Surrounding Sea". Only about one-quarter of the sphere was considered to contain inhabitable land, leaving three-quarters covered with water. Since so much of the sphere was water, there was no need to try to map it in any detail.[20] The sea that covered three-quarters of the globe is represented on this map by the ring of the "Surrounding Sea".

The landmasses are all depicted in the central circle. There is an enormous eastward extension of Africa so that nearly the entire southern quadrant (top) is land. The Mediterranean comes in from the right (West), and the Indian Ocean from the left (East). The vertical blue band of the River Nile comes from the top (South) into the Mediterranean. There are three large circular islands in the Mediterranean (Cyprus, Crete, and Sicily), and three in the Indian Ocean (the islands of Awāl, Ḥarak, and Lāfit in the Persian Gulf). Muslim Spain (al-Andalus) is clearly labelled at the top of the triangular landmass representing Europe. Neither Italy nor Greece is indicated in any way. The "Gulf (*khalīj*) of Constantinople" appears to connect with the "Surrounding Sea". In the lower half, the Caspian and Aral Seas are indicated by two nearly contiguous circles.

---

[19] Bodleian Library, MS Ouseley 373, folios 3b-4a, copied in 1297 (696 H), a Persian translation of *Kitāb al-Masālik wa-al-mamālik* (Book of Routes and Provinces) by al-Iṣṭakhrī.

[20] This contrasts with the numerous attempts to map the sphere of the heavens; see E. Savage-Smith, "Celestial Mapping," in *The History of Cartography* vol. 2, part 1, 12-70.

Between the dark blue band of water nearby these circular seas and the "Gulf of Constantinople" lies al-Rūs (north-eastern Europe) and the realm of Gog and Magog, indicated by an enclosed semi-circle. India, Tibet and China occupy the north-eastern coast of the Indian Ocean, while Baghdad lies at the triangle of blue water near the map's centre.[21]

By the author's own description of the map, the rectangular map of the world in the *Book of Curiosities* is based on Ptolemaic material. At the end of Book I (on the stars), the author states:

> This is the end of the Book I, with the blessings of God and His support. It is followed by the second book, consisting of twenty-five chapters. The first chapter [of Book II] is on the mensuration of the Earth and its map (*ṣūrah*), including the seas [?], from the Equator to the farthest limit of the inhabited world, which is at 66 degrees, as related by Ptolemy *al-qalūdhī* (Claudius) in his book known as *Geography*.[22]

The author here appears to conflate the first two chapters of Book II. While the mensuration of the Earth is indeed discussed in the first chapter of Book II, the author's reference to the map (*ṣūrah*) of the Earth appears to be to the rectangular map of the world, which makes up the entire second chapter of the same book.

The title of the rectangular world map is to be found at the end of the first chapter of the book, where the author says: "There follows the second chapter on the map of the earth (*fī ṣūrat al-arḍ*)."[23] The map focuses upon the lands known to be well inhabited—unlike the Balkhī world maps of the day that tried to show all the land-masses in the hemisphere and the edges of the inhabitable worlds. While the author claims that the map extends to "the farthest limit of the inhabited world, which is at 66 degrees," the rectangular world map does not in fact even represent the whole of the "inhabited world." Arab geographers customarily divided the inhabited area north of the Equator into seven zones, or climes, according to hours of maximum daylight. In this map, however, it is evident that only

[21] For an interesting analysis of this general type of circular world map with surrounding sea, see the doctoral dissertation by Karen C. Pinto, "Ways of Seeing. 3: Scenarios of the World in the Medieval Islamic Cartographic Imagination" (diss., Columbia University, 2002).

[22] Book I, chapter 10, fol. 22a.

[23] Fol. 23a, last line.

the first through sixth climes are illustrated in their entirety. The semicircular "Mountain of the Moon" lies on the Equator where the scale is located, while the boundary between the third and fourth climes would come, were it indicated, approximately halfway up the map. Almost no land south of the Equator is shown at the top of the map, and only select elements of the seventh clime—which included the lands of Gog and Magog and much of northern Europe—are illustrated.

On the map itself, the anonymous author incorporated both Ptolemaic and non-Ptolemaic elements derived from al-Khwārazmī's treatise of mathematical geography. This influence can most clearly be seen in the brown land mass on the lefthand side of the rectangular map. The "Island of the Jewel" is distinctive to al-Khwārazmī. It is a non-Ptolemaic island that, according to the illustration in al-Khwārazmī, lies close to the Equator, and is surrounded by the Sea of Darkness and a nearly encircling mountain range.[24] On the rectangular world map, a narrow portion extends upward and touches what would be the top of the scale, possibly past the Equator, while there is a northern (lower) extension that seems to run into an unidentified brown area filling the now damaged lower left corner of the map. Al-Khwārazmī also indicates the longitude and latitude location of the Island of the Jewel in the text of his treatise.[25]

Reconstructions of the relevant sections of al-Khwārazmī's world map, based on the coordinates he supplies in his *Depiction of the Earth* (*ṣūrat al-arḍ*), show the Island of the Jewel to be placed in a very similar position to that on the rectangular world map.[26] In the recon-

[24] For this island al-Khwārazmī generally used the name Jazīrat al-Jawhar (Island of the Jewel), as can be seen on his map of the island; occasionally, however, in the tables he also used Jazīrat al-Yāqūt (Island of Sapphires). For an illustration, see al-Khwārazmī, *Das Kitāb ṣūrat al-arḍ*, Tafel 1; or G.R. Tibbetts, "The Beginnings of a Cartographic Tradition," 105 Fig. 4.8.

[25] At various places in his geographic tables al-Khwārazmī provides coordinates for various coast lines of the island, for three localities on it, for two internal mountains, and for the surrounding mountain range. See al-Khwārazmī, *Das Kitāb ṣūrat al-arḍ*, 7-8, 40-42, and 83.

[26] Several reconstructions have been published; for example, Sezgin, *Mathematische Geographie*, Kartenband 4, no. 1b (using a modern projection, and shown with North at the top); and H. Daunicht, *Der Osten nach der Erdkarte al-Ḫuwārizmīs. Beitrage zur Historischen Geographie und Geschichte Asiens. Bd. I: Rekonstruktion der Karte, Interpretation der Karte: Südasien*, Bonner Orientalistische Studien, Neue Serie 19 (Bonn: Selbstverlag des Orientalischen Seminars der Universität, 1968). See also G.R. Tibbetts, *A Study*

structions of al-Khwārazmī's maps, the Island of the Jewel lies to the east of a large peninsula protruding southwards. This large peninsula is distinctively Ptolemaic, as can be seen in any of the late medieval reconstructions of Ptolemy's maps, as well as later printed Ptolemaic maps.[27] It is not found on the world maps of the "Balkhī school," which depict the northern coasts of the Indian Ocean as a straight line. Both in the reconstructions of al-Khwārazmī's world map and the Ptolemaic maps, as well as in the rectangular world map, the labels for this peninsula indicate localities in India and China. The prominence of the island of Ceylon (Sri Lanka) nearby this large peninsula, which is distinctive to both the Ptolemaic maps of this region and the al-Khwārazmī reconstructions, provides an explanation for the unnamed circular island highlighted by bits of reflective gold glitter, nearby the peninsula on the rectangular world map.

Another distinctive Ptolemaic feature incorporated into al-Khwārazmī's work and then reflected in our map is the Mountain of the Moon at the source of the Nile, along the Equator. Al-Khwārazmī not only lists the Mountain of the Moon in his tables, but also depicts it in his map of the Nile—one of his four surviving maps. The similarity between the depiction of the Mountain of the Moon by al-Khwārazmī and its rendering in the rectangular world map is remarkable. Although most of the place-names on al-Khwārazmī's map are contemporary to him, the name of the Mountain of the Moon is taken from Ptolemy. The parachute-like depiction of the Mountain of the Moon is not found in the "Balkhī school" circular world maps, although it is later incorporated into the world maps accompanying copies of al-Idrīsī's *Entertainment for One Who Longs to Travel the World* (*Nuzhat al-Mushtāq fī ikhtirāq al-āfāq*) composed about 1154 for Roger II, the Norman king of Sicily.[28]

In addition, two general features of the rectangular world map strongly suggest that its prototype was mathematically plotted. First,

---

of the Arabic Texts Containing Material on South-East Asia (Leiden: Brill and London: Royal Asiatic Society, 1979), Fig. 1b.

[27] For example, Milan, Biblioteca Ambrosiana, MS gr. 997, D 527 inf. (early 14th century), where the large peninsula can be seen in the world map, fols. 94v-95r, and the separate map of Asia and the Indian Ocean, fols. 99v-100r; for reproductions, see Sezgin, *Mathematische Geographie*, Kartenband, 63 no. 31 and 66 no. 31c.

[28] It is also to found on the circular world map in the *Book of Curiosities*, fols. 27b-28a.

the vast majority of labels refer to localities, specifically cities, and cities were the focus of tables of coordinates forming the basis of mathematical geography. In contrast, on surviving "Balkhī school" world maps the labels refer almost exclusively to regions or provinces. The world maps of the Balkhī school, as noted by the historian of Islamic cartography Gerald Tibbetts, are "an arm-chair attempt to see all the provinces set down relative to each other."[29] For this reason, the surviving world maps of al-Iṣṭakhrī and Ibn Ḥawqal carry labels designating the Islamic provinces, which are then illustrated in detail in the accompanying regional maps. Moreover, in the "Balkhī school," the delineation of the non-Islamic kingdoms is restricted to the world map, and plays no role in the rest of the treatise or maps. In this kind of non-plotted world map, which aims to give only a general overview of the landmasses, there is no place for individual localities such as cities.[30] World maps with labels primarily designating regions rather than cities are also typical of later non-plotted maps, such as those of al-Idrīsī and others. Such maps differ substantially from plotted maps, where the emphasis is on the location of individual cities.

The final indicative feature of the *Book of Curiosities* map is the rectangular shape itself. As early Islamic scholars knew that the world is a sphere, and as they assumed that only one hemisphere was inhabited, they usually chose to present the Earth as a circle or disc. This was the most obvious way of depicting the globe on the flat surface of a map, given that the other hemisphere was nothing but water. The world maps of the "Balkhī school" are shaped as circles, and so are practically all the non-plotted world maps in the later Islamic tradition. A notable exception is an oval world map in a late recension of Ibn Ḥawqal copied in 1445.[31] This unique oval world map differs markedly from other "Balkhī school" maps. The Ptolemaic Nile system in the south and al-Khwārazmī's "Island of the Jewel" in the east are both present, as in the *Book of Curiosities* rect-

[29] Tibbetts, "The Balkhī School of Geographers," 120.

[30] An exception that proves the rule is a late fifteenth-century Timurid map, which is categorized by Tibbetts as being of the Balkhī school in spite of possessing climate boundaries: see "The Balkhī School of Geographers," 127-28, fig. 5.25.

[31] Paris, Bibliothèque nationale de France, MS arabe 2214, fols. 52b-53a, copied 1445 (847 H); for an illustration see G.R. Tibbetts, "Later Cartographic Developments," 140, Fig. 6.3.

angular world map, although they are not found in other "Balkhī school" world maps. The oval shape and the presence of elements from both Ptolemy and al-Khwārazmī led Tibbetts to conclude that "the editor of this revision, whether or not it was Ibn Ḥawqal, must have used a copy of a world map on a Ptolemaic projection rather than on the rectangular projection recommended by Suhrāb."[32] In other words, Tibbetts felt that the oval format strongly suggested a plotted prototype, but one employing Ptolemy's second projection that yielded a curved map rather than a rectangular one.

In any case, a deviation from the convention of representing the world as a circle has to be explained, and the most obvious explanation is an attempt at mathematical projection, which meant an oval in the case of a Ptolemaic projection and a rectangle in the case of the projection suggested by Suhrāb.

The influence of al-Khwārazmī upon the *Book of Curiosities* is particularly evident in the separate map of the River Nile, one of five rivermaps that form the eighteenth chapter of the second book (see Plate VI). The separate map of the River Nile in the *Book of Curiosities* shows remarkable similarity to the surviving map of the Nile made by al-Khwārazmī. It is also similar to another separate map of the Nile found in the same late recension of Ibn Ḥawqal's work that contained the oval world map, a version clearly influenced by al-Khwārazmī.[33] As in our rectangular world map, the "Mountain of the Moon" forming the source of the River Nile in this Ibn Ḥawqal manuscript is indicated by the parachute-like formation that was also employed in al-Khwārazmī's Nile map. The general layout of the river map, together with the division into climes, in both the *Book of Curiosities*' separate Nile map and that in the Ibn Ḥawqal manuscript, is also similar to Khwārazmī's.

In the delineation of the Nile in the *Book of Curiosities*, both in its separate map as well as in its rectangular world map, an additional tributary emerges from "white dunes" in west Africa. This feature is not found in either al-Khwārazmī's or Ibn Ḥawqal's maps of the Nile. Moreover, the maker of the rectangular world map has placed Fazān and Zaghāwah correctly on a tributary coming from the west,

---

[32] G.R. Tibbetts, "Later Cartographic Developments," 138.

[33] Paris, Bibliothèque nationale de France, MS arabe 2214, fol. 13b, copied 1445 (847 H); for an illustration see G.R. Tibbetts, "Later Cartographic Developments," 139, Fig. 6.2.

rather than maintaining the misplaced positions on Khwārazmī's map, where they are on a tributary coming from the east.

Most unusually, the labels on the separate Nile map in the *Book of Curiosities* include some longitude and latitude coordinates. Thus, the Mountain of the Moon is noted to lie between 46° and 59° longtitude. The diameters of the marshes, and the distances between the tributaries of the Nile are indicated by celestial (*falakīyah*) degrees, which are then converted into miles. For example, the diameter of the eastern marsh is given as "five celestial degrees, equivalent to 284 miles." The centre of the largest marsh is given as 58° (possibly a mistake for 53°) longitude and 2° latitude, and the location of al-Fayyūm is given as longitude 48° 5' and latitude 30°. Labels with indications of coordinates are not found on the surviving river map of al-Khwārazmī, but they may have been on other now lost copies. The text of al-Khwārazmī's work provides longitude and latitude values for all these localities, but with slightly different values.

All in all, it is clear that the author of the *Book of Curiosities* used a work of mathematical geography for the compilation of his work and the production of at least some of his maps. On the map of the River Nile, which is remarkably similar to the extant map of the Nile made by al-Khwārazmī, there are indications of longitudes and latitudes that strongly suggest the author was using a work of mathematical geography which contained maps as well as tables. Specifically, the general features of the rectangular world map in the *Book of Curiosities* suggest that the author had before him a mathematically plotted world map. The rectangular shape of the map—unlike the circular forms common to almost all other Islamic world maps—suggests an orthogonal projection similar to that advocated by Suhrāb. The abundant indication of cities, rather than merely regions or provinces, is also highly unusual for a world map of this period. These features, together with the presence of the scale, even in its corrupt form, point to a model or prototype that was mathematically plotted.

Further evidence for reliance upon treatises of mathematical geography is the fact that the edges of the inhabited landmasses were illustrated with features distinctive to al-Khwārazmī and not used by the "Balkhī school" nor by the later al-Idrīsī. The "Island of the Jewel" at the easternmost limits of the inhabitable world is shown as it appears in al-Khwārazmī's separate map of the island. The large peninsula combining Persia and India and parts of China was a Ptolemaic feature incorporated by al-Khwārazmī into his tables, and reflected here in a similar peninsula on the rectangular world map. The very prominent Mountain of the Moon and its marshes forming the source of the Nile were also derived from the maps known to be associated with al-Khwārazmī.

It seems likely that the author used a prototype world map that was either based directly upon the tables of al-Khwārazmī or on similar tables, employing an orthogonal projection that resulted in a rectangular shape. Such a map would have included a scale, numerous labels for individual localities, and some distinctive "Khwārazmian" features. Although no such maps survive, there are several references to their circulation. It seems, however, that the author of the rectangular world map had an inadequate understanding of mathematical geography, and did not fully comprehend either the purpose of the scale or the tables of longitude and latitude from which he was copying. Nonetheless, this map, with its unique scale, remains the earliest preserved one providing evidence of a plotted prototype.

NEW PERSPECTIVES ON PARADISE— THE LEVELS OF REALITY IN BYZANTINE AND LATIN MEDIEVAL MAPS

Maja Kominko

This paper concentrates on certain aspects of the depiction and description of Paradise during Late Antiquity and the Early Middle Ages which seem to be illustrative of a particular perception of reality and space. The principal focus is on the map of the world in the *Christian Topography*[1] and in the *Commentary on the Apocalypse* of Beatus of Liebana.[2]

The *Christian Topography* was written anonymously.[3] Sources from the eleventh century onwards attribute it to a certain Cosmas Indicopleustes, that is "Cosmas who sailed to India", but the authenticity of this name has been contested.[4] Though not revealing his name, the author provides some information concerning himself and the circumstances in which his treatise was written. On the basis of the internal references we can establish that he wrote around 547-549.[5] We know that by then he was already quite old and had retired from his mercantile career, which earlier in life had carried him far in commercial pursuits. He claims to have sailed on the Mediterranean, Red Sea and Persian Gulf.[6] Although it is not clear whether he ever

[1] See W. Wolska-Conus, *Cosmas Indicoplèustes. Topographie Chrétienne, Sources Chrétiennes,* (Paris: Éditions du Cerf, 1968, 1970, 1973), 141, 159, 197. My numbering of the paragraphs of the text follows from this edition.

[2] See J. Williams, *The Illustrated Beatus. A Corpus of the Illustrations of the Commentary on the Apocalypse.* 5 vols. (London: Harvey Miller, 1994-2000).

[3] He only refers to himself as a *Christian*: see *Christian Topography* V:257; VII:1; VII:96-97; VIII:31.

[4] The name Cosmas accompanied by the epithet *Monk* appears in one of the eleventh century manuscripts of the treatise, Laur.Plut.IX.28, and in the manuscripts of Gospels and Psalters which quote the *Christian Topography* in the commentary. Wolska-Conus, *Topographie Chrétienne,* v.1, 107-115; See also W. Wolska-Conus, *La Topographie Chrétienne de Cosmas Indicoplèustes. Théologie et Science au VIe siècle* (Paris: PUF, 1962), 1.

[5] For dating of the *Christian Topography*, see Wolska-Conus, *Topographie Chrétienne,* v.1, 16.

[6] *Christian Topography* II:54-56; VI:1-2.

went as far as India in the modern meaning of this word, we should bear in mind that in Byzantine sources the term *India* designated not only the Indian subcontinent, but also Arabia and Ethiopia.[7] Thus, because it is clear that the author of the *Christian Topography* travelled at least as far as Axum, there is no need to challenge the appropriateness of his nickname. It is interesting to note that a sixth century inscription found in 'Abu Sha'ar on the Red Sea mentions a certain Andreas Indicopleustes, attesting the use of such an epithet during Late Antiquity.[8] As to the name itself, Cosmas was very common in Egypt, but it is suspiciously appropriate for the author of the work on the cosmos, and therefore its authenticity should be treated with caution.[9]

Whether the author of the *Christian Topography* was an Egyptian has also been disputed. Snippets of information scattered throughout the treatise make it fairly clear that it was written in Alexandria, and that the author was an inhabitant of that city rather than a visitor.[10] However, his theological views and his exegetical method appear to be as foreign to those of other Alexandrian writers as they are close to those of Antiochene and East Syrian exegetes. It was largely on account of this affinity that an attempt to identify Cosmas with a certain Constantinus of Antioch has been greeted with great enthusiasm by modern scholars.[11] In my view this identification is doubtful. It is based only on a short passage of the seventh century Armenian Geography which mentions a certain Constantinus of Antioch as the author of the *Christian Topography*. Unlike Alexandria, Antioch is

---

[7] P. Mayerson, "A Confusion of Indias: Asian India and African India in the Byzantine Sources," *Journal of the American Oriental Society*, 113, 2 (1993), 169-74.

[8] R.S. Bagnall, J.A. Sheridan, "Greek and Latin Documents from 'Abu Sha'ar, 1992-1993," *Bulletin of the American Society of Papyrologists* 31 (1994), 109-20.

[9] It has been suggested that this name was invented by a later scribe in much the same way as the epithet "*Climacus*" given to John, the author of the *Heavenly Ladder*. See C. R. Beazley, *The Dawn of the Modern Geography: A History of Exploration and Geographical Science from the Conversion of the Roman Empire to AD 900*, v.1 (London: J. Murray, 1897), 277. Even so, on this basis the author of the *Christian Topography* should have been called Cosmicus rather than Cosmas. See E.O. Winstedt, *The Christian Topography of Cosmas Indicopleustes, edited with geographical notes* (Cambridge: Cambridge University Press, 1909), 2.

[10] For an extensive discussion of the problem, see M. V. Anastos, "The Alexandrian Origin of the Christian Topography of Cosmas Indicopleustes," *Dumbarton Oaks Papers* 3 (1946), 73-80.

[11] W. Wolska-Conus, "Stephanos d'Athènes et Stephanos d'Alexandrie: Essai d'identification et de biographie," *Revue des Études Byzantines* 47 (1989), 28-31.

barely mentioned in the *Christian Topography,* and the link with Syrian theology is in fact explained in the passage where Cosmas proclaims himself to be a devoted student of Patrikios, identified with Mar Aba, a Nestorian *katholikos* of Persia, who had visited Alexandria a few years previously.[12] It thus appears that it was through the intermediation of Mar Aba, a student and later teacher at the school in Nisibis, that the influence of the East Syrian and Antiochene exegetes was transmitted into the *Christian Topography.* In fact Cosmas goes as far as claiming that all his knowledge derives from the Bible and his master Mar Aba, since he had not received any formal education. The claim to lack a proper schooling, and in particular the claim to lack any rhetorical ability, is a common literary *topos* usually attesting to the contrary.[13] In this case, however, it does ring true. In short, the literary style of the *Christian Topography* is not very sophisticated.[14] But dearth of rhetorical skills does not imply dearth of any sort of knowledge. Cosmas may be a bad writer, but he is certainly well read, both in theology as well as in ancient science and philosophy. It should be noted that he engaged in scholarly discussion with John Philoponus, his contemporary fellow citizen of Alexandria, whose scholarship helped to fuel the Renaissance breakaway from Aristotle.[15] Cosmas was an ingenious compiler, and in an effort to understand the world around him he brought together and sought to reconcile a variety of sources which reflect the diverse and complicated milieu in which he was writing.

The original manuscript of the *Christian Topography* is not preserved. There are three Byzantine copies extant, dated to the ninth (Biblioteca Apostolica Vaticana, Cod. Gr. 699) and eleventh centuries

[12] *Christian Topography* II:2; On the identification, see P. Peeters, "Observations sur la vie syriaque de Mar Aba, catholicos de l'église Perse (540-552)," in *Recherches d'Histoire et de Philologie Orientales, Subsidia Hagiographica* 27, v. I (Brussels: Soc. des Bollandistes, 1951).

[13] Such a claim appears already in the second Pauline epistle to the Corinthians (2 Cor. 11:6). It can be found later, for example, in the works of Theophilos of Antioch, *Ad Autolycum* 1:1, and of Orosius, who at the beginning of his *History* adopts an extended metaphor in which he compares himself to a dog (Preface, 1-8).

[14] See criticism by Photius, who quotes the *Christian Topography* in codex 36 of his *Bibliotheca.*

[15] See R. Sorabji, "John Philoponus," in R. Sorabji (ed.), *Philoponus and the Rejection of Aristotelian Science* (London: Duckworth, 1987), 5-18; also B. Elweskiöld, *Cosmas Indicopleustes and John Philoponus. A Christian controversy on the structure of the world in sixth-century Alexandria* (diss., Lund University, 2005).

(Sinai, Saint Catherine Monastery, Cod. Gr. 1186; Biblioteca Medicea Laurenziana, Cod. Plut. IX.28). All three are lavishly illustrated, and most of the illustrations in all three are the same. This uniformity, and the close relationship between the text and the scientific illustrations, suggest that the miniatures in the preserved manuscripts copy those in the original, sixth century codex. The case of the Biblical illustrations is slightly more complex, but that is a subject for another paper.[16]

Cosmas imagines the universe in the form of an oblong, vaulted building, based on the flat earth and divided into two superimposed spaces by the firmament placed in the middle of its height.[17] These two spaces correspond to two states of existence: the state of mortality in the space below where people currently live, and the state of immortality and perfection in the space above, which will open only at the end of times.[18] The revealed image of the world is the tabernacle built by Moses in the wilderness, described in Exodus and depicted in the miniatures of the *Christian Topography*. The Holy of the tabernacle corresponds to the lower space, while the Holy of Holies reflects the upper, heavenly space. [19]

According to Cosmas, the earth is flat (Fig. 1). The inhabited earth, placed in the middle, is surrounded by the Ocean, which in turn is encircled by another earth, where in the east Paradise is located.[20] Four seas—four gulfs of the ocean—break the outline of the oikoumene.[21] The rivers of Paradise travel under the Ocean and reappear in the inhabited earth.[22] For Cosmas, the revealed image of the earth is the table of the tabernacle. Its description and depiction in the *Christian Topography* (Fig. 2)[23] are based on the Biblical

[16] For the relationship between the text and illustrations, see L. Brubaker, "The Relationship between Text and Image in the Byzantine MSS of Cosmas Indicopleustes," *Byzantinische Zeitschrift* 70 (1977), 42-57. My recently completed Oxford doctoral thesis addresses the question of the sources for the miniatures in the Byzantine manuscripts of the *Christian Topography*.

[17] *Christian Topography* II:20; II:21; III:14; IV:4; IV:9; V:227.

[18] See Wolska, *Théologie et science*, 87-112.

[19] *Christian Topography* II:35-36; III:16; III:51-52; III:55; IV:25; V:1; V:20; V:27. See also Wolska, *Théologie et science*, 113-46.

[20] *Christian Topography* II:17; II:19; II:24; II:36; III:52; IV:11.

[21] *Christian Topography* II:29; II:30.

[22] *Christian Topography* II:81.

[23] *Christian Topography* II:19; II:36; V:33-34; Sin.gr.1186, fol. 81r; Laur.Plut. IX.28, fol. 111v. See also: L. Brubaker, "The Tabernacle Miniatures of the Byzantine

Fig. 1. The Earth according to Cosmas, Laur.Plut.IX.28, fol. 92v, eleventh century.

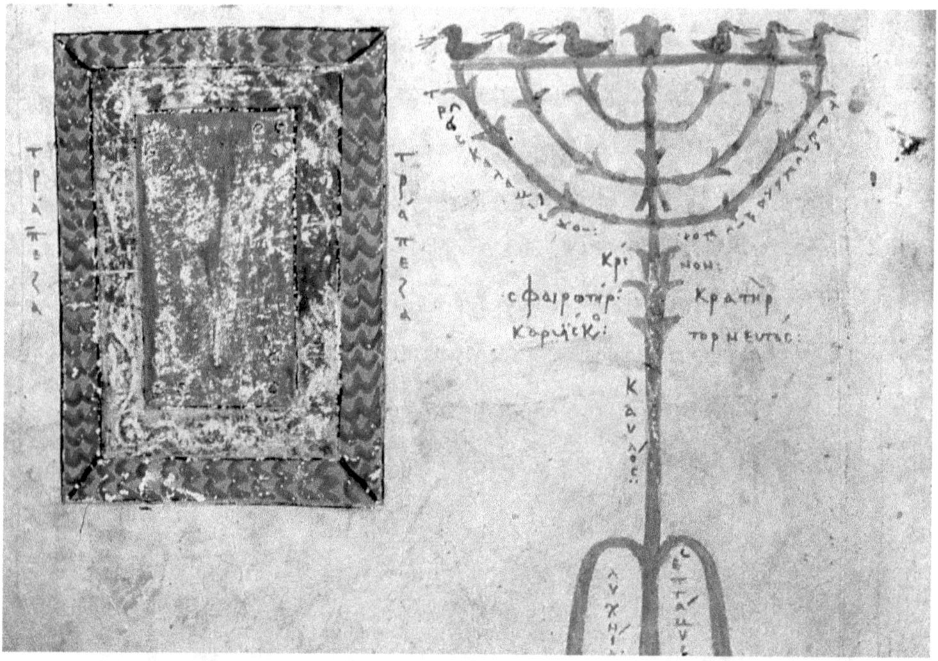

Fig. 2. Table of the Tabernacle and Menorah, Laur.Plut.IX.28, fol. 111v.

account in Exodus 25:23-24. Original to Cosmas is the minute correspondence he draws between the structure of the table and that of the earth. According to him, both the table and the earth are twice as long as wide. The oikoumene corresponds to the middle of the table, the Ocean to the crown around it, and the earth beyond the Ocean to the outer border.[24] No other source describes such close correspondence between the structure of the earth and that of the table. Moreover, the description of the table in the *Christian Topography* differs a little from that in the Bible. Both Exodus and the *Christian Topography* describe the middle part of the table as surrounded by a wreath, which in turn is surrounded by a crown. In Exodus, however, another wreath is described surrounding the crown. Cosmas omits this from his description. The discrepancy is difficult to explain, but the divergence from the Biblical text suggests

Octateuchs," *Actes du XVe Congrès International d'Etudes Byzantines* II: *Art et Archéologie* (Athens: Association Internationale des Etudes Byzantines, 1981), 73-92.

[24] *Christian Topography* II:19; II:43; V:34.

that Cosmas followed that text a little less slavishly than has hitherto been believed.[25]

A very similar structure of the earth, consisting of the oikoumene, surrounded by the Ocean, which in turn is surrounded by the other earth where Paradise is located, is described by Ephrem the Syrian (306-373)[26] and Narsai (died 502).[27] Furthermore, to illustrate the structure of the earth, Ephrem compares it with the altar built by Moses described in Exodus 27:1.[28] The similarity of this comparison to that in the *Christian Topography* suggests that Cosmas was probably inspired by an East Syrian source. Yet the fact that Cosmas compares the earth to the table rather than to the altar is striking; the altar is square, while the table is twice as long as wide, a proportion given to the earth by many ancient geographers.[29] This is not the only characteristic of Cosmas' earth which tallies with the accounts of ancient geographers. The notion of the oikoumene as an island was accepted by the majority of ancient geographers.[30] The seas, including the Caspian, were usually described as the gulfs of the Ocean.[31] Moreover, the Black Sea on Cosmas' map is depicted in the form of two semicircles aligned on top of a straight line. The text of Cosmas' treatise offers no explanation for such a shape. However, Strabo and Pliny the Elder both compare the shape of the Black Sea to a Scythian bow made of two semicircular horns joined to one another.[32] On

[25] However, this is not the only way in which Cosmas' description of the table deviates from the Biblical text. For him, the table is not only the image of the earth, but also a symbol of a year, with the twelve loaves, placed on the table, symbolising the twelve months of the year. For this reason, according to Cosmas, the loaves were divided into four batches of three, one batch in every corner, symbolising four seasons of three months each. This description runs counter to the Biblical text, where in Leviticus 24,5-6 the loaves are described as distributed in two rows of six. See *Christian Topography* III:51; V:20; V:34; IX:8.

[26] Ephrem, *Hymns on Paradise,* 1, 8-9; 2,6.

[27] Narsai, *Homilies on Creation,* I:396-398; I:404; III:308-325.

[28] Ephrem, *Hymns on Paradise,* 1, 8-9.

[29] See Agathemerus, *Geographiae informatio,* 1,2; Geminus, *Introduction to Phenomena,* 16,3-5.

[30] Exceptions are Herodotus, *History,* 4,45; Hipparchus, quoted by Strabo, *Geography,* 1,1,9; Ptolemy, *Geography,* 7,5,2.

[31] A very widespread error, opposed by Herodotus, *History,* 1,203; Aristotle, *Meteorology,* 2,1,354a; Hipparchus, quoted in Agathemerus, *Geographiae informatio,*1,2; Ptolemy, *Geography,* 7,5.

[32] Strabo, *Geography,* 2,5,22; Pliny, *Natural History,* 4,7. See also M. Kominko, "The Map of Cosmas. The Albi Map and the Tradition of Ancient Geography," *Mediterranean Historical Review* 20,2 (2005), 169.

the whole, Cosmas gives an image of the oikoumene which essentially corresponds to that described by ancient geographers. He correctly locates the sources of the Blue Nile in Ethiopia.[33] Although at some point he confuses the Indus and the Ganges,[34] his geography is unaffected by the miraculous origin of the rivers of Paradise; rather, he believes that their sources in the oikoumene are merely the points where they resurface after travelling through the Ocean. In effect he maintains a division between what could be explored and known, that is the oikoumene, and the miraculous, that is Paradise, placed beyond impassable Ocean.

This clear separation is unusual in Late Antiquity. According to the Greek version of Genesis 2:8, Paradise was planted in the east.[35] Although some of the Christian writers took Paradise to be allegorical,[36] most understood it literally and believed that it was "under this very heaven in the east".[37] The fact that Paradise was hidden and inaccessible to men made any discussion of its location quite problematic. The issue was further complicated by the existence of the four rivers of Paradise (Genesis 2:10-14), which connected it with the oikoumene. Three of them—Tigris, Euphrates and Gheon—were easily identified. The first two were called by their own names, while the identification of Gheon as the Nile was prompted by the fact that Greek translators rendered "mouth of Horus" in Jeremiah 2:18 as "waters of Gheon."[38] Identification of

---

[33] *Christian Topography* II:81.

[34] *Christian Topography* XI:24.

[35] Septuagint renders by *kata anatolas* the Hebrew expression *miqqedem,* which can be taken to mean both *in the beginning* and *in the east.* Aquila, Symmachus and Theodotion all take it to have temporal meaning (Aquila *apo archethes,* Symmachus *ek protes,* Theosotion *en protois),* and so does Jerome, rendering it as *in principio*; but Vetus Latina has *in orientem* .

[36] Origen, *Homily on Numbers*, 16,5; *Dialektikos,* 8. See also J. Daniélou, "Terre et Paradis chez les Pères de l'Église," *Eranos Jahrbuch* 22 (1954), 433-72; M. Alexandre, "Entre ciel et terre: les premiers débats sur le site du Paradis (Gen. 2,8-15 et ses réceptions)," in B. Deforge (ed.), *Peuples et pays mythiques* (Paris: Les Belles Lettres, 1988), 187-224 ; A. Scafi, "Mapping Eden: Cartographies of the Earthly Paradise," in D. Cosgrove (ed.), *Mappings* (London: Reaktion Books, Ltd., 1999), 50-70.

[37] Theophilus of Antioch, *Ad Autolycum,* 2, 24; Epiphanius, *Ancoratus, Patrologia Graeca* 58, col. 120 BC. On the medieval and early modern theories concerning the geographical location of Paradise, see M. Alexandre, "Entre ciel et terre," 194-98.

[38] This verse from Jeremiah is quoted by Severianus of Gabala to support the

Phison brought more difficulty. Some of the Christian writers believed it to be the Danube,[39] others identified it with the Ganges.[40] Others still made double-identifications: Pseudo-Caesarius believed Phison to be the Danube, which was the same as the Indus,[41] and Epiphanius thought Phison to be the Indus, which was the same as the Ganges.[42] At the same time it was generally understood that the Tigris and Euphrates have their sources in Armenia, while the Nile flowed from somewhere in Ethiopia. In response to the discrepancy between this information and the theory that the rivers originated in Paradise, writers took various attitudes. Some, like Isidore, disregarded geographical knowledge where it contradicted the Scriptures, or suggested that despite identity of names Euphrates and Tigris are not the real rivers of Paradise.[43] But many writers opted for a solution close to that described by Cosmas, and they pictured the rivers of Paradise as flowing underground for the initial part of their course.[44]

We should bear in mind that the idea of a river flowing underground and resurfacing again appears very frequently in ancient sources.[45] Some rivers are described as travelling not only underground, but also through a sea without mingling their waters with it.[46] Probably the most cited was the case of Alpheus, which was

identification (*Homilies on Creation, Patrologia Graeca* 56, col. 479). Epiphanius, *Ancoratus, Patrologia Graeca* 43, col. 117 C. It should be perhaps noted that this identification of the river appears earlier in Jewish sources (Book of Jubilees 2,18; Josephus, *Jewish Antiquities,* 1,38-39).

[39] St. Ephrem, *The Commentary on Genesis* 2,4; Severianus of Gabala, *Homilies on Creation, Patrologia Graeca* 56, col. 478.

[40] Avitus, *De Mosaicae Historiae Gestis, Patrologia Latina* 59, col. 329-30 ; also Josephus, *Jewish Antiquities,* 1,38; Jerome, *Onomastica Sacra,* 155.

[41] Caesarius, *Dialogi, Patrologia Graeca* 37, col. 852.

[42] Epiphanius, *Ancoratus, Patrologia Graeca* 43, 117 C.

[43] Isidore, *Etymologies*, 13, 21, 10.

[44] Philo, *Questions and Answers on Genesis,* 8; Augustine, *De Genesi ad Litteram,* 8,7; Severianus of Gabala, *Homilies on Creation,* 5,6, *Patrologia Graeca* 56, col. 478-79; Theodoret of Cyr, *Questions on Genesis*, 29.

[45] Pliny, *Natural History,* 5,51 (theory of Juba II concerning the course of the Nile); Strabo, *Geography*, 6,2,9 (Orontes in Syria, the Tigris in Mesopotamia and the Nile in Libya); Procopius, *Buildings,* 2,2,16 (a river that went underground in Dara and re-emerged in Theodosiopolis).

[46] For example Inopos: see Strabo, *Geography*, 6,2,4.

believed to travel thus from the Peloponnese to Sicily.[47] The story was so widely known that it was used as a rhetorical figure by Libanius and by Gregory of Nazianzus. [48] There is in fact an indication that Ephrem, the first writer to picture the rivers of Paradise as travelling through the Ocean and underground, does so under the inspiration of Greek sources; to describe the way in which the rivers reach the oikoumene he uses a Greek loan word *kantharos*.[49] Ephrem's inspiration for placing Paradise in the earth surrounding the Ocean appears to originate from quite a different, eastern tradition, which we find reflected on the Babylonian world map, where the land encompassing the Ocean is associated with mythical creatures.[50] It is quite striking that similar land surrounding the Ocean appears on some much later Arabic maps, where it is also associated with a fantastic dimension. Although no direct link can be demonstrated, it appears possible that all these placements reflect a similar tradition. The appearance of the land surrounding the Ocean on the maps is particularly fascinating as these accompany texts which describe a spherical earth, and therefore are difficult to reconcile with the concept of the all-encompassing land.[51] We should also not forget that the eschatological dimension was associated with lands beyond the Ocean in many ancient sources. Circe instructs Odysseus to "cross through the Ocean" on his way to the underworld.[52] Hesiod locates

[47] Polybius, *Historiae,* 12,4d; Virgil, *Eclogue* 10,1-5;21-22;47-49 and *Aeneid* 3, 694-696; Ovid, *Metamorphoses* 5,600-41; Strabo, *Geography*, 6,2,4; Seneca, *Quaestiones Naturales,* 3,26,5; 6,8,2; Pliny, *Natural History*, 2,225; 31,55; Pausanias, *Description of Greece*, 5,7,3; Lucian, *Dialogi Marini*, 3,1,1-3,2,11; Ausonius XXIX,91-96; Claudian IV,9,12; Nonnus, *Dionysiaca,* 13,322; 37,172; 42,98; John of Lydus, *De Mensibus*, 4,78,1; Suidas, *Lexicon*, "Adrias"; "Arethusa" (10th century A.D).

[48] Libanius, *Oratio* 11,68; Gregory in the Funeral oration to Basil (c. 380); Sermon 43,2; and *On Virtue*, 1,2,10. See also J. Nimmo Smith, "The River Alpheus in Greek, Christian and Byzantine Thought," *Byzantion* 74,2 (2004), 416-32.

[49] See *St. Ephrem Selected Prose Works,* trans. E. Matthews, J.P. Amar, ed. K. McVey (Washington: Catholic University of America Press, 1994), 101, n. 128.

[50] A.R. Millard, "Cartography in the Ancient Near East," in J.B. Harley and D. Woodward (eds.), *The History of Cartography*, vol. 1 (Chicago: University of Chicago Press, 1987), 111-12.

[51] See G.R. Tibbets, "Later Cartographical Developments," in *The History of Cartography,* vol. 2.1, 149-55; S. Maqbul Ahud, "Cartography of al Sharif al idrisi," ibid. 160-63; 170-71.

[52] Homer, *Odyssey*, 10,508.

the Isles of the Blessed "beyond glorious Ocean,"[53] and likewise beyond it lies his island of the heroes.[54]

Neither the association of the mythical or eschatological dimension with the land beyond the Ocean, nor the concept of rivers travelling underground or under the sea, were foreign to Greek and Latin sources. And yet Christian writers did not use them the way Cosmas did, to detach Paradise from the oikoumene. One conceivable reason for this difference could be simply that they did not believe the oikoumene to be surrounded by Ocean. But such reasoning would seem to be false. Although the Ocean surrounding the inhabited earth is not mentioned in the Biblical text, the Biblical description of the gathering of waters below the firmament in one place was sometimes taken to indicate that only a single body of waters can exist, and thus all seas must be interconnected. This concept probably corroborated the idea of the existence of one surrounding Ocean, described by many exegetes.[55] Widespread acceptance of this idea is confirmed by a passage of the *Soghita* written for the dedication of the cathedral church in Edessa; here the church, located between a river and a pond, is described as follows: "Indeed, it is an admirable thing that in its smallness it should resemble the great world, not in size but in type; waters surround it like the sea."[56]

In consequence it is difficult to understand why the Christian writers did not take the chance to solve the difficult issue of the relationship between the oikoumene and Paradise by placing the latter beyond the Ocean, as Cosmas did. No doubt such a placement could have been facilitated by a perception of the Orient as fabulous and fantastic, one stemming largely from a tradition inspired by the accounts of Ktesias and the Alexander historians.[57] This is not to say that the Orient was invariably perceived as fantastic and mythical.

[53] Hesiod, *Theogony,* 215, 274, 294.

[54] Hesiod, *Works and Days*, 170-174.

[55] The Ocean surrounding the earth is explicitly mentioned by Ephrem, *Hymns on Paradise,* 1,10; Basil, *Hexamaemeron,* 35E—36E; Eusebius of Caesarea, *In Praise of Constantine,* 6,6; Narsai, *Homilies on Creation,* 2,34-56.

[56] Translation: K.E. McVey, "The Domed Church as Microcosm: Literary Roots of an Architectural Symbol," *Dumbarton Oaks Papers* 37 (1983), 98.

[57] J.S. Romm, *The Edges of the Earth in Ancient Thought. Geography, Exploration and Fiction* (Princeton: Princeton University Press, 1992), 86-91; 93-94.

There are sources which describe it without making any references at all to miraculous elements, but they either ignore the issue of Paradise altogether, like Orosius;[58] or, like Cosmas, they locate it beyond the Ocean. Nonetheless, the tradition of the miraculous Orient was clearly strong enough to make the location of Paradise there a viable theory. For Cosmas, the combination of the placement of Paradise beyond Ocean, and the description of the rivers of Paradise as travelling underground, effectively allowed him to maintain a scientific image of the oikoumene. Consequently, in contrast to many other Christian writers, he does not provoke a clash of science and religion, but merges them in a quite harmonious way. His oikoumene is unaffected by the miraculous origins of the rivers of Paradise, and the presence of Paradise does not change the Far East into a fantastic land. In fact Cosmas has a particular interest in the eastern confines of the earth, describing not only Sri Lanka and the Indian subcontinent, but also, very briefly, China.[59]

The *Commentary on the Apocalypse* by Beatus of Liebana, composed around 776, provides an interesting comparison to the treatise of Cosmas. Both the *Christian Topography* and the *Commentary* of Beatus were created with accompanying illustrations. Both are essentially works of Biblical interpretation; and both, although to different degrees, show interest in a physical world, as is attested by the presence of maps. Moreover both use a variety of sources, although Beatus' personal input seems to be much smaller, for he himself declares that his *Commentary* is essentially a collection of excerpts from works of holy fathers.[60] As in the case of the *Christian Topography,* the original manuscript of Beatus' *Commentary* is lost. There are, however, 26 entire or fragmentary illustrated copies known. Only 14 of them contain a map, but the text makes it clear that a map to illustrate the mission of the apostles was an integral part of the work.[61]

[58] See Y. Janvier, *La Géographie d'Orose* (Paris: Belles Lettres, 1982), 153-64.

[59] M. Kordosis, "The Limits of the Known Land (Ecumene) in the East According to Cosmas Indicopleustes. Tzinista (China) and the Ocean," *Byzantion* 69.1 (1999), 99-106; Zhang Xu-Shan, "The Name of China and its Geography in Cosmas Indicopleustes," *Byzantion* 74.2 (2004), 452-62.

[60] J. Williams, *The Illustrated Beatus,* v. I, 31 ff. See also E. Edson, *Mapping Time and Space. How Medieval Mapmakers Viewed their World* (London: The British Library, 1997), 149-50.

[61] See: J. Williams, "Isidore, Orosius and the Beatus Map," *Imago Mundi* 49 (1997), 8.

Unlike the case of the *Christian Topography,* not all the maps preserved in Beatus' manuscript are the same. All share some characteristics. East is always at the top, the inhabited earth is surrounded by Ocean, and Paradise appears on the eastern side of oikoumene.[62] Nevertheless, some maps show the inhabited earth as rectangular, while in others it is oval.[63] It has been argued that the rectangular maps are closer to the archetype. Not only does the oldest manuscript of the *Commentary* contain a rectangular map, but also in two cases where we know the tenth century model for a later map, the rectangular shape of the original gave way to an oval in the copy.[64] Whether or not the archetype map was rectangular, an important point is that Beatus' maps are always elongated on a south-north axis, which constitutes a break from the tradition of depicting the earth either as round, or as stretching from east to west.

All Beatus maps share one more element: an additional land south of the oikoumene, divided from it by a narrow strip of waters identified on some of the maps as *Internal Ocean* or as the *Red Sea.* The land itself is also variously labelled. On the map in the oldest manuscript (Pierpont Morgan Library), it is inscribed as "A neighbouring desert land unknown to us because of the heat of the sun." In later copies, however, an inscription identifies it as the fourth part of the world, unknown on account of great heat, where legendary Antipodes live. In one of the copies (Beatus, Burgo de Osma, Arch. De la Catedral, Cod. 1, ff. 34v-35), a figure of a *sciapod* accompanies the inscription.

Interpretation of this land has been much debated, but if we suppose any continuation between Beatus maps and earlier geographical traditions, we can only assume that this is indeed the land of antipodes, since no other land was ever located south of the oikoumene. However, as has been already pointed out, the depiction of the antipodean continent is against the very purpose of the Beatus' map, which was to illustrate the mission of the Apostles. According to the Gospel of Mathew 28:19 and to the Acts of Apostles 2:4-5, they were to carry the Gospel to all people. The antipodes were believed to be separated from the oikoumene by the Torrid Zone, impassable on

[62] D. Woodward, "Medieval Mappaemundi," in *The History of Cartography,* vol. 1, 303-304; E. Edson, *Mapping Time and Space,* 151.

[63] D. Woodward, "Medieval Mappaemundi," 303-305.

[64] J. Williams, "Isidore, Orosius and the Beatus Map," 23-26.

account of great heat; so had they existed, this mission would be impossible to fulfil.[65] Furthermore, beginning with Augustine, Christian sources rejected the idea of antipodes for the reason that their inhabitants could not possibly be descendents of Adam, and therefore were not redeemed by Christ.[66] It was only Isidore who reclaimed antipodes for the Christian world, although he subjected them to a curious change. At one point in his *Etymologies* (IX,2,133) he writes that, according to some sources, antipodes are the people "on the other side of the earth", adding that there is no reason to believe this. In another passage (XI,3,24) he describes antipodes as the inhabitants of Libya, who have their feet facing backwards and eight toes; this is a meaning of the term which he clearly finds much more credible. The last passage of the *Etymologies* (XIV,5) to be noted in this connection is that from which the inscription identifying the southern land on some of the Beatus' maps is taken. It follows on from a description of the marvels and fantastic beasts of Ethiopia: "There are two Ethiopias: one near the rising sun, and the other toward the west in the area of Mauritania. In addition to these three parts of the earth there is a fourth, situated on the other side of the inner ocean, to the south, which is unknown to us because of the heat of the sun. It is said that the legendary antipodes live there."[67] It is not entirely clear how this passage is to be interpreted, for the antipodes mentioned here are not described as inhabitants of Libya, nor do they seem to be the antipodes of the ancient geographies separated from the oikoumene by the impassable outer Ocean. Rather, these antipodes have moved closer to the oikoumene, and have started to inhabit its edges, but at the same time they retain their legendary and fantastic status. Antipodes inherited from ancient geography were not acceptable from the point of view of Christian dogma. Curiously, instead of denying their existence, Isidore and creators of Byzantine maps reinterpret them, and make the inhabit-

---

[65] E. Edson, *Mapping Time and Space*, 153-55; J. Williams, "Isidore, Orosius and the Beatus Map," 17-23.

[66] Augustine, *De Civitate Dei* 16,9. See also Procopius of Gaza, *Commentary on Genesis, Patrologia Graeca* 87, col. 69 B. On the problem of the attitude of later Latin sources to the antipodes, see V. I. J. Flint, "Monsters and Antipodes in the Early Middle Ages and Enlightenment," *Viator* 15 (1984), 65-80.

[67] J. Williams, "Isidore, Orosius and the Beatus Map," 21-23. On the perception of southern confines of the inhabited earth as fantastic, see J.S. Romm, *The Edges of the Earth in Ancient Thought*, 49-59.

ants of the other hemisphere into fantastic dwellers of the edge of the known world.

The maps of both Cosmas and Beatus seem to be based on a combination of ancient geographical traditions and Biblical interpretation. For Cosmas, a clear separation between these two sets of sources is apparent. The image of the oikoumene, which can be subjected to empirical research and scrutiny, is based on ancient sources. By contrast, everything beyond the impassable Ocean is a subject of revelation, and remains beyond the reach of human science and understanding. On the Beatus map there is no such division, however. His maps are something of a paradox. On one hand the continent of the antipodes, which for ancient geographers was inaccessible, becomes the fourth part of the inhabited world and thus open to access; meantime on the other hand the presence of Paradise clearly indicates limitations of human knowledge even within the oikoumene.

RASHI'S MAP OF THE LAND OF CANAAN, CA. 1100, AND ITS CARTOGRAPHIC BACKGROUND[1]

Benjamin Z. Kedar

The presence of maps in medieval Hebrew manuscripts has been almost totally overlooked by students of medieval cartography. For instance, in the multivolume *History of Cartography* edited by J. B. Harley and David Woodward, which surveys and discusses the development of cartography from antiquity onwards, from pre-Columbian America to the realm of Islam to aboriginal Australia, there is not even an allusion to the existence of medieval Hebrew maps.

Students of Jewish history, on the other hand, have repeatedly dealt with maps or diagrams that appear in the most influential Jewish commentary on the Pentateuch, that by Rashi (acronym for Rabbi Shlomo Yitzhaqi), who lived mainly in Troyes, France, between 1040 and 1105. Abraham Berliner, who edited Rashi's commentary in 1866, observed that three of the manuscripts on which he based his edition—namely, those of the Leiden and Munich libraries and of the Saraval collection—contain, within the framework of the discussion of the Book of Numbers, chapter 34, "drawings of the Land [of Israel] considered necessary for Rashi's commentary." (Fig. 1) Berliner went on to say that since Rashi's grandson and pupil Rabbi Shmuel ben Meir (ca. 1080—after 1158) had stated that "my grandfather wrote a commentary and drew borders," it would appear that Rashi himself, too, made drawings of the Land's borders to elucidate his commentary with their help.[2] Berliner's remark that Rashi, *too*, may have drawn borders suggests that Berliner did not believe the drawings in the three manuscripts necessarily originated with Rashi. In any case, he did not reproduce these drawings in his edition.

[1] My thanks to the participants of the Vancouver conference, and especially to Professor Patrick Gautier Dalché, for their comments and criticism.

[2] Abraham Berliner, ed., *Rashi on the Pentateuch* (Berlin, 1866), 300, n. 5 at Num. 34:3 [in Hebrew].

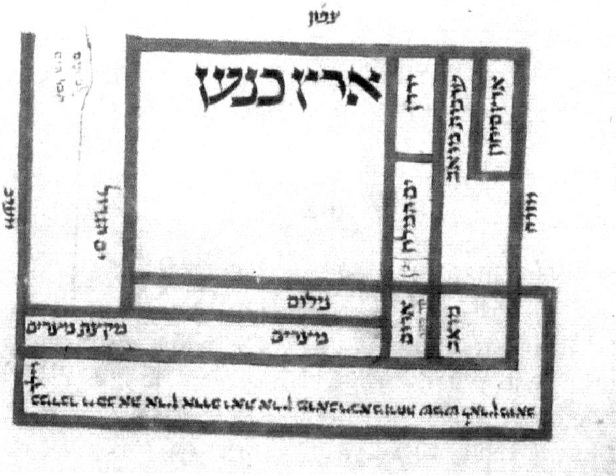

Fig. 1. Map of Canaan (north at top). Munich, Bayerische Staatsbibliothek, Cod. hebr. 5 [Π], fol. 139v.

Nine years later, in 1875, Moritz Steinschneider observed, in his catalogue of the Hebrew manuscripts of the Royal and State Library of Munich, that MS 5, copied in 1213, is one of the most important manuscripts of Rashi's commentary, and that it contains, on folio 140, "a plan of [the Land of] Canaan."[3] In the catalogue's second edition, published in 1895, Steinschneider gave the manuscript's correct date, 1233.[4]

In 1944, Zev Vilnay in his book *The Hebrew Map of the Land of Israel*, written in Hebrew, presented the "diagrams" in Rashi's commentary as the earliest attempt in Hebrew writings to visually represent the Land of Israel. From the statement by Rashi's grandson quoted above, Vilnay assumed that Rashi himself had drawn these diagrams and, relying on Steinschneider's catalogue, he drew atten-

[3] Moritz Steinschneider, *Die hebräischen Handschriften der K. Hof- und Staatsbibliothek in München* (Munich, 1875), 2.

[4] Moritz Steinschneider, *Die hebräischen Handschriften der K. Hof- und Staatsbibliothek in München*, 2nd ed. (Munich: Palm in Comm., 1895), 2. The correct date had been already given by [Max] Lilienthal, "Bibliographische Notizen über die hebräischen Manuscripte der Königlichen Bibliothek zu München," *Allgemeine Zeitung des Judenthums* 2 (1838), Beiblatt 13, 50.

tion to the diagram appearing in MS 5 of the Munich library.[5] In the second edition of his book, published in 1968, Vilnay spoke of "two maps of the Land of Israel, 'the land of Canaan'," that appear in the Munich manuscript, and he published black-and-white photographs of them for the first time.[6]

The maps of the Munich manuscript were discussed in detail in an article written in Hebrew by Bezalel Narkiss, a leading expert on medieval Jewish art, published in a festschrift in honor of Vilnay. This appeared in 1984. Narkiss, like Vilnay, regarded the Munich maps as reproducing Rashi's work. He was also of the opinion that no better maps existed in Rashi's time, and that Rashi's maps were superior to several Hebrew maps of the fourteenth to sixteenth centuries. Focusing on the relationship between the maps and the parts of Rashi's commentary adjacent to them, Narkiss concluded that the maps depended totally on Scripture and on statements by later Jewish sages, and do not reflect geographical knowledge. Narkiss dealt at some length with the Nile, which appears in one of the maps along the *southern* border of Canaan. He regarded this as a mistake on the part of Rashi, who erroneously identified the Nile with the Torrent of Egypt (Nahal Mizrayim), the present-day Wadi al-ʿArish, without realizing that the Nile flows from south to north.[7]

In the 1990s, Mayer Gruber devoted a number of articles in English to Rashi's maps. One of them he co-authored with Catherine Delano Smith, the well-known historian of cartography, especially of maps appearing in Biblical exegesis. Gruber, too, adopts the view that the maps were originally drawn by Rashi himself, both on account of his grandson's testimony and because the maps appearing in French, German, Italian and Syrian manuscripts of Rashi's commentary are largely identical. He published the maps appearing in Rashi's commentary on the Pentateuch as preserved in Bodleian MS. Opp. 34, an Ashkenazi manuscript of the early thirteenth century.

[5] Zev Vilnay, *The Hebrew Maps of Palestine: A Research in Hebrew Cartography* (Jerusalem, 1944), 7, 39 [in Hebrew].

[6] Zev Vilnay, *The Hebrew Maps of the Holy Land: A Research in Hebrew Cartography*, 2nd ed. (Jerusalem, 1968), 5 [in Hebrew]. For a study of the illuminations appearing in the Munich manuscript, see Thérèse Metzger, "Le manuscrit illuminé Cod. Hebr. 5 de la Bibliothèque d'Etat à Munich," in *Études de civilisation médiévale (IXe-XIIe siècles). Mélanges offerts à Edmond-René Labande* (Poitiers: CESCM, 1974), 537-52.

[7] Bezalel Narkiss, "Rashi's Maps," in Ely Schiller (ed.), *Zev Vilnay's Jubilee Volume* (Jerusalem: Ariel, 1984), 345-49.

Gruber also convincingly explains how, in the manuscript tradition, the maps gradually disappeared. In the first stage, such indicators by the copyist as "and now comes the drawing," or "like this," were followed by the map. In the second stage, these words were followed by an empty space. In the third, no space follows the words. In the fourth stage, the copyist did not comprehend the meaning of the words "and now comes the drawing," which were no longer followed by a blank space, and so decided to omit them. As all early printed editions happened to rely on fourth-stage manuscripts, the maps were not reproduced in them, and all later editions down to the present followed suit.[8]

The article co-authored by Gruber and Delano Smith contains black-and-white reproductions of the two maps of the Munich manuscript as well as an English translation of their legends. The authors point out that the first map mainly illustrates Rashi's commentary, whereas the second illustrates the scriptural text itself. They observe that while early Christian maps highlight the division of the country among the Twelve Tribes, Rashi's maps concentrate on its borders, because these borders delimit the area in which a Jew must keep the precepts applying solely to the Land of Israel. The authors also maintain that Rashi's maps attest to a cartographic sophistication and they wonder whether Judaism, too, had a tradition of using diagrams for the elucidation of texts.[9]

In a later article, Gruber offered the rather startling hypothesis that Rashi's map unwittingly reproduces a cuneiform map of the ancient Egyptian province of Canaan that has yet to be discovered

[8] Mayer I. Gruber, "What Happened to Rashi's Pictures?," *The Bodleian Record* 14.2 (April 1992), 111-24. For the provenance and date of MS. Opp. 34 (whose new number is 186), see Malachi Beit-Arié, comp., *Catalogue of the Hebrew Manuscripts of the Bodleian Library*, ed. R.A. May (Oxford: Clarendon Press, 1994), 26.

[9] Catherine Delano-Smith and Mayer I. Gruber, "Rashi's Legacy: Maps of the Holy Land," *The Map Collector* 59 (Summer 1992), 30-35. The maps, with an English translation of the legends, appear likewise in Eva and Gimpel Wajntraub, *Hebrew Maps of the Holy Land* (Vienna: Hollinek, 1992), 2-3. The authors also reproduce the maps appearing in a fourteenth-century Provençal manuscript of Rashi's commentary (Florence, Biblioteca Laurentiana, MS. Plaut. 3.10). For arguments for Rashi's authorship of the maps, see Mayer I. Gruber, "Light on Rashi's Diagrams from the Asher Library of Spertus College of Judaica," *The Solomon Goldman Lectures* 6 (1993), 73-85, and Gabrielle Sed-Rajna, "Some Further Data on Rashi's Diagrams to his Commentary on the Bible," *Jewish Studies Quarterly* 1 (1993-94), 149-53.

but may still come to light as a result of some future excavation, in the archive of a Canaanite city such as Hatzor![10]

A new phase in the study of Rashi's map was inaugurated by a series of articles published between 1991 and 2003 by Avraham Grossman, a leading expert on Rashi and his work.[11] Grossman persuasively argued that MS B.H. 1 of the University Library of Leipzig, dated to the first half of the thirteenth century, constitutes the most important witness to the original text of Rashi's commentary. At the same time, he maintained that the maps of the Leipzig manuscript are closer to Rashi's original maps than the maps of the Munich manuscript. This is so, claimed Grossman, because the winding border lines of the maps of the Leipzig manuscript are consistent with Rashi's assertion, in his commentary, that the border veers to the left or to the right. In other words, the straight reddish border lines of the maps of the Munich manuscript—the only ones hitherto discussed, whether by Vilnay, Narkiss or Gruber—amount to a schematization of Rashi's original drawing. It should be noted in addition that winding border lines appear also on the map of the Bodleian manuscript published by Gruber,[12] as well as on two thirteenth-century Paris manuscripts more recently reproduced by Gabrielle Sed-Rajna.[13] Grossman further strengthened the case for Rashi's authorship of the maps by pointing out that the remark of his grandson, Rabbi Shmuel ben Meir, that Rashi "wrote a commentary and drew borders," appears in the grandson's commentary on the self-

---

[10] Mayer I. Gruber, "The Sources of Rashi's Cartography," in Norman Simms (ed.), *Letters and Texts of Jewish History* (Hamilton, New Zealand: Outrigger Publ., 1998), 61-67. Gruber also reproduces maps appearing in two manuscripts of Rashi's commentary, one of the thirteenth century, the other of 1305 (Parma, Biblioteca Palatina, MSS de Rossi 543 and 175).

[11] Avraham Grossman, "Marginal Notes and Addenda of R. Shemaiah and the Text of Rashi's Commentary," *Tarbiz* 60 (1990-91), 93-95; "Eretz-Israel in Rashi's Teaching," *Shalem* 7 (2002) 30-31; id. and Benjamin Z. Kedar, "Rashi's Maps of the Land of Israel and their Historical Significance," *Newsletter of the Israel Academy of Sciences and Humanities* 25 (November 2003), 26-27 [all three publications are in Hebrew].

[12] Gruber, "What Happened to Rashi's Pictures?," 118.

[13] Sed-Rajna, "Some Further Data," Figs. 1 and 2, reproducing maps appearing in Paris, Bibliothèque nationale de France, MSS Hébreu 154 and 155; *Les manuscrits hébreux enluminés des bibliothèques de France* (Leuven and Paris: Peeters, 1994), 225, 228, with the maps of MS Hébreu 156 (first quarter of the 14th century) appearing on 231-32.

same verses of the Book of Numbers that Rashi's maps attempt to elucidate.[14]

A closer look at the two drawings of Canaan reveals that they differ substantially from each other. In the first (Fig. 2), we see a line, partly single, partly double, that delimits Canaan's boundaries. The text written on both sides of the single line or between the double lines is literally lifted from Numbers 34:3-12. For instance, along the line on the right margin, which designates the country's southern border, the text reads from the top down: "from the end of the Dead Sea which is on its eastern side." These words, with the exception of "which is," are taken from verse 3. Then follows the sentence: "It [that is, the border] shall then turn for you." To the left of this sentence, inside the single line, is: "from the south up the Ascent of Scorpions;" and underneath that: "and pass by Tzin." To the left of these words, inside the single line, the text is: "and its southern limit shall be Qadesh Barne'a." All four of these groups of words appear, exactly in this order, in Numbers 34:4. In addition, we see, to the left of the place names "Ascent of Scorpions" and "Qadesh Barne'a," two elongated rectangles within which these place-names are repeated. Likewise, the phrase "by Tzin" is repeated, with the single border line encircling it on three sides to the left of its original appearance in the third group of words. Underneath the encircled "by Tzin" is, outside the line designating the border, first: "and it shall proceed to," and then, inside the line "Hatzar Addar." Underneath "Hatzar Addar" another place name appears "'Atzmon." Evidently, the intention is a continuation of Numbers 34:4, namely: "and it [that is, the border] shall proceed to Hatzar Addar and continue to 'Atzmon."

This drawing plainly represents an attempt to break down the written description of the borders into a number of segments and to arrange them along the continuous line that represents the borders. The interior of the country is not depicted at all, and much of the space encompassed by the borders is taken up by an excursus on the southern and northern mountains named Mount Hor. The countries bordering on Canaan are not marked either. Thus we have here a

[14] For the crucial influence of Rashi's maps on rabbinical literature down to the nineteenth century, see the drawings reproduced in Wajntraub, *Hebrew Maps of the Holy Land*, 6-11, 15-20, 25-28, 39, 42, 65, 71, 74, 79, 97, 104-12, 121, 154.

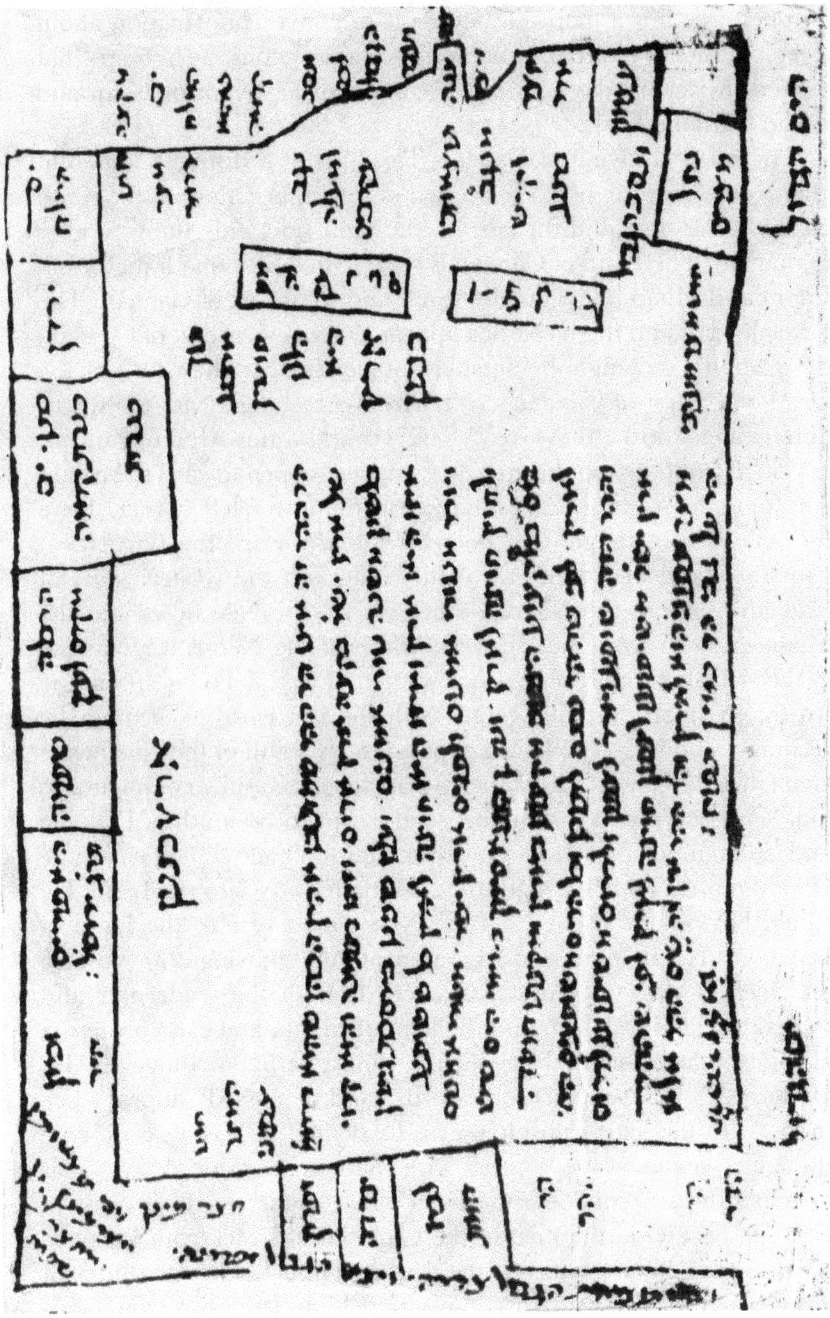

Fig. 2. Diagram of Canaan's borders (north at left). Leipzig, Universitätsbibliothek, B.H. 1, fol. 161r.

diagram that attempts to elucidate verses describing the country's borders; it is not a map that attempts to convey information about a region, or the relationship among several regions, as in maps that existed in Rashi's day and much earlier in the realm of Islam and in the Catholic West.

The other drawing of Canaan (Fig. 3) is of a distinctly different character. Here the north is marked on the upper margin, the south on the lower one, and the east on the right side; only the west goes unmarked. The Land of Canaan is represented by a quadrangle, and Edom and Moab by rectangles south and southeast of Canaan. The River Jordan and the Dead Sea appear as rectangles east of Canaan, the Jordan's rectangle being substantially smaller than that of the Dead Sea. West of Canaan's rectangle is a rectangle with the words "Philistines" and "the Great Sea," that is, the Mediterranean, inscribed from the top down. The rectangle is open toward the north. The most interesting details appear in the lower left corner. Here the Nile is a rectangle that opens *westward* toward the Great Sea, with the Land of Edom east of the Nile and the western part of Canaan north of the Nile. In other words, the Nile flows into the Mediterranean from east to west. South of the Nile is a rectangle, in the right-hand part of which is written "Egypt." To its left are the words "A small part of Egypt," with the last word exceeding the rectangle's boundaries. Thus Egypt is largely south of the Nile, while Canaan is largely north of the river. Indeed, Rashi's commentary, too, leaves no doubt that he believed Egypt to be south of the Nile and Canaan north of it, for he wrote: "If God had wished, when the Children of Israel left Egypt, to hasten their entry into the Land, He would have made them cross the Nile *northwards* into the Land of Israel; but He did not do so." Underneath the drawing is a rectangle in which is written: "And he passed through the wilderness and skirted the land of Edom and the land of Moab, and came from the east to the land of Moab," which is almost verbatim Judges 11:18. Underneath that sentence, the words "and he passed" appear once more, with the letters upside down. Evidently, in Rashi's prototype the entire sentence was written with the letters upside down, so as to make the sentence follow—from left to right and from west to east—the route on the map of the Children of Israel from Egypt to the Land of Moab. Indeed, the scribe of the Leipzig manuscript clarifies that this was the case in his model, but he desists from emu-

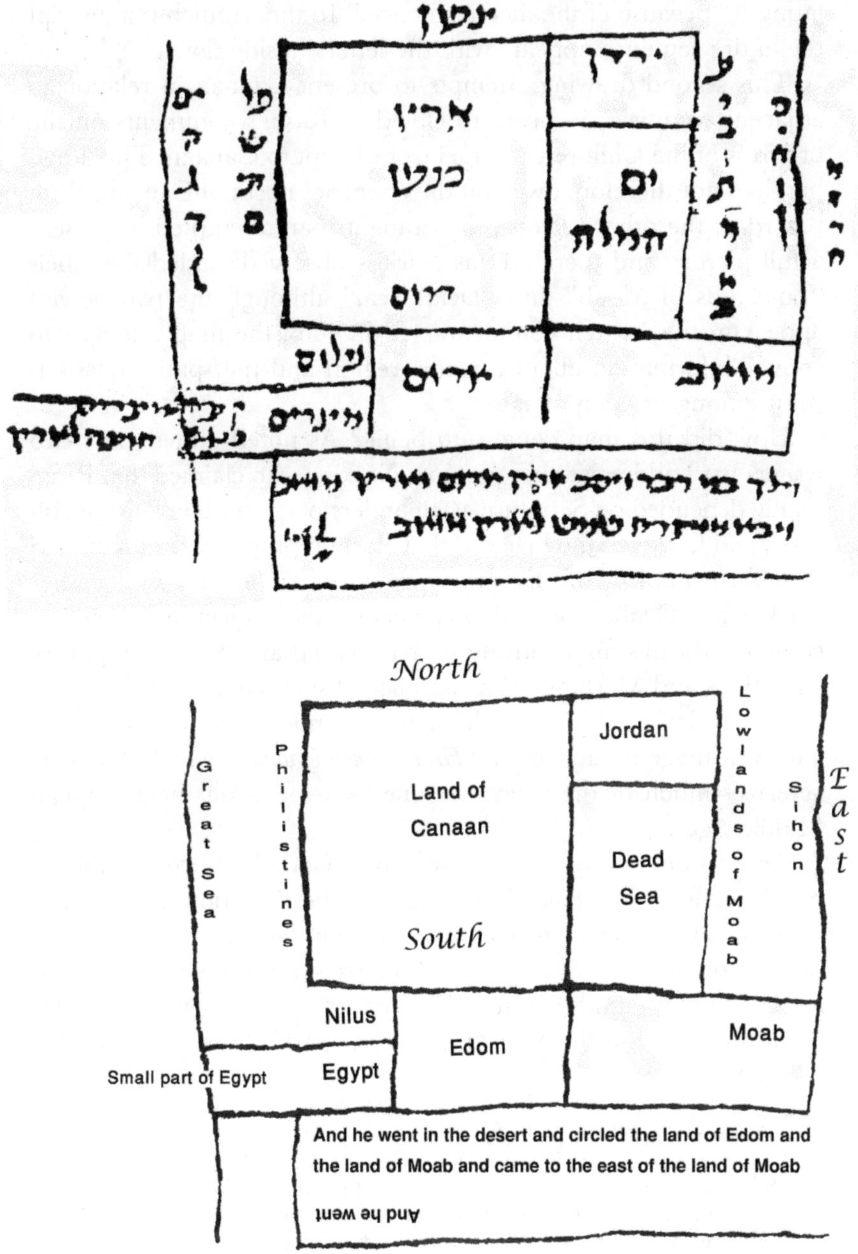

Fig. 3. Map of Canaan (north at top). Leipzig, Universitätsbibliothek, B.H. 1, fol. 160v.

lating it "because of the inconvenience." In the Munich manuscript the entire sentence appears with the letters upside down.

This second drawing attempts to present Canaan in relation to adjacent countries, in order to elucidate Rashi's comments on the progress of the Children of Israel from Egypt to Canaan. The drawing does not therefore dwell on the countries north of Canaan. With regard to the other regions, the draughtsman attempted to present a full picture and therefore (as Narkiss observed) added the labels "Lowlands of Moab" and "Dead Sea," although the two do not appear in the written commentary.[15] In sum, the map attempts to convey information about a whole region and the spatial relationships among its components.

How did this map come into being? As noted above, only two writers have addressed this question: Narkiss, who claimed that Rashi totally depended on Scripture as he understood it and on later Jewish texts; and Gruber, who believed that Rashi reproduced an unattested cuneiform map of the Egyptian province of Canaan. But neither Narkiss nor Gruber, nor other writers on the subject, attempted to compare Rashi's map with those that existed among contemporary Christians and Muslims.[16] The absence of such an attempt is typical of the insularity which still characterizes much of Jewish studies—a mirror image of the dictum "*Hebraica non leguntur*," which still characterizes much of the research done by experts on the European Middle Ages.

An examination of maps drawn up in Catholic Europe until the twelfth century immediately reveals that the depiction of a country in the form of a rectangle was quite conventional, as Konrad Miller pointed out as far back as 1895.[17] Countries are represented by rectangles on the eighth-century Albi map,[18] the late eighth-century world map now in the Vatican (Vat. Lat. 6018),[19] as well as on the

[15] Narkiss, "Rashi's Maps," 436-37.

[16] For a comparison of European and Islamic maps available around 1100, see my "Some Reflections on Maps, Crusading and Logistics," in J. Pryor (ed.), *Logistics of Warfare in the Age of the Crusades* (Aldershot: Ashgate, 2006), 159-83.

[17] Konrad Miller, *Mappaemundi: Die ältesten Weltkarten*, 3. Heft (Stuttgart: Roth, 1895), 59.

[18] See for instance Evelyn Edson, *Mapping Time and Space: How Medieval Mapmakers Viewed Their World* (London: The British Library, 1999), 32.

[19] On this map see Catherine Delano-Smith, "Geography or Christianity? Maps

so-called Anglo-Saxon map of the tenth century and the Ripoll map of 1055 or 1056.[20] Closest in conception to Rashi's map is the eleventh- or twelfth-century map now in the Bibliothèque nationale de France (Fig. 4), appearing in a manuscript of Sallust's *De Bello Jugurthino*—which, like that by Rashi, endeavors to elucidate a specific text. In the middle of the lower part is a large quadrangle marked *Mare rubrum*; to the left and right of it are two rectangles of equal size marked *Hierusalem* and *Egiptus* respectively. Italy, Europe and Asia are also represented by rectangles, of different sizes. Thus the rectangles and the quadrangle on Rashi's map, which designate countries and seas, have many parallels, some very close, in the cartography of Catholic Europe. Countries designated by rectangles or parallelograms appear also on some maps produced by the Balkhi School of the realm of Islam.[21]

The same is true of Rashi's Nile, situated south of Canaan and flowing into the Mediterranean from east to west. The Nile follows the same course in the late eighth-century world map now in the Vatican. In the Anglo-Saxon and Ripoll maps the Nile also flows from east to west, but toward the conclusion of its course it turns northwest to the Mediterranean.[22] On the map dated to the eleventh or twelfth century (Fig. 4), the Nile figures south of Egypt and flows roughly from east to west. On Guido of Pisa's map of 1119, the Nile flows from east to west, and its emptying into the Mediterranean is already marked south of Judea and Jerusalem. But Guido appears to have changed his mind and let the Nile flow into the Red Sea.[23] In a world map in a Munich manuscript of Isidore's *Etymologies* now dated to the first half of the twelfth century,[24] the Nile appears south

of the Holy Land before AD 1000," *Journal of Theological Studies* 42 (1991), 150-52; Kedar, "Some Reflections," 173, n. 28.

[20] Reproductions of these two maps, too, can be consulted in Edson, *Mapping Time and Space*, 8, 83; a large scale reproduction of the second is appended to Alexandre Vidier, "La mappemonde de Théodulphe et la mappemonde de Ripoll (IXe-XIe siècles)," *Bulletin de géographie historique et descriptive* 26 (1911).

[21] See for instance the maps reproduced in J.B. Harley and D. Woodward (eds.), *The History of Cartography*, Vol. 2.1 (Chicago: University of Chicago Press, 1992), 121, 123; Evelyn Edson and Emilie Savage-Smith, *Medieval Views of the Cosmos* (Oxford: Bodleian Library, 2004), 76.

[22] For reproductions, see note 20.

[23] See for instance Edson, *Mapping Time and Space*, 117.

[24] Elisabeth Remak-Honnef and Hermann Hauke, *Katalog der lateinischen*

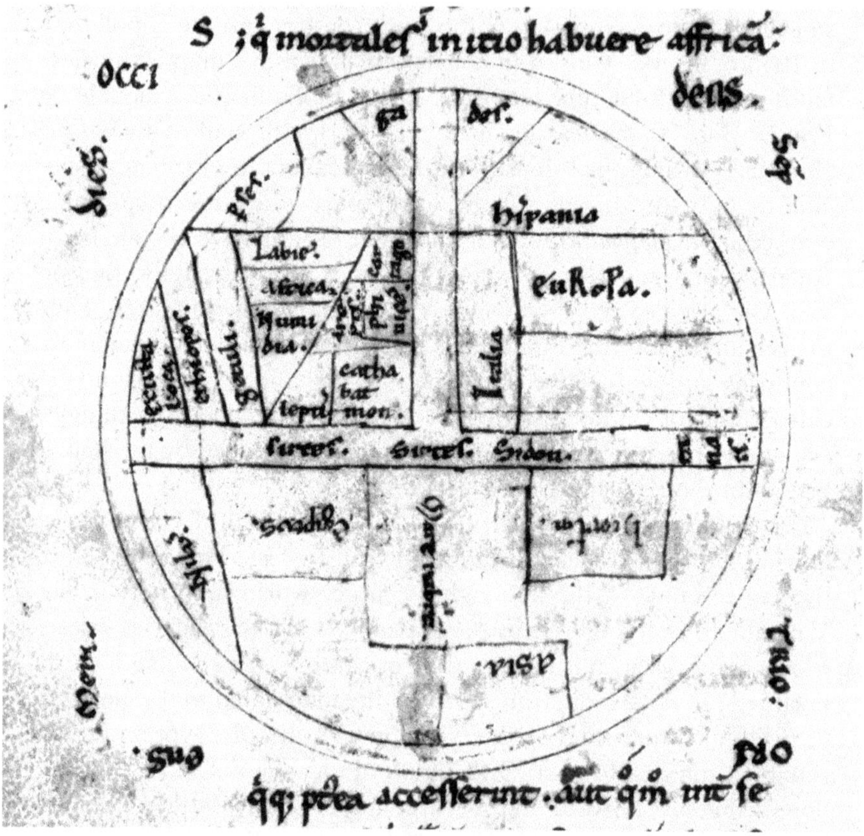

Fig. 4. World map, 11th or 12th century (north at right). Paris, Bibliothèque Nationale de France, Lat. 6253, fol. 52v.

of the Holy Land, flowing from east to west and emptying into the Mediterranean by two branches, one turning north, the other northwest.[25] And a twelfth-century manuscript of three works of Jerome now in the British Library—the '*Liber hebraicarum questionum*', the '*Liber de interpretationibus nominum*', and (from Eusebius) the '*Liber locorum*'—contains a map, probably based on a much earlier prototype, that also shows the Nile south of Palestine, flowing from east to west and reaching the Mediterranean by two branches; but there is also an

---

Handschriften der Bayerischen Staatsbibliothek München, Clm 10001-10930 (Wiesbaden: Harrossowitz, 1991), 36-37.

[25] See for instance Edson, *Mapping Time and Space*, 161.

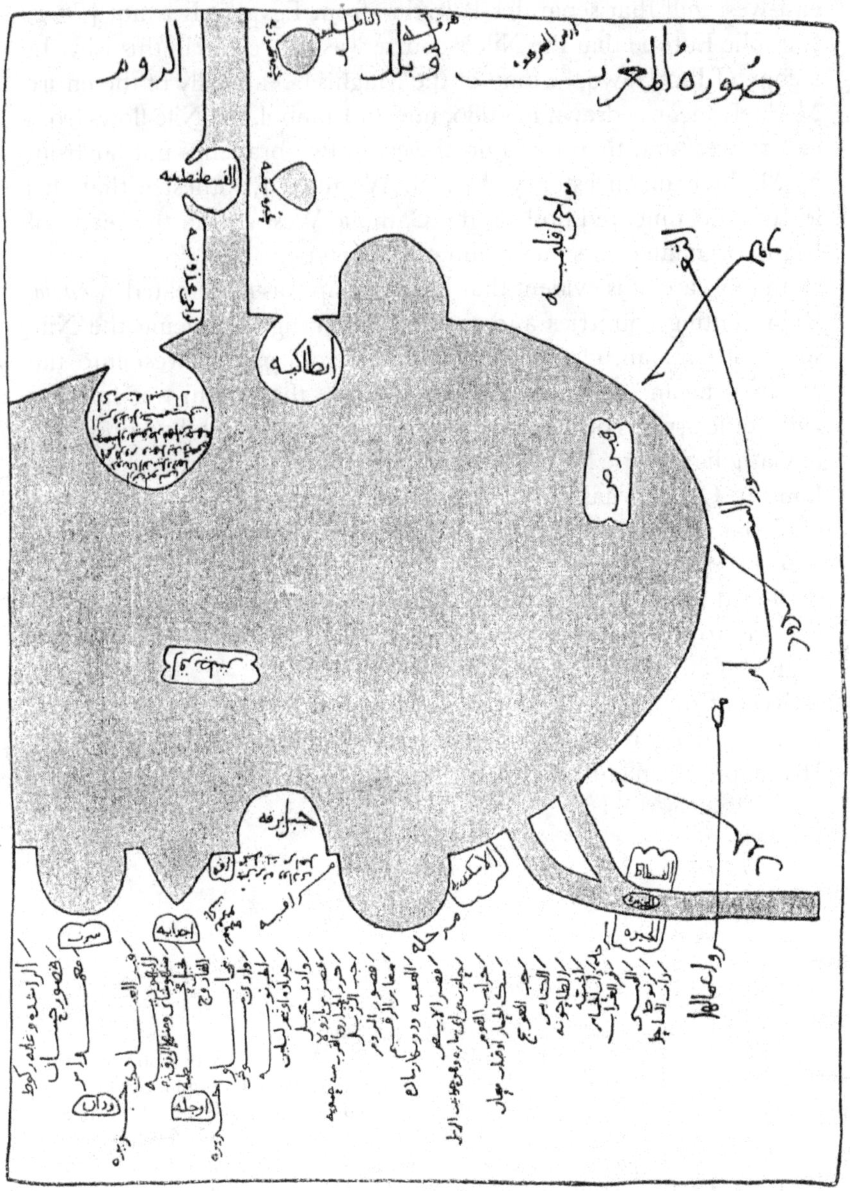

Fig. 5. Ibn Hawqal's map of the Maghreb, eastern part (north at top). Muhammad Ibn Hauqal, *Configuration de la Terre (Kitab surat al-ard)*, trans. J.H. Kramers and G. Wiet, 2 vols. (Beirut and Paris: Aubenas, 1964), vol. 1, after p. 60.

east-west gulf that separates Palestine from Egypt.[26] It is not just in Catholic Europe that the Nile's course was conceived in this way. In a copy of Ibn Hawqal's map of the Maghreb—actually of the entire Mediterranean—drawn in 1086, now in Istanbul, the Nile flows from east to west and then turns northwest in two branches not far from its Mediterranean estuary (Fig. 5). We may hypothesize that this feature, documented both in the Catholic West and in the realm of Islam, originated in some common prototype.

In any case, it is evident that Rashi's map was not created *ex nihilo*. Representing countries and seas by quadrangles, placing the Nile south of Canaan, charting its progress from east to west into the Mediterranean—all these features indicate that Rashi was familiar with some aspects of the cartographic traditions which had prevailed in Catholic Europe for a long time. Moreover, Rashi's quadrangular Land of Canaan has parallels in Christian maps: in the world map of the late eighth century, the map of the eleventh or twelfth century, and the map in Jerome's *Onomasticon*. It is also possible that Rashi preferred the term *Nilus* (quite uncommon in medieval Hebrew literature) to the Hebrew biblical term *Ye'or*,[27] because of the impact of these Christian cartographic traditions.

Recent studies of Rashi's work have demonstrated his familiarity with very many technical facets of the surrounding Christian world.[28] His map of Canaan is further testimony to that familiarity.

---

[26] See for instance Edson, *Mapping Time and Space*, 29.

[27] Exodus 1:22; the Vulgate translates *ye'or* with *flumen*. It may be also noted that Rashi, commenting on Num. 34:8, mentions Antioch. The spelling he uses, "Antochia," appears also in the Scroll of Ovadyah the Proselyte and the Itinerary of Benjamin of Tudela: see Antonio de Rosa and Mauro Perani (eds.), *Giovanni-Ovadiah da Oppido, proselito, viaggiatore e musicista dell'età normanna. Atti del convegno internazionale Oppido Lucano, 28-30 marzo 2004* (Florence: Giuntina, 2005), 297; Marcus N. Adler (ed. and trans.), *The Itinerary of Benjamin of Tudela* (London: Frowde, 1907), 18.

[28] See Avraham Grossman, *Rashi: R. Shlomo Yitzhaqi* (Jerusalem: Merkaz Zalman Shazar le-toldot Yiśra'el, 2006), 116-118 [in Hebrew]. An interesting example is Rashi's observation about the garment that Christian ladies wore when riding horses.

MAPS AND PANEGYRICS: ROMAN GEO-ETHNOGRAPHICAL RHETORIC IN LATE ANTIQUITY AND THE MIDDLE AGES

Natalia Lozovsky

In a public speech delivered in the late 290s, Eumenius, a professor at Autun, described what appears to be a rather detailed map of the Roman Empire. In order that the students might "learn more clearly with their eyes what they comprehend less readily by their ears," the map depicted "the sites of all locations with their names, their extent, and the distances between them, the sources and terminations of all the rivers, the curves of all the shores, and the Ocean, both where its circuit girds the earth and where its pressure breaks into it."[1] The map did not limit itself to instruction in geography, for its ultimate purpose was to demonstrate "the most noble accomplishments of the bravest Emperors through representations of separate regions."[2] Because it has not survived, we can only guess at the exact appearance of the map and of the *diuersa regionum argumenta.*[3] Eumenius' description, however, indicates that by listing places which historically belonged to the Empire or were recently reconquered, the map

---

[1] *Eumenii pro Instaurandis Scholis* in *In Praise of Later Roman Emperors:* The Panegyrici Latini, ed. R.A.B. Mynors, trans. C. E. V. Nixon and B. S. Rodgers (Berkeley: University of California Press, 1994), 20.2, text at 563, trans. at 171: *Siquidem illic, ut ipse uidisti, credo, instruendae pueritiae causa, quo manifestius oculis discerentur quae difficilius percipiuntur auditu, omnium cum nominibus suis locorum situs spatia interualla descripta sunt, quidquid ubique fluminum oritur et conditur, quacumque se litorum sinus flectunt, qua uel ambitu cingit orbem uel impetus inrumpit Oceanus.*

[2] Eumenius 21.1, text at 563, trans. at 172: *Ibi fortissimorum imperatorum pulcherrimae res gestae per diuersa regionum argumenta recolantur, dum calentibus semperque uenientibus uictoriarum nuntiis reuisuntur gemina Persidos flumina et Libyae arua sitientia et conuexa Rheni cornua et Nili ora multifida....*

[3] As Patrick Gautier Dalché and Richard Talbert suggested during the discussion of this paper at the workshop, the regions may have been represented by personifications, a tradition well established in Rome. For discussion and examples, see R. R. R. Smith, "*Simulacra Gentium*: The *Ethne* from the Sebasteion at Aphrodisias," *Journal of Roman Studies* 78 (1988), 50-77; Claude Nicolet, *Space, Geography, and Politics in the Early Roman Empire* (Ann Arbor: University of Michigan Press, 1991), esp. 29-56; Gudrun Bühl, *Constantinopolis und Roma: Stadtpersonifikationen der Spätantike* (Kilchberg and Zurich: Akanthus, 1995).

reaffirmed traditional Roman ideas about conquest and domination over geographical space. It celebrated the emperors for once again expanding Roman territories and bringing glory to Rome. Thus both the map itself, which seems to have existed as a physical object, and its description by Eumenius were meant to serve as vehicles for imperial propaganda.

Eumenius' map, both described in a panegyric and serving as one, presents a convenient starting point for an exploration of the intersection between descriptions of lands and peoples and imperial rhetoric in late antiquity and the Middle Ages. Delivered in the period when Roman emperors still ruled in the West and East, Eumenius' speech, uniquely among the surviving panegyrics, explicitly connects a map with imperial praise. It captures the traditional Roman attitudes toward conquest, maps, and propaganda, the same attitudes that were to survive the breakdown of the Roman Empire and to inspire later generations of rulers and scholars.

Connections between rulers and maps in antiquity have attracted considerable attention of late.[4] Recent scholarship has also highlighted the role that maps and geographical texts played in supporting imperial power in Carolingian Europe and in twelfth-century England.[5] Here, building upon recent trends, the focus will be on

[4] Nicolet, *Space*; C. R. Whittaker, *Frontiers of the Roman Empire. A Social and Economic Study* (Baltimore: John Hopkins University Press, 1994); Susan P. Mattern, *Rome and the Enemy: Imperial Strategy in the Principate* (Berkeley: University of California Press, 1999); Daniela Dueck, "The Augustan Concept of 'an Empire without Limits'," in Michael Dickhardt and Vera Dorofeeva-Lichtmann (eds.), *Creating and Representing Sacred Spaces, Göttinger Beiträge zur Asienforschung* 2-3 (Göttingen: Peust und Gutschmidt, 2003), 211-27; Trevor M. Murphy, *Pliny the Elder's* Natural History*: The Empire in the Encyclopedia* (Oxford: Oxford University Press, 2004).

[5] Patrick Gautier Dalché, "Principes et modes de la représentation de l'espace géographique durant le haut moyen âge," in *Uomo e spazio nell'alto medioevo 4-8 aprile 2002*, Settimane de studio del Centro italiano di studi sull'alto medioevo 50 (Spoleto, 2003), 117-50; id., "L'Heritage Antique de la Cartographie Médiévale: Les Problèmes et les Acquis ," in this volume; Marcia Kupfer, "Medieval World Maps: Embedded Images, Interpretive Frames," *Word and Image* 10 (1994), 262-88, esp. 268-69. On Charlemagne's maps: Emily Albu, "Imperial Geography and the Medieval Peutinger Map," *Imago Mundi* 57 (2005), 136-48; Natalia Lozovsky, "Roman Geography and Ethnography in the Carolingian Empire," *Speculum* 81 (2006), 325-64; Nathalie Bouloux, "Les usages de la géographie à la cour des Plantagenêts dans la seconde moitié du XIIe siècle," *Médiévales* 24 (1993), 131-48; and Daniel Birkholz, *The King's Two Maps: Cartography and Culture in Thirteenth-Century England*, Studies in Medieval History and Culture, 22 (New York: Routledge, 2004).

how Roman perception and use of maps and geo-ethnographical descriptions as vehicles of imperial propaganda influenced the scholars and rulers of the Latin West.

Eumenius' description of a map as an object meant both for instruction and propaganda has no exact parallel in the formal panegyrics surviving from antiquity. However, his technique of using geographical and ethnographical references in order to glorify emperors and the might of Rome is common to the praise of rulers.[6] Other contemporary panegyrists create verbal mini-maps, flash pictures meant to highlight the military triumphs of the emperor in question and to indicate the extent of Roman power. They use names of rivers as geographical markers that signify the four cardinal directions, and give catalogues of conquered places and peoples reduced to subjugation. All these conventional features emphasize the magnitude and invincibility of the Empire.[7]

The authors of panegyrics used these techniques because classical rhetoric recommended them. What rendered such techniques particularly effective, however, was the background knowledge that panegyrists shared with their audience and considered to be common. The authors assumed that the addressees of the praise, as well as other listeners and readers, knew how classical geography described lands and peoples, as well as what representations of lands and peoples meant in the ideology of Roman conquests and expansion. Pliny and Strabo in their geographical works, and other Roman writers and poets in more openly ideological contexts, proudly catalogued places and peoples that belonged to the sphere of Roman power or were just beyond it. As Trevor Murphy has recently demonstrated, imperial geography thus paralleled Roman triumphs with their displays of trophies and slaves from different parts of the world or representations of conquered nations.[8] The idea of Rome as an "empire without end," so powerfully expressed by Virgil in his *Aeneid*,

[6] Ulrike Asche, *Roms Weltherrschaftsidee und Aussenpolitik in der Spätantike im Spiegel der Panegyrici Latini* (Bonn: Habelt, 1983); Sabine MacCormack, "Latin Prose Panegyrics," in Thomas Alan Dorey (ed.), *Empire and Aftermath: Silver Latin II* (London: Routledge and Kegan Paul, 1975), 143-205; Roger Rees, *Layers of Loyalty in Latin Panegyric, AD 289-307* (Oxford: Oxford University Press, 2002).

[7] For examples, see the following panegyrics in *In Praise of Later Roman Emperors:* The Panegyrici Latini: 8.3.3, 8.5.1-3, 10.2.6, 10.7.3-7, 11.6.6, 11.16.3-5, and the editors' comments on these passages.

[8] Murphy, *Pliny the Elder's* Natural History, 129-64.

persisted even when Rome was no longer expanding. Late antique geographical texts written in the fourth and fifth centuries go even further than earlier authors by including places like India under the heading of "provinces" (*provinciae*).[9] Like earlier works, these texts served not only to inform the audience, but also to celebrate the extent of Roman power. Triumphal geography, both earlier imperial and late antique, thus shared the ideology of contemporary panegyrics and provided support for their geo-ethnographical rhetoric.

One more instance of late antique testimony, a poetic one, combines the praise of an emperor and a reference to a map. Unlike Eumenius' description, this poem influenced the Middle Ages, judging from its survival in several manuscripts from the ninth to the twelfth century. [10] According to the poem, Emperor Theodosius ordered two men to produce a map and a text, which were to describe "the sum of all the world" with its "oceans and mountains, rivers and harbours, gulfs and cities." The poem glorifies Theodosius as the ruler whose greatness exceeds earthly limits and whose wisdom has initiated the description of the world.[11] This praise expresses the

[9] Virgil *Aeneid* 1.278-79: *imperium sine fine*. Nicolet, *Space*; Whittaker, *Frontiers*, 14-16; Julius Honorius, *Cosmographia,* in A. Riese (ed.), *Geographi Latini Minores* (Heilbronn: apud Henningeros fratres, 1878), 25.

[10] Eumenius' report of a map apparently left no traces of a direct influence on medieval thought. The poem, on the other hand, is preserved in the late antique geographical text *Diuisio orbis* in the following manuscripts: Vatican, Vat. lat. 642, s. XII, and Leiden, Bibliotheek der Rijksuniversiteit, Scaliger 39, s. XVII. It is also found in the ninth-century text of Dicuil's *De mensura orbis terrae*, because he used the *Diuisio orbis* as a source. The only surviving medieval manuscript is Paris, BN lat. 4806, s. IX. The separate text of the *Diuisio orbis*, with the poem, is edited in P. Schnabel, "Die Weltkarte des Agrippa als wissenschaftliche Mittelglied zwischen Hipparch und Ptolemaeus," *Philologus* 90 (1935), 405-40 (the poem at 440), and in Ludwig Bieler, "The Text Tradition of Dicuil's *Liber de mensura orbis terrae*," *Proceedings of the Royal Irish Academy*, 64 C 1 (1965), 1-31 (the poem at p. 29). For Dicuil, see *Dicuili liber de mensura orbis terrae*, ed. J. J. Tierney and L. Bieler (Dublin: Dublin Institute for Advanced Studies, 1967), 5.4, 56-58. For manuscripts, see also Patrick Gautier Dalché, "Notes sur la 'carte de Théodose II' et sur la 'mappemonde de Théodulf d'Orléans'," *Geographia Antiqua* 3 (1994), 91-106, esp. 95.

[11] Dicuil, *De mensura* 5.4, pp. 56-58: *....totus quem uix capit orbis,/ Theodosius princeps ..../ Sed tamen hoc tua nos docuit sapientia, princeps.* Dicuil used two works that may go back to this enterprise: *Demensuratio prouintiarum* and *Diuisio orbis*, the latter of which contains the above verses. On the fifth-century circumstances of the production of these works, see the interesting hypothesis in Wanda Wolska-Conus, "Deux contributions à l'histoire de la géographie. II. La carte de Theodose II: sa destination ?," *Centre de recherché d'histoire et civilization de Byzance. Travaux et memoirs* 5 (1973), 274-79.

traditional Roman notion that the power to map the empire, or to put its descriptions in order, equals the supreme power to command spaces and peoples. It may have reminded the audience of earlier rulers who had ordered a survey of the imperial lands, such as the emperor Augustus.[12]

The poem does not specify whether it means Theodosius the Great (379-395) or Theodosius II (408-450). Unfortunately, no surviving contemporary chronicle hints at this event, so it is not entirely clear which Theodosius commissioned the map. Most scholars who identify the addressee of the poem as Theodosius II cite as one of the reasons this emperor's interest in geo-ethnography. According to a subscription found in some manuscripts of Solinus' *Collectanea rerum memorabilium*, this text, composed in the third century, was copied "by the diligent efforts of our lord Theodosius, the invincible emperor." The copying of Solinus, as well as the map and the compilation mentioned in the poem, were parts of Theodosius' broader interest in classical texts. [13] At the same time, the author who wrote the poem and the emperor who ordered the compilation both surely realized that maps and geographical texts expressed imperial power over space, and could be used to extol that power.

Roman geo-ethnography, steeped in imperial ideology, continued to provide imperial models even after the Roman Empire in the West ceased to exist in a political sense. Throughout late antiquity and

---

[12] For possible connections between the *Diuisio orbis*, the late antique geographical text that contains the poem, and the so called "map of Agrippa," which may have resulted from Augustus' enterprise, see Tierney, "Introduction" in *Dicuili liber*, 17-26. For imperial possession of maps as an expression of power in Rome, see Pascal Arnaud, "L'affaire Mettius Pompusianus ou la crime de cartographie," *Mélanges de l'École française de Rome: Antiquité* 95 (1983), 677-99.

[13] Most scholars take the poem to refer to Theodosius II: Tierney, "Introduction" in *Dicuili liber*, 24; Wolska-Conus, "Deux contributions"; O. A. W. Dilke, "Cartography in the Byzantine Empire," in *The History of Cartography* vol. 1, 259; Ekkehard Weber, "Zur Datierung der Tabula Peutingeriana," in Heinz E. Herzig and Regula Frei-Stolba (eds.), *Labor omnibus unus: Gerold Walser zum 70. Geburtstag* (Stuttgart: Steiner-Verl.-Wiesbaden-GmbH, 1989), 117. Some, however, allow that it might as well refer to Theodosius I: D. Detlefsen, *Ursprung, Einrichtung und Bedeutung der Erdkarte Agrippas* (Berlin: Weidmann, 1906), 15, and Mario Esposito, "An Irish Teacher at the Carolingian Court: Dicuil," *Studies* 3 (1914), 651-76, reprinted in Mario Esposito, *Irish Books and Learning in Mediaeval Europe* (Aldershot: Ashgate, 1990), 663. For the subscription, see Theodor Mommsen, "Introduction," in his Solinus, *Collectanea rerum memorabilium* (Berlin: Weidmann 1895), p. XCVIII: *studio et diligentia d[omini] n[ostri] Theodosii invictissimi principis.*

the Middle Ages scholars diligently studied Roman imperial geography and made use of the ideas of power that it conveyed. They read, copied, and made notes, which testify to the use of Roman geography in medieval schools. They also compiled their own texts that relied on the Roman tradition. Thus in 825, Dicuil, an Irish scholar at the courts of Charlemagne and his son Louis the Pious, wrote a treatise *On the Measurement of the World.* The work, as Dicuil informed his readers, was based on the description of lands ordered by the emperor Theodosius. He also included in his treatise the late antique poem, already discussed, which praised Theodosius as the great and wise ruler who had initiated the description of the world. Dicuil, who had dedicated his earlier work to the emperor Louis the Pious, may have used this glorification of Theodosius and his enterprise as subtle praise of the present emperor. Such praise would conform to contemporary imperial rhetoric because both Louis the Pious' legislation and the allusions by his court poets strove to present him as the second Theodosius. Dicuil was drawing on the tradition of praise established in late antique panegyrics. He also used Roman imperial geography in order to present his picture of the world. Far from being a mindless copyist, he chose and combined his material in such a way as to reflect the shift of his interest. Instead of the Mediterranean focus traditional for Roman geography, Dicuil's treatise subtly privileges the lands located to the north. Combined with the praise of a Roman emperor, his geographical survey of the world may have served to remind the Carolingian emperor of Roman might. This survey may have also imitated its Roman imperial sources, as well as asserting the might and extent of the new Christian empire. [14]

Carolingian imperial rhetoric, building on the prestige of the Roman Empire, presented the Franks as successors and legitimate heirs to the Romans. Frankish sources not only claimed that the Franks were descended from the Trojans like the Romans, but they also asserted that the Franks were superior to the Romans in their conquests. Relying on the legacy of Roman imperial praise, Caro-

[14] For changes in the Carolingian picture of the world, see Patrick Gautier Dalché, "Tradition et renouvellement dans la représentation de l'espace géographique au IXe siècle," *Studi Medievali*, 3rd ser. 24 (1983), 121-65; for imperial contexts of Carolingian geography, see Albu, "Imperial Geography," and Lozovsky, "Roman Geography."

lingian poets used geo-ethnographical references in order to celebrate the might of their emperors. Like Roman panegyrists, they spoke of various nations conquered by the Frankish army and brought to their knees before the emperor. Charlemagne emerges from one poem as an emperor who has subjected savage barbarians to his laws, and who now rules the whole world.[15] Emperor Louis the Pious, according to another poem, had made savage and dangerous barbarians living to the north and south fear his power.[16] Later in the ninth century, nostalgically looking back at the time of Charlemagne, an anonymous poet proclaimed that this emperor had conquered peoples whose names the Romans did not even know.[17] In its assimilation of Roman triumphal imagery, Carolingian imperial rhetoric reveals its reliance on Roman poetic, ideological and geographical models, presumably familiar to learned audiences.

The three maps that Charlemagne owned according to Einhard's report have not survived, and only speculation is possible about their appearance or exact contents. However, the very fact that the emperor had the maps at his court appears significant. Charlemagne and his milieu inherited Roman ideas about the imperial control of

[15] Modoin, *Ecloga* 2. 78-81, ed. Ernst Dümmler, in *MGH Poetae* 1 (Hannover, 1881), 390:

Aurea lux terris caelo demissa relucet,
Quae mare, quae et terras, quae totum mitigat orbem,
Quae saevos populos subigit gentesque refrenat
Legibus innumeras, totumque coerceat orbem....

Ibid. 2.115-17, 391:

Caesareo populum Carolus gentesque coercet
Tegmine, cuncta regit terrarum regna per orbem,
Imperioque pio toto dominabitur orbi.

[16] Walahfrid Strabo, *De imagine Tetrici* 253-55, ed. Michael W. Herren, "The 'De imagine Tetrici' of Walahfrid Strabo: Edition and Translation," *Journal of Medieval Latin* 1 (1991), 118-39:

Sic Vulgar Sarraeque cynus, malus hospes Hiberis,
Brutus Britto, Danus uersutus et horridus Afer
Subdat honorandis sua colla exterrita dextris.

[17] Poeta Saxo, ed. Paul von Winterfeld, in *MGH Poetae* 4/1 (Berlin, 1899), 5. 647-52, 70:

Romani multis ducibus multisque sub annis
Italiae populos vix sibi subdiderant:
Unus hic in spacio perpauci temporis omnem
Subiecit victor, disposuit dominus:
Adde tot Europae populos, quos ipse subegit,
Quorum Romani nomina nescierant.

space and probably of its representations as well. The first two maps depicted Constantinople and Rome, both imperial and Christian capitals, and may have presented the new Christian emperor with a visual reminder of the succession of power. The third, a map of the world, may have reinforced the message of poetic panegyrics, which glorified the rule of the emperor over the entire world. [18]

Carolingian scholars adapted Roman models to the ideology of the new Christian empire. By their use of imperial rhetoric that traditionally relied on geo-ethnography, they and their rulers in their turn set examples for later generations. A series of famous images that depict rulers—in luxury manuscripts produced approximately between 870 and 1014—reveal one more way in which Roman triumphal vocabulary provided medieval artists with visual means to express contemporary imperial ideas and to create pictorial panegyrics to their rulers. Varying in emphasis and details, these images use a common motif ultimately going back to the Roman tradition. All of them depict women, personifying provinces or *gentes*, who pay homage or bring gifts to the ruler.[19] In Roman art and poetry, female

[18] Einhard, *Vita Karoli Magni* 33, ed. O. Holder-Egger, *MGH SS rer. Germ.* [25] (Hannover, 1911), p. 40. For a reconstruction of the third map, see F. N. Estey, "Charlemagne's Silver Celestial Table," *Speculum* 18 (1943), 112-17. For an analysis of the imperial context of the maps see M. Kupfer, "Medieval World Maps," 268-69. For a more detailed treatment of Roman geography in the Carolingian Empire, see Lozovsky, "Roman Geography." The recent hypothesis by Emily Albu about the Carolingian provenance of the famous Peutinger Map, traditionally dated to late antiquity, appears very attractive; see her contribution to the present volume. For late antique imperial models of Charlemagne's tables, see Deborah Deliyannis, "Charlemagne's Silver Tables: The Ideology of an Imperial Capital," *Early Medieval Europe* 12 (2003), 159-77.

[19] Seven manuscript images contain this motif: Munich. Bayerische Staatsbibliothek, Clm. 14000, fol. 5v; Chantilly, Musée Condé, ms. 14bis; Bamberg, Staatsbibliothek, Clas. 79; Munich, Staatsbibliothek, Clm. 4453, fols. 23v and 24r; Bamberg, Staatsbibliothek, Bibl. 140, fol. 59v; Munich, Staatsbibliothek, Clm 4452, fol. 2r; Munich, Staatsbibliothek, Clm. 4456, fol. 11v. The modern literature focusing on these *Herrscherbilder* is vast, and sharp debates surround almost every aspect, from dating the manuscripts and identifying the rulers to the ideas which the images express and reflect. For discussion with particular attention to the personifications, see especially Percy Ernst Schramm, "Das Herrscherbild in der Kunst des frühen Mittelalters," *Vorträge der Bibliothek Warburg* 2/1 (1922-23), 145-224; Wilhelm Weizsäcker, "Imperator und huldigende Frauen," in *Festschrift für Karl Gottfried Hugelmann zum 80. Geburtstag am 26. September 1959 dargebracht von Freunden, Kollegen und Schülern*, vol. 2 (Aalen: Scientia Verlag, 1959), 815-31; Konrad Hoffmann, "Das Herrscherbild im 'Evangeliar Ottos III.' (clm 4453)," *Frühmittelalterliche Studien* 7 (1973); Percy Ernst Schramm, *Die deutschen Kaiser und Könige in Bildern ihrer Zeit*

personifications of places and nations expressed and emphasized imperial power over space and peoples.[20] When the creators of the Codex Aureus, produced ca. 870, depicted Charles the Bald sitting on his throne, flanked by two female figures with cornucopiae, they were drawing on and paying tribute to Roman and Byzantine traditions of representing imperial might. The portrayal of the two personifications, identified by an inscription as *Francia* and *Gotia*, was also politically relevant, signifying the two *regna* of Charles' realm, chosen among others for their importance.[21] The power over space projected by this image mirrored the messages of imperial domination conveyed by Carolingian poetic panegyrics.

The rulers who succeeded Charlemagne's dynasty in the eastern part of the former Carolingian Empire—the state that in the twelfth century would be called the Holy Roman Empire—continued the appropriation of the imperial legacy. The image occupying a double leaf in the so-called *Gospels of Otto III*, produced at the end of the tenth century at Reichenau, on the right leaf shows a ruler, most likely Otto III, holding an orb inscribed with a cross in his left hand

751-1190, ed. Florentine Mütherich (Munich: Prestel, 1983); Gerhart Ladner, *L'immagine dell'imperatore Ottone III* (Rome: Unione Internazionale degli Istituti di Archeologia, Storia e Storia dell'Arte, 1988); Henry Mayr-Harting, *Ottonian Book Illumination: An Historical Study*, 2 vols. (London and New York: Harvey Miller, 1991); Joachim Ott, "*regi nostro se subdit Roma benigno*—Die Stiftung des Perikopenbuche Heinrichs II. (Clm 4452) für den Bamberger Dom vor dem Hintergrund der bevorstehenden Kaiserkrönung," *Jahrbuch für fränkische Landesforschung* 54 (1994), 347-70; Rainer Kahsnitz, "Herrscherbilder der Ottonen," in Mario Kramp (ed.), *Krönungen: Könige in Aachen—Geschichte und Mythos: Katalog der Ausstellung in zwei Bänden* (Mainz: von Zabern, 2000), vol. 1, 283-93; Florentine Mütherich, "Das Kaiserbild," in Florentine Müterich and Karl Dachs (eds.), *Das Evangeliar Ottos III.: Clm 4453 Bayerischen Staatsbibliothek München* (Munich: Prestel, 2001), 31-35; Ursula Nilgen, "Blonde Roma? Zum Sinn des Blondhaars in der Buchmalerei der Reichenau," *Zeitschrift für Kunstgeschichte* 66 (2003), 19-32; Eliza Garrison, "Henry II's *Renovatio* in the Pericope Book and Regensburg Sacramentary," in Nigel Hiscock (ed.), *The White Mantle of Churches: Architecture, Liturgy, and Art Around the Millennium* (Turnhout: Brepols, 2003), 57-79. For an alternative to the prevailing datings and identifications, see Ulrich Kuder, "Die Ottonen in der ottonischen Buchmalerei: Identification und Ikonographie," in Gerd Althoff and Ernst Schubert (eds.), *Herrschaftsrepräsentation im ottonischen Sachsen* (Sigmaringen: Thorbecke, 1998), 137-218.

[20] For the Roman tradition, see Smith, "*Simulacra*"; Nicolet, *Space*; Bühl, *Constantinopolis*. For the medieval use of the tradition, see Schramm, "Herrscherbild," 195-97; Weizsäcker, "Imperator"; Hoffmann, "Herrscherbild."

[21] Janet L. Nelson, "The Frankish Kingdoms, 814-898: the West," in Rosamond McKitterick (ed.), *New Cambridge Medieval History*, vol. 2, c. 700-c. 900 (Cambridge: Cambridge University Press, 1995), 131-33.

and a scepter with a bird on top of it (Plate VII). On the left leaf, four women form a procession, slightly bowing before the emperor and offering him gifts. Inscriptions above their heads identify the women as *Roma*, *Gallia*, *Germania* and *Sclavinia*.[22] According to the interpretation elaborated by Percy Ernst Schramm and supported by many modern scholars, this image presents in pictorial form the ideology of Otto III's *Renovatio Imperii Romanorum*.[23] Rome as an imperial and cultural ideal was especially dear to Otto III (983-1002), the son of the German emperor Otto II and a Byzantine princess Theophanu. Otto III even made Rome his residence in the last years of his life. Rome also inspired Otto's advisers, among them Gerbert of Aurillac, an outstanding scholar, who as pope in 999-1003 took the name of Sylvester II in memory of the pope contemporary to Constantine the Great. His ideas about Otto's *renovatio* closely correspond to the pictorial image of the emperor, complete with the names of the provinces: "Ours, ours is the Roman Empire! Italy fertile in fruit and Gaul and Germany fertile in warriors give strength to the empire, and we do not lack [the support of] the most powerful kingdoms of Scythia [Slavic lands]. You are our Caesar, emperor of the Romans, and Augustus. Born of the highest blood of the Greeks, you surpass the Greeks in imperial power, you rule the Romans by hereditary

[22] Most scholars identify the ruler as Otto III: see Weizsäcker, "Imperator," 816-17; Schramm, *Deutschen Kaiser*, 205-207, and other works listed in note 19 above. However, Kuder in "Die Ottonen," 140 and 193-96, suggests the date of "1004 oder etwas später, somit um 1005-10," and identifies the ruler as Henry II.

[23] The ideas that Schramm expressed in *Kaiser, Rom und Renovatio: Studien und Texte zur Geschichte des römischen Erneuerungsgedankens vom Ende des karolingischen Reiches bis zum Investiturstreit* (Leipzig: Teubner, 1929) have also inspired his analysis of the imperial images in "Herrscherbild" and *Deutschen Kaiser*. Recently Schramm's notion of the Ottonian *Renovatio Imperii Romanorum* and his interpretation of evidence, including the images of emperors, have come under attack. See Knut Görich, *Otto III., Romanus, Saxonicus, et Italicus: Kaiserliche Rompolitik und Sachsische Historiographie* (Sigmaringen: Thorbecke, 1993). This book has provoked heated debates. Most relevantly to the present discussion, for a general support of Görich's critique, see Gerd Althoff, *Otto III*, trans. Phyllis G. Jestice (University Park: Pennsylvania State University Press, 2003); against Görich, see Heinrich Dormeier, "Die Renovatio Imperii Romanorum und die "Außenpolitik" Ottos III. und seiner Berater," in Benjamin Scheller (ed.), *Polen und Deutschland vor 1000 Jahren : die Berliner Tagung über den "Akt von Gnesen"* (Berlin: Akad.-Verlag, 2001), 163-91, both with references to earlier literature. For analysis of this image in the framework of Schramm's ideas, see most recently Mayr-Harting, *Ottonian Book Illumination*, 159-62, and Mütherich, "Das Kaiserbild."

right, and you exceed both in intelligence and eloquence."[24] As in the Codex Aureus, the names of the lands in this proclamation and in the manuscript image are politically relevant. In response to the ideology of the renewal of the Roman Empire, *Roma* heads the procession of personified lands. *Gallia* and *Germania* stand for the two core parts of Otto's empire, signifying the lands, respectively, to the west and east of the Rhine. Friendly relations between Otto III and the Poles may explain the presence of *Sclavinia*.[25]

In addition to late antique iconographic examples, this image could have been directly inspired by Roman triumphal geography. As scholars have long noted, the depiction of the four provinces here resembles that of the personified lands in the illuminated copies of the *Notitia Dignitatum*, a list of dignitaries and their areas of responsibility going back to the late Roman Empire (probably ca. the fifth century A.D.). The *Notitia Dignitatum* survives today in later manuscripts, but the copy known to Ottonian scholars may also have contained Roman geographical works, such as the *Cosmography* of Pseudo-Aethicus and the *Itinerarium Antonini*. It may also have included Dicuil's *De mensura*.[26] Furthermore, the artists at Reichenau where

[24] Gerbert of Aurillac, *Opera*, ed. A. Olleris, Paris 1867, p. 298: *Nostrum, nostrum est Romanum imperium; dant vires ferax frugum Italia, ferax militum Gallia et Germania, nec Scythae nobis desunt fortissima regna. Noster es, Caesar, Romanorum imperator et Auguste, qui summo Graecorum sanguine ortus, Graecos imperio superas, Romanis haereditario jure imperas, utrosque ingenio eloquentia praevenis.* For analysis of the Ottonian imperial ideology, in addition to Schramm see Carl Erdmann, "Das ottonische Reich als Imperium Romanum," *Deutsches Archiv für Erforschung des Mittelalters* 6 (1943), 412-41.

[25] Schramm, *Deutschen Kaiser*, 205-207; Mayr-Harting, *Ottonian Book Illumination*, 159-61; Mütherich, "Das Kaiserbild."

[26] For a detailed treatment of Roman influences see Schramm, *Deutschen Kaiser*; Weizsäcker, "Imperator"; Hoffmann, "Herrscherbild"; Mütherich, "Das Kaiserbild." According to reconstructions based on surviving manuscripts, the so called Speyer Codex—which disappeared after it was copied in the sixteenth century—contained the illustrated *Notitia Dignitatum* and a number of Roman and Roman-inspired geographical works. It was either this manuscript, or a very similar one, that was used by Ottonian scholars. See J. J. G. Alexander, "The Illustrated Manuscripts of the Notitia Dignitatum," in R. Goldburn and P. Bartholomew (eds.), *Aspects of the Notitia Dignitatum: Papers presented to the conference in Oxford December 13 to 15, 1974* (Oxford: British Archaeological Reports, 1976), 11-12, 25; Bieler, "The Text Tradition." Another manuscript very likely to have been used by the Ottonians is Dresden, Landesbibliothek Dc 182, destroyed during the Second World War; it contained the same geographical works as the Codex Spirensis, but did not include the *Notitia Dignitatum*. Some scholars believe that it was given to Otto III by Gerbert of Aurillac: see Uta Lindgren, *Gerbert von Aurillac und das Quadrivium: Untersuchungen zur Bildung im Zeitalter der Ottonen* (Wiesbaden: Steiner, 1976), 93. Later, under Henry

this and other imperial images were produced may have been familiar with a map that depicted a world essentially going back to late antiquity and centered on Rome, a map that served as one of the models for the famous Peutinger Map.[27] The enthroned figures personifying Rome and other cities on this map may have provided inspiration for the imperial images in manuscripts.[28]

In their iconographic details, the image of Otto III in the *Gospels*, as well as other similar imperial images, was meant to recall the emperor Augustus.[29] Since the time of Orosius, he was known as the emperor whose rule providentially coincided with the birth of Christ. According to the Gospel of Luke and to an independent tradition, Augustus issued the edict to describe the world that was his dominion. References to Augustus would thus cast the new emperors in the same glorious light, as the Christian heirs to the Roman power over the world.[30]

In this image and others, the personified provinces along with other details borrowed from the Roman vocabulary of power form an important component of the visual rhetoric expressing imperial ideology.[31] Roman geo-ethnographical models, adapted to the changing political situations, provided ideological support for the new medieval empire.

Traces of a similar inspiration can be found in a less spectacular manuscript. Paris, Bibliothèque Nationale de France MS lat. 4806, produced in the third quarter of the ninth century at Reims, contains a collection of geographical texts, most of them late antique, such as the *Cosmographia* of Pseudo-Aethicus and *Itinerarium Antonini*.[32] This codex also preserves the treatise of the Carolingian scholar Dicuil,

II, it became part of the library at Bamberg: see Hartmut Hoffmann, *Bamberger Handschriften des 10. und des 11. Jahrhunderts* (Hannover: Hahn, 1995), 75 and 176.

[27] Albu, "Imperial Geography."

[28] Alexander, "Illustrated Manuscripts," 17.

[29] For instance, the bird sitting on the top of the emperor's Otto's scepter closely resembles the eagle on the scepter of Augustus from the Roman cameo attached to the processional cross of German emperors: see Schramm, *Deutschen Kaiser*, esp. 206.

[30] For an analysis of Orosius' influence, see Hoffmann, "Herrscherbild."

[31] The rhetoric of contemporary panegyrics, continuing both Roman and Carolingian traditions, also strongly relies on triumphal geo-ethnographical imagery: see for instance Hoffmann, "Herrscherbild."

[32] The date and place were identified by Bernhard Bischoff: see Bieler, "The Text Tradition," 3.

discussed above. On the verso leaf—which the ninth-century scribe had left blank—between the list of the seven hills of Rome and the beginning of Dicuil's *On the Measurement of the Earth*, a later, eleventh-century hand wrote three poems: an epitaph for pope Leo IX (1049-54); an epitaph for emperor Henry III (1017-56); and a poem entitled "The Exhortation to the Noblemen of the Realm." The author of the last, most likely an Italian, calls upon his countrymen—and especially upon the Romans—to support the young king Henry IV and to repel the Normans, Saracens and Hungarians.[33] Then, the poet continues, the nations that once belonged to the Roman Empire will again be subject to the Romans. The names of the lands and peoples deliberately evoke classical texts. According to the poet, Ethiopians and Phrygians, Gauls and Britons will all come back into the fold of the new Roman Empire.[34] He names Julius Caesar and Augustus as model emperors; he also mentions Charlemagne as an emperor equal to both Roman rulers.[35] The poet envisions the time when the young scion of Henry III will assume power as the rightful heir of all these emperors.

It seems significant that the scribe has chosen to place this poem between the text that lists Rome's hills and aqueducts and Dicuil's treatise, which is essentially based on Roman geography. Awareness of connections between Roman geographical information and imperial ideology may have triggered the choice. Poems celebrating the

[33] For a discussion of imperial ideas in the eleventh century, including those of this poem, see Tilman Struve, "Kaisertum und Romgedanke in salischer Zeit," *Deutsches Archiv für Erforschung des Mittelalters* 44 (1988), 424-54. The poem is edited in Ernst Dümmler, "Gedichte aus dem elften Jahrhundert," *Neues Archiv* 1 (1876), 177.

[34] *Exhortatio ad proceres regni* 13-22:
Et cum iusticia subdita erit Libia.
Grecia iuncta aderit, nec erunt orientia bella,
Et cum muneribus curret Arabs timidus.
Subdita erit uobis reuerenter Hiberia fortis,
Romanas leges Cantaber excipiet.
Querite nunc Calabros, pelleo zemate Parthos,
Quascumque et gentes pars Orientis habet.
Memphis cum Tyriis uenerabitur arma Quiritis,
Tellus Aethiopum, gens simul atque Frigum.
Gallus erit famulus, subiectus necne Britannus,
Francus in auxiliis pronior obsequiis.

[35] Ibid., 26:
Iulius et Caesar [above the line: *i.e. Octauianus*], *Karolus his quoque par.*

German emperors as successors to the Carolingians and the Romans find an appropriate place among Roman and Carolingian geographical works. The manuscript thus reveals a direct intersection between learned geography and medieval imperial rhetoric, both grounded in the Roman tradition.

Roman imperial and geographical rhetoric exercised its influence not merely in those states that officially called themselves empires. Imperial legacy—sometimes claimed openly, sometimes implied—became a stock part of the language of power in medieval Europe. Around 1100, Baudri, abbot of Bourgeuil and later archbishop of Dol, wrote a long poem praising Adela, the countess of Blois. Adela belonged to one of the most important families in Europe. She was a daughter of William the Conqueror and the wife of Stephen, the count of Blois. Her son Stephen was later to become King of England. The poem describes the chamber of the countess in great detail, and modern scholars disagree about whether it is real or imaginary.[36] According to Baudri, the wall tapestries in the room depicted scenes from ancient, biblical and contemporary history. A map of the sky occupied the ceiling, and a map of the world decorated the floor.[37]

Baudri was very well educated in classical models, and he skillfully used allusions and direct quotations from Roman poets, such as Virgil, Lucan, and Ovid.[38] Throughout his poem, Baudri exploits Roman imperial motifs. He begins by addressing the countess as deserving the name of a queen, and by giving praise to her father as a great king. Baudri declares that William has conquered unconquerable Angles with his sword, submitted Normans to his yoke by iron, and

[36] Not enough evidence exists to draw a firm conclusion one way or another; for the purpose of this discussion, the question is not of primary importance. For an imaginary chamber, see Christine Ratkowitsch, *Descriptio picturae. Die literarische Funktion der Beschreibung von Kunstwerken in der lateinischen Großdichtung des 12. Jahrhunderts* (Vienna: Verlag der Österreichischen Akademie der Wissenschaften, 1991); Patrick Gautier Dalché and Jean-Yves Tilliette, "Un nouveau document sur la tradition du poème de Baudri de Bourgueil à la comtesse Adèle," *Bibliothèque de l'École des Chartes*, 144/2 (1986), 241-57. For a real chamber, see David Woodward, "Medieval *Mappaemundi*," in *The History of Cartography*, vol. 1, 339; Benjamin Z. Kedar, "Some Reflections on Maps, Crusading and Logistics," in John H. Pryor (ed.), *Logistics of Warfare in the Age of the Crusades* (Aldershot: Ashgate, 2006), 159-84. I thank Professor Kedar for allowing me to see this valuable work before publication.

[37] Baudri de Bourgueil, *Poèmes*, ed. Jean-Yves Tilliette, vol. 2 (Paris: Les Belles Lettres, 2002), *Adelae comitissae*, v. 723:

Quippe pavimentum mundi fuit altera mappa....

[38] Ratkowitsch, *Descriptio picturae*, 25-107.

surpassed emperors by his generosity. William the Conqueror, according to Baudri, has overpowered not only peoples but geography itself: "over the tempestuous abyss of the sea, he joined two shores and made them one kingdom." The entire earth trembles before his imperial power.[39] Baudri draws on the traditional Roman triumphal images, such as the conquest of geographical space and the entire world trembling before the conqueror. To complete the picture, he even casts the English as ferocious savages who shamefully run away from the victorious William.[40]

Baudri goes on to describe the tapestry placed near the countess' bed. Depicting the conquest of England, the tapestry shows "the wealth of the king, his glory, his wars and triumphs."[41] Baudri draws on Virgil for details describing William's life, such as his flight from his native land, his speech to his armies, and especially the battle scenes. This evocation of Aeneas allows Baudri to raise William to the pedestal of an epic hero.[42] Praise escalates even further when Baudri cites ancient examples of military and royal glory. He declares that William exceeded all ancient kings and heroes as a ruler of men and a general, and that even Xerxes could not have assembled such

---

[39] Baudri, *Adelae comitissae* 6-18:
..."*Comitissa, ualeto,*
Reginae potius nomine digna, uale."
Haec est illius, si nescis, filia regis
Anglos indomitos qui domuit gladio,
Qui sibi Normannos tollentes iura paterna
Perculit ense fero subposuitque iugo.
Iste procellosas pro littore duxit abyssos
Et quasi conduxit littora littoribus.
Iste, licet sumptus superauerit imperiales.
....
Denique tantus erat ut solus fecerit orbem
Numen ad imperii subtrepidare sui.

[40] Ibid., 417-18:
Tunc prae tristitia gens effera praeque pudore
Egreditur palans insequiturque uagos.

[41] Ibid., 561-62:
Regis diuitiae, sua gloria, bella, triumphi
In uelo poterant singula uisa legi.
For a discussion of Baudri's tapestry and its possible relation to the Bayeux Tapestry, see O. K. Werckmeister, "The Political Ideology of the Bayeux Tapestry," *Studi medievali*, 3rd ser. 17.2 (1976), 535-95, esp. 550-51 and 562-63.

[42] Tilliette notes the significance of Baudri's use of the *Aeneid* in his commentary on the poem. See also Ratkowitsch, *Descriptio picturae.*

an army and fleet.[43] The poets and writers of the generation previous to Baudri had already established the tradition of favorably comparing William to Julius Caesar, and Baudri follows in their footsteps. According to this tradition, William is superior to Julius Caesar, for whereas Caesar conquered Britain when supported by the might of the Roman Empire, William did the same by relying only on his own strength.[44] Baudri proclaims that William is greater than all dukes and Caesars, and that he will soon be crowned with imperial honor.[45]

Not only does Baudri praise Adela's father as the new emperor, he celebrates her as a ruler in her own right. If her room in the poem symbolizes the world, it is Adela herself who, by ordering the decoration of the room, initiates the depiction of the world. Baudri describes Adela using her ruler to indicate what her serving women should do. This description evokes a passage from Martianus Capella's fifth-century encyclopedia of the liberal arts. Baudri relies on Martianus' encyclopedia later in his poem when he describes the contents of Adela's map and talks about the liberal arts. Geometry appears in Book VI of Martianus' encyclopedia as "a distinguished-looking woman holding a ruler in her right hand and a solid globe in her left." Geometry also appears as the mistress of geographical knowledge, because Book VI is largely dedicated to a description of the world. In an eleventh-century manuscript copy of Martianus' encyclopedia produced in France, Geometry is shown as a regal woman, enthroned with a ruler in her hands (Fig. 1).[46] Thus Baudri's allusion

---

[43] Baudri, *Adelae comitissae* 349-50:
Ille potens Serses non tantam iungere classem
Nec tantum potuit iungere nauigium.

[44] For a discussion of the Caesar theme in the Anglo-Norman poetic tradition celebrating William, see Elisabeth M. C. Van Houts, "Latin Poetry and the Anglo-Norman Court 1066-1135: The *Carmen de Hastingae Proelio*," *Journal of Medieval History* 15 (1989), 39-62, esp. 41 and 57. Norman historians, too, often cast Norman dukes as successors to Roman heroes.

[45] Baudri, *Adelae comitissae* 555-58:
Optinuit regnum rex optinuitque ducatum
Dux, et sic nomen Caesaris optinuit.
Solus et ipse duos, dum uixit, rexit honores,
Cunctis Caesaribus altior et ducibus.
Ibid., 242:
Mox extra seriem de duce Cesar erit.

[46] Ibid., 103-104:
Astiterant dictans operantibus ipsa puellis
Signaratque suo quid facerent radio.

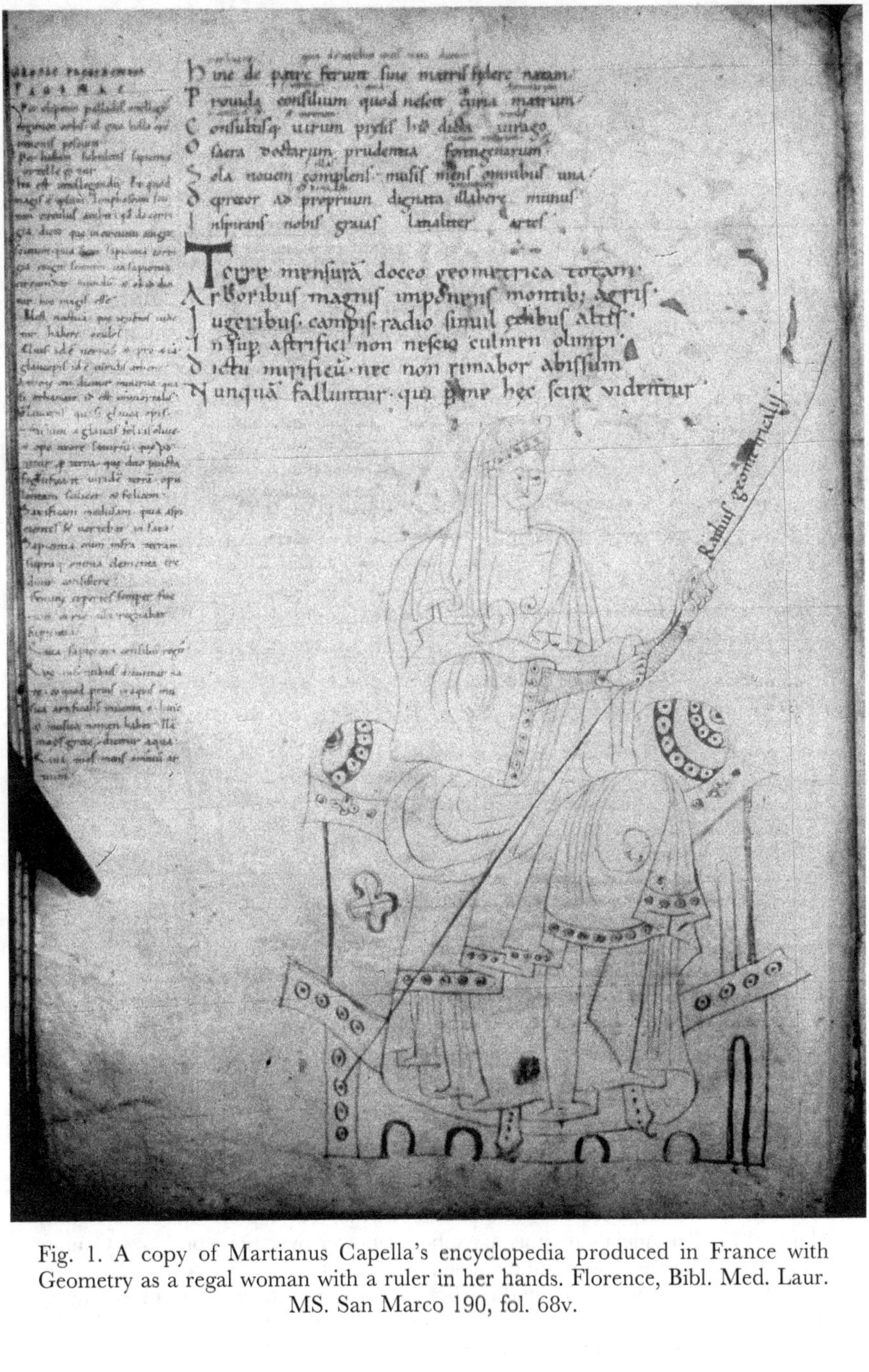

Fig. 1. A copy of Martianus Capella's encyclopedia produced in France with Geometry as a regal woman with a ruler in her hands. Florence, Bibl. Med. Laur. MS. San Marco 190, fol. 68v.

places Adela in the same category as Geometry, who has the power to describe the earth often assigned to emperors, and as Roman emperors such as Augustus and Theodosius, who were praised for measuring and organizing space.

The extended praise of Adela and her father, which contains many deliberate imperial connotations, prepares the reader for the appearance of Adela's world map, as does the long-standing tradition of associating world maps with power. World maps decorated the rooms of rulers such as Pope Zacharias in the eighth century and Henry III of England in the thirteenth. Medieval monarchs inherited Roman imperial regalia, including the orb, which represented the world in miniature. Some medieval images even show an emperor holding the orb with a tripartite map inscribed on it.[47] Thus it appears that Baudri and his learned audience would perceive Adela's map not only as a symbol of earthly vanity or the wealth of human learning. They would be likely to understand it as a symbol of power.[48] In describing the map, Baudri relied on classical information about the world, borrowed mainly from Martianus Capella. Baudri talks about the three continents, and about their great rivers, mountains, and wonders. The description of the map is essentially the sum of Roman knowledge. However, the focus of attention is shifted toward the parts of Europe lying to the north of Italy and, more specifically, to the Loire region.[49]

"The countess herself gave orders to the girls who performed the work and indicated with her ruler what they should do." Cf. Martianus Capella, *De nuptiis Philologiae et Mercurii*, ed. J. Willis (Leipzig: Teubner, 1983), 6.580: ... *prospicio quandam feminam luculentam radium dextera, altera sphaeram solidam gestitantem.* For the depiction of Geometry, see Fig. 1.

[47] Woodward, "Medieval *Mappaemundi*," 337; Kupfer, "Medieval World Maps," 267-68 and 276-79; Marcia Kupfer, "The lost wheel map of Ambrogio Lorenzetti," *The Art Bulletin* 78-2 (1996), 286-310, esp. 304; Anna-Dorothee von den Brincken, "'...ut describeretur universus orbis.' Zur Universalkartographie des Mittelalters," *Miscellanea Mediaevalia* 7 (1970), 249-78, esp. 249-50.

[48] For Adela's map as a symbol of vanity, see Woodward, "Medieval *Mappaemundi*," 339. For the wealth of learning and princely virtue, see Kupfer, "Medieval World Maps," 277. For Adela's map as a symbol of power over the world, see Gautier Dalché and Tilliette, "Nouveau document," 243 n. 4.

[49] The Loire is singled out and celebrated as the most pleasant and healthful of rivers, and as one of the four great rivers of Europe (the other three being the Po, Rhine and Danube): Baudri, *Adelae comitissae* 877 and 897-98.

The map presented Roman imperial geography, was made by Adela's order, and placed on the floor under her feet. Baudri's description of the map can thus be seen as an extension of the praise given to Adela and her father earlier in the poem. William the Conqueror has just appeared before the reader as the second Aeneas and the second Julius Caesar, as a mighty ruler destined to build an empire more powerful than all older empires. The *Aeneid*, as Baudri and his learned audience well knew, glorified Rome and contained the promise of an "empire without end." This shared knowledge lent depth and meaning to Baudri's praise of Adela and her family, and allowed subtle nuances to be added at will. Readers familiar with the traditional vocabulary of power were welcome to understand the map as a symbol and promise of imperial domination that extended from the past into the future.

The several cases chosen for the discussion above demonstrate how Roman ideas about geo-ethnography, conquest and power provided imperial rhetoric with useful tools. Maps, geo-ethnographic descriptions and personifications of lands and peoples came to be used as symbolic expressions of triumph and domination. From antiquity to the Middle Ages and beyond, scholars and rulers relied on the Roman vocabulary of power, adding and changing terms as needed, in order to celebrate imperial might and to reinforce political and ideological messages. Thus late antique panegyrists used geo-ethnographical sketches to glorify even modest military successes by emperors. In a parallel development, compilers of late antique geographical texts, continuing the triumphalist tradition of earlier writers, encouraged the audience even as the late Roman Empire was experiencing defeats.

As new European rulers were contending for the imperial title, the idea of Rome, the center of an empire that once dominated the world, came to exercise powerful influence over political theory, ideology and imagination.[50] Models provided by Roman geo-ethnogra-

[50] For an overview, see Arturo Graf, *Roma nella memoria e nelle immaginazioni del medio evo*, 2 vols. (Turin: Ermanno Loescher, 1882-83); Fedor Schneider, *Rom und Romgedanke im Mittelalter: die geistigen Grundlagen der Renaissance* (Munich: Drei Masken Verlag, 1926); Bernhard Kytzler (ed.), *Rom als Idee* (Darmstadt: Wissenschaftliche Buchgesellschaft, 1993), especially essays by William Hammer, "The Concept of the New or Second Rome in the Middle Ages," and Michael Seidlmayer, "Rom und Romgedanke im Mittelalter," both reprinted from earlier publications.

phy, together with models drawn from history and literature, proved useful as rhetorical tools. Borrowing the Roman vocabulary of domination, medieval poets and scholars celebrated new emperors and their achievements. Measured against the eternal Roman example, new empires were proclaimed to have succeeded and to have superseded the old Roman one. Roman geo-ethnography supported this quest in the Carolingian Empire, where geographical studies informed the audience and geo-ethnographical imagery was used to glorify the emperors. In Ottonian manuscripts, the image of German emperors, accompanied by the traditional personifications of provinces, asserted the might of the self-proclaimed successors to both the Romans and the Carolingians. At a direct intersection of Roman geography and imperial praise, an eleventh-century copyist added poems filled with triumphal imperial rhetoric to a collection of Roman geographical texts. In the twelfth century, the celebration of Anglo-Norman rulers followed the same pattern. Baudri of Bourgeuil glorified Adela and her father by using Roman triumphal vocabulary, which included geo-ethnographical references and a map.

What made such continuous use of geo-ethnographical rhetoric effective, at least in the minds of medieval writers, was knowledge of Roman geography and ethnography as part of a broader education in classical literature shared by learned audiences. Geo-ethnographical references, like allusions to Roman poets and historians, were expected to be understood. A study of geo-ethnographical rhetoric thus provides an instructive, and to date insufficiently appreciated, perspective on such important questions as the reception of classical knowledge and its use in ideology. However, the influence of maps and texts which creatively developed Roman models was not limited to scholarly education. Rather, in the medieval period geo-ethnography also contributed to ideology and politics by enriching the understanding and justification of *translatio imperii.*

"*USQUE AD ULTIMUM TERRAE*": MAPPING THE ENDS OF THE EARTH IN TWO MEDIEVAL FLOOR MOSAICS

Lucy E.G. Donkin

In the eleventh, twelfth and early thirteenth centuries, figurative floor mosaics were laid in a number of churches in northern Italy, part of a wider revival of the art in western Europe.[1] The pavements feature a broad range of imagery, including subjects of a geographical, and even cartographical, nature.[2] Two examples from Piedmont, in which such imagery is accompanied by unusually full and specific inscriptions, are of particular interest here: the well-known mosaic from S. Salvatore in Turin, which depicts the oceanic islands and winds familiar from medieval *mappaemundi*; and the presbytery pavement of Asti cathedral, which includes personifications of the Rivers of Paradise.[3] It is revealing to consider both works in the light of the definition of maps proposed in the first volume of J. B. Harley and David Woodward's *History of Cartography*, that is to say, "graphic representations that facilitate a spatial understanding of things, concepts,

---

[1] For an overview of the mosaics in northern Italy, see Arturo Carlo Quintavalle, *Wiligelmo e Matilde: L'officina romanica* (Milan: Electa, 1991), 97-111, 280-304, 397-408, 495-98; Elena Pianea, "I mosaici pavimentali," in Giovanni Romano (ed.), *Piemonte Romanico* (Turin: Cassa di Risparmio di Torino, 1994), 393-420.

[2] Studies that consider this aspect of medieval floor mosaics include Chiara Frugoni, "La figurazione basso-medioevale dell'imago mundi," in *"Imago Mundi": La conoscenza scientifica nel pensiero bassomedioevale* (Todi: presso l'Accademia Tudertina, 1983), 225-69; Xavier Barral I Altet, "Poésie et iconographie: un pavement du XIIe siècle décrit par Baudri de Bourgueil," in William Tronzo et al. (eds.), *Studies on art and archaeology in honour of Ernst Kitzinger on his seventy-fifth birthday, Dumbarton Oaks Papers* 41 (1987), 41-54; Jeanne Fox-Friedman, "Sacred and Secular: Modena Cathedral and Monumental World Maps," *Arte Medievale*, 2nd ser., 10 (1996), 39-55; Jeanne Fox-Friedman, "Vision of the World: Romanesque Art of Northern Italy and the Hereford Mappamundi," in P. D. A. Harvey (ed.), *The Hereford World Map: Medieval World Maps and their Context* (London: The British Library, 2006), 137-51.

[3] This article draws on Chapters 7 and 8 of my doctoral thesis: "*Ornata decenter.* Figurative Ecclesiastical Floor Mosaics in Northern Italy, 1030-1213" (diss., University of London, 2005). The research was carried out at the Courtauld Institute of Art and funded by the Arts and Humanities Research Board, to whom I am most grateful.

conditions, processes, or events in the human world."[4] From this perspective, not only can comparisons be drawn with both world maps and schematic diagrams in other media, but the mosaics themselves can also be seen to have possessed a mapping function appropriate to their location within the church.

The mosaic of S. Salvatore in Turin is the more explicitly cartographic of the two examples, and this aspect of its iconography has provoked scholarly interest since the time of its discovery in 1909.[5] A central design of a Wheel of Fortune surrounded by animal roundels is encircled by a band representing the ocean, with personifications of the twelve winds at the corners of the square composition (Fig. 1). S. Salvatore was a palaeochristian foundation, which may have formed part of the early episcopal complex of Turin and was used by a college of canons from the Carolingian period onwards.[6] Recent excavations have indicated that the church was rebuilt at some point during the eleventh century.[7] At a later date, the presbytery was raised and enlarged in works that may well have entailed the abandonment of the crypt. The final arcades of the nave were closed off with low walls, and filled in to form a platform reached by a flight of seven steps and probably enclosed by plain marble slabs.[8] It is this modified presbytery that was decorated with the mosaic pavement, generally dated to the end of the twelfth century.[9] The mosaic covered all the area between the steps and the altar; but it was badly damaged in the 1490s when S. Salvatore was demolished to make way for the new cathedral, and it was further disturbed by burials over the following centuries. The surviving fragments, which were taken up shortly after discovery, have recently been restored to

[4] *The History of Cartography*, vol. 1 (Chicago: University of Chicago Press, 1987), xvi.

[5] Pietro Toesca, "Vicende di un'antica chiesa di Torino: Scavi e scoperte," *Bollettino d'Arte* 4 (1910), 1-16; Pietro Gribaudi, "Di un mosaico cosmografico medioevale scoperto a Torino," *Bollettino della Società Geografica Italiana*, 4th ser., 12 (1911), 619-37; Federico Patetta, "A proposito del mosaico medioevale scoperto a Torino nel Marzo del 1909," *Atti della Società Piemontese di Archeologia e Belle Arti* 8 (1910/17), 318-40.

[6] Luisella Pejrani Baricco, "La Basilica del Salvatore e la Cattedrale di Torino: Considerazioni su uno scavo in corso," in L. Mercando and E. Micheletto (eds.), *Archeologia in Piemonte*, vol. 3, *Il Medioevo* (Turin: Allemandi, 1998), 132-49.

[7] For the results of the 1996-97 excavations, see ibid., 140-44.

[8] Ibid., 142, fig. 97, 143, fig. 98, 144-47.

[9] Pianea, "I mosaici pavimentali," 413-16, esp. 415 n. 79.

Fig. 1. Presbytery pavement of S. Salvatore, Turin, late twelfth century, photographed in situ at the time of its discovery in 1909.

their original position.[10] The floor is best known to scholars through an influential article by Ernst Kitzinger.[11] Positing that the geographical elements were taken from a single cartographic source, he drew comparisons with the form of an early twelfth-century Beatus map now in the Biblioteca Nazionale in Turin, but he considered the model for the floor to have been closer to something resembling the Hereford map in its use of inscriptions.[12] In Kitzinger's view, the combination of such elements with a Wheel of Fortune was to "characterize the earth as a scene of vain pursuits."[13] This interpretation was followed in the brief discussion of the mosaic in the first volume of *The History of Cartography*, where it was the only medieval floor to receive attention.[14] My intention is to focus on the representation of the edges of the earth, exploring their wider associations on the premise that the geography of the mosaic does not necessarily depend on a single model.

With its circular form and wavy infill, the mosaic ocean displays a notable affinity with *mappaemundi*. Two blocks of text within the ocean represent islands in the north-west of the known world (Fig. 2). The larger of the two refers to "*Britania*", "*Orcades*" and "*Tile*": *BRITANIA/ INSULA INTER/ FUSA MARI/ ORCADES INSULE/ TILE ULTIMA INSULA*. The smaller panel describes "*Scocia*": *SCOCIA INSU/LA P(ro)XIMA BRI/TANIE UBI/ NULLA ANGU(is)*.[15] Although the islands are mentioned in a number of geographical

[10] The mosaics were initially removed to the Museo Civico d' Arte Antica. They entered the collection of the Museo di Antichità in 1989, and were reassembled in their original position in 2005. I am most grateful to Dottoressa Pejrani Baricco of the Soprintendenza Archeologica del Piemonte for the opportunity to examine some of the fragments in storage.

[11] Ernst Kitzinger, "World Map and Fortune's Wheel: A Medieval Mosaic Floor in Turin," *Proceedings of the American Philosophical Society* 117, no. 5 (1973), 344-73.

[12] Ibid., 356-61. Turin, Biblioteca Nazionale Universitaria, Sgn. I.II.1, fols. 38v-39r (fols. 45v-46r according to the post-restoration foliation); John Williams, *The Illustrated Beatus: A Corpus of the Illustrations of the Commentary on the Apocalypse*, vol. 4, *The Eleventh and Twelfth Centuries* (London: Harvey Miller, 2002), 26-30, fig. 119a-b.

[13] Kitzinger, "World Map and Fortune's Wheel," 367.

[14] David Woodward, "Medieval *Mappaemundi*," in *The History of Cartography*, vol. 1, 339, fig. 18.41. Antique and Byzantine floor mosaics were discussed in the chapters by O. A. W. Dilke: "Roman Large-Scale Mapping in the Early Empire," 230-31; "Itineraries and Geographical Maps in the Early and Late Roman Empires," 246-48; "Cartography in the Byzantine Empire," 263-67.

[15] "*Scocia*' is used here to mean Ireland.

Fig. 2. Details of the winds and islands from the presbytery pavement of S. Salvatore, Turin, late twelfth century.

texts, Kitzinger demonstrated that the wording here most closely resembles a passage in Book Fourteen of Isidore's *Etymologiae*.[16] These and other oceanic islands are common features of contemporary world maps. The detailed zonal map included in a late twelfth-century copy of the *Liber Floridus* by Lambert of St. Omer depicts several, including "*Anglia*", "*Hybernia*", and "*Thyle*" in the north west (Fig. 3).[17] It is rare, however, that islands are identified by anything more than a simple name-label, at least in the eleventh and twelfth centuries. Descriptive text is only found with any regularity on larger scale *mappaemundi* from the thirteenth century, such as the Ebstorf and Hereford maps. The fuller inscriptions on the S. Salvatore pavement may, therefore, have been taken from an early example of this kind of world map. Alternatively, the designer of the mosaic may have taken advantage of the space afforded by the floor surface to create an original design with descriptive inscriptions.

In either case, but particularly the latter, it is significant that the wording of the inscriptions emphasises the remote and marvellous nature of the islands. "*Tile*", or Thule, for example, is described as the "*ultima insula,*" a position reflected in the *Liber Floridus* map, where it is shown at the western end of the row of islands, closest to the Antipodes.[18] The first reports of its existence are traditionally ascribed to Pytheas of Massilia, whose circumnavigation of Britain in the fourth century BC was recorded by Polybius, Strabo and Pliny the Elder.[19] Under the Romans, to quote James Romm, Thule possessed "symbolic rather than geographical resonance," coming to "represent all the most distant regions to which exploration and conquest could aspire."[20] In the Middle Ages, beginning with Dicuil in 825, the name appears to have generally referred to Iceland.[21] Thus, although the inscription on the pavement draws on Isidore's text, it

[16] Isidore, *Etymologiae* 14.6.2-6 (Migne *PL* 82.513a-b); Kitzinger, "World Map and Fortune's Wheel," 350.

[17] Wolfenbüttel, Herzog August Bibliothek, Cod. Guelf. 1 Gud. lat., fols. 69v-70r. Danielle Lecoq, "La mappemonde du *Liber Floridus* ou la vision du monde de Lambert de Saint-Omer," *Imago Mundi* 39 (1987), 17, 43.

[18] Ibid.

[19] Polybius, *Histories* 34.5.3-4; Strabo, *Geography* 2.4.1; Pliny, *Natural History* 2.187; 4.104.

[20] James S. Romm, *The Edges of the Earth in Ancient Thought: Geography, Exploration and Fiction* (Princeton: Princeton University Press, 1992), 158.

[21] Ibid., 157 n. 82; John Kirtland Wright, *The Geographical Lore of the Time of the Crusades: A Study in the History of Medieval Science and Tradition in Western Europe* (New

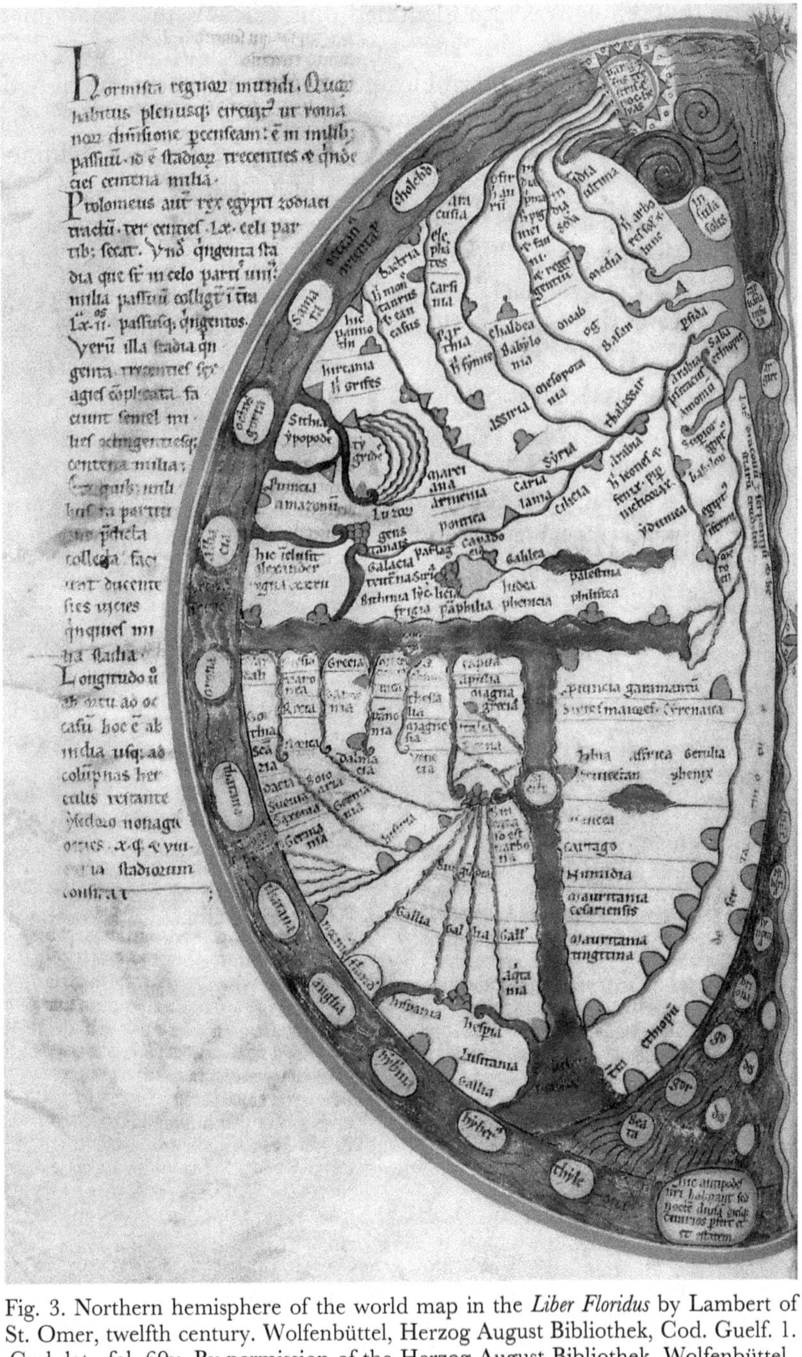

Fig. 3. Northern hemisphere of the world map in the *Liber Floridus* by Lambert of St. Omer, twelfth century. Wolfenbüttel, Herzog August Bibliothek, Cod. Guelf. 1. Gud. lat., fol. 69v. By permission of the Herzog August Bibliothek, Wolfenbüttel.

may be that Thule was also identified thus here. At the same time, its characterisation as the "last" or "farthest" island suggests that remoteness was considered significant. Islands with better-established identities are also described in exotic terms on the pavement. "*Britania*" is said to be located in the "*fusa mari*", another feature with its roots in ancient geographical thought; both Pytheas and Agricola are said to have observed the sluggish, viscous quality of the waters north of the island.[22] The northern reaches of the ocean were described as unnavigable by Pytheas on account of the frozen sea which lay beyond Thule and, like it, entered the geographical imagination of the Middle Ages. In his description of the area in the *Imago Mundi*, Honorius Augustodunensis refers to an area of perpetual cold where the sea is frozen: "*Ultra hanc, versus aquilonem, est mare congelatum et frigus perpetuum.*"[23] "*Scocia*" is described on the S. Salvatore pavement as a place devoid of serpents, a characteristic ascribed to various north-western islands by a number of commentators: Bede writes of the lack of snakes in Hibernia; Honorius specifies the island of Tanatos.[24]

The fragmentary nature of the S. Salvatore mosaic makes it impossible to tell whether it originally featured more islands, but it is possible that those which have survived were the only ones depicted. "*Britannia*", "*Hibernia*" and "*Thule*" alone appear on a world map in an early twelfth-century computus manuscript from Thorney Abbey (Fig. 4).[25] An elaboration of a basic T-O structure, this map is embellished with inscriptions marking places of primarily biblical importance. The horizontal bar of the T is prominently labelled "*Hierusalem,*" with an additional inscription "*crux xpi*" perhaps referring to the Tau cross. At the centre of the map is a circle with a cross identified as "*Mons Syon.*" The islands are among the few non-biblical elements on the map. Britannia is shown within the ocean, while the other

York: Dover Publications, 1965), 75; Dicuil, *Liber de mensura orbis terrae* 7.6-13, ed. J. J. Tierney (Dublin: Dublin Institute for Advanced Studies, 1967), 72-75.

[22] Strabo, *Geography* 2.4.1; Tacitus, *Agricola* 10; Romm, *The Edges of the Earth,* 22-23, 157-58.

[23] Honorius Augustodunensis, *De Imagine Mundi* 1.31 (Migne *PL* 172.130c).

[24] Ibid. (Migne *PL* 172.130b); Bede, *Ecclesiastical History of the English People* 1.1.

[25] Oxford, St. John's College Library, MS 17, fol. 6r. Faith E. Wallis, "MS Oxford, St John's 17: a Medieval Manuscript in its Context" (diss., University of Toronto, 1985), 219-23; Evelyn Edson, *Mapping Time and Space: How Medieval Mapmakers Viewed Their World* (London: The British Library, 1997), 86-90, fig. 5.4.

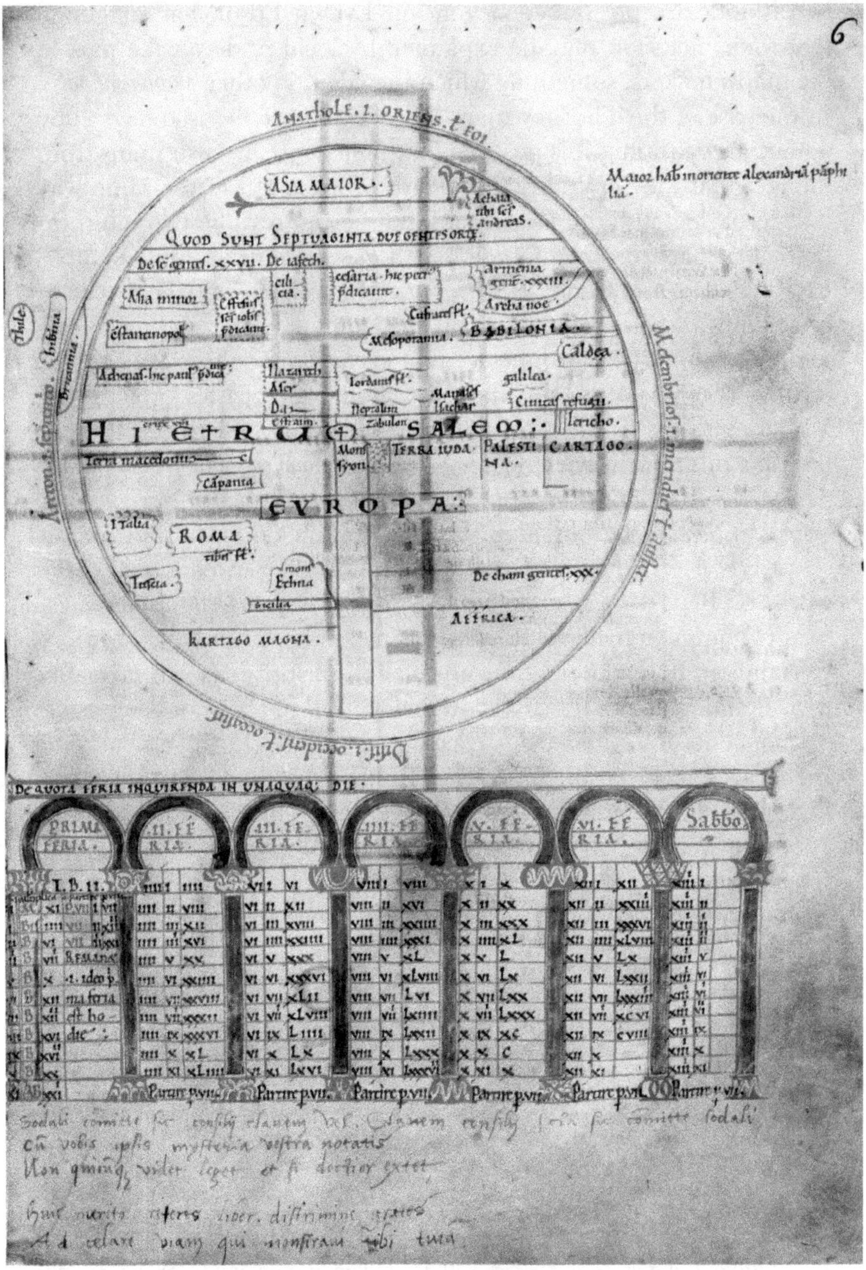

Fig. 4. World map in a computus manuscript from Thorney Abbey, 1110. Oxford, St. John's College Library, MS 17, fol. 6r. By permission of the President and Scholars of St John Baptist College in the University of Oxford.

two islands are positioned beyond it. Evelyn Edson has suggested that their inclusion may be explained by local pride on the part of the map's makers, something which is evident in other *mappaemundi*.[26] In the case of the Thorney map, however, where the islands occupy a marginal position, it is possible that their presence also stems from the fact that they epitomised the ends of the earth in the same way that Jerusalem and the Holy Land did the centre. The two interpretations are far from incompatible. An exegetical tradition, exemplified by Bede's *Ecclesiastical History*, regarded both the initial conversion of Britain and its acceptance of the Roman Easter dating as the fulfilment of biblical prophecies that salvation would extend to the farthest reaches of the earth and its "multitude of isles."[27] Boniface V's letter to Justus of Canterbury, for example, refers to the archbishop's ministry in such terms: "*Et vere 'per omnem terram exisse sonum eorum, et in fines orbis terrae verba ipsorum' universalis gentium confessio, suscepto Christianae sacramento fidei, protestetur*."[28] Since the Thorney manuscript is concerned with the calculation of Easter and contains three works by Bede—*De natura rerum*, *De temporibus* and *De temporum ratione*—its compiler is likely to have been familiar with this tradition. Certainly in Italy, the liminal nature of the British Isles was probably their defining characteristic, partly conceived in terms of geographical concepts inherited from classical antiquity. In an embassy to Frederick I in 1155, the citizens of Rome proclaimed that their city had not only extended her empire to the ends of the earth, but had added to it islands lying outside the world: "*quin etiam insulas extra orbem positas orbi adiciens*."[29]

Mappaemundi do not, however, provide the only parallels for the geographical elements at S. Salvatore. Depictions of the ocean and winds surrounding a non-cartographic central scene are found in

[26] Edson, *Mapping Time and Space*, 89.

[27] Ps. 96(97):1 AV. Jennifer O'Reilly, "Islands and idols at the ends of the earth: exegesis and conversion in Bede's *Historia Ecclesiastica*," in Stéphane Lebecq et al. (eds.), *Bède le Vénérable entre tradition et postérité / The Venerable Bede. Tradition and posterity* (Lille: CEGES, Université Charles-de-Gaulle, Lille 3, 2005), 119-45.

[28] Bede, *Ecclesiastical History* 2.8. The references are to Ps. 18(19):5 and Rom. 10:18.

[29] Otto of Freising, *Gesta Friderici* 2.29, ed. B. von Simson and G. Waitz, 3rd ed. (Hannover; Leipzig: Hahn, 1912), 135. The passage, usually taken to refer to Britain, is cited by Richard Southern in connection with the Thorney map: *Medieval Humanism and Other Studies* (Oxford: Basil Blackwell, 1970), 133 n. 1.

manuscripts of Beatus of Liébana's *Commentary on the Apocalypse*. The composition is often used to illustrate Chapter Seven of the Book of Revelation, in which four angels standing at the corners of the earth restrain the winds, and a fifth ascends from the east to mark the elect of the tribes of Israel.[30] In the mid tenth-century Morgan Beatus, the tribes are shown in a landscape that takes up almost the whole of the earth, which is surrounded by an oval ocean. At the four corners of the page are winged figures of the angels blowing out streams of air identified as "*ventus.*"[31] The correspondence with the floor is even closer in the case of the mid eleventh-century Saint-Sever Beatus, where the earth is round and the angels hold winged heads representing the winds (Fig. 5).[32] The same type of composition was also used in the Turin Beatus manuscript.[33] Kitzinger noted such illustrations in connection with the personification of the winds in the mosaic without considering the more fundamental similarity, that of representing the outer ocean independently from the central elements of a *mappamundi*.[34] If the representation of the winds at the corners of the composition reflects the text's "*quattuor angulos terrae*" understood in the sense of parts or regions of the world, the inclusion of the ocean may express a corresponding interest in the confines or ends of the earth.[35] These too had apocalyptic associations, being employed in the Old Testament to denote the extent of God's rule and of the Messianic Kingdom. Psalm Seventy-One describes the Messiah as ruling from sea to sea, and from the river to the ends of the earth: "*Et dominabitur a mari usque ad mare, et a flumine usque ad terminos terrae.*"[36] Similar language permeates the prophecies of Isaiah,

---

[30] Rev. 7:1-3 Vulg.

[31] New York, Pierpont Morgan Library, MS M. 644, fol. 115v; Williams, *The Illustrated Beatus*, vol. 2, *The Ninth and Tenth Centuries* (London: Harvey Miller, 1994), 21-33, fig. 49.

[32] Paris, Bibliothèque Nationale, MS lat. 8878, fol. 119r; Williams, *The Illustrated Beatus*, vol. 3, *The Tenth and Eleventh Centuries* (London: Harvey Miller, 1998), 44-57, fig. 406.

[33] Turin, Biblioteca Nazionale Universitaria, MS I.II.1, fol. 103v. Williams, *The Illustrated Beatus*, vol. 4, *The Eleventh and Twelfth Centuries* (London: Harvey Miller, 2002), fig. 143.

[34] Kitzinger, "World Map," 359-60.

[35] For a detailed discussion of these terms, see Anna-Dorothee von den Brincken, *Fines Terrae: Die Enden der Erde und der vierte Kontinent auf mittelalterlichen Weltkarten* (Hannover: Hahnsche Buchhandlung, 1992), 14-18.

[36] Ps. 71(72):8 Vulg.

Fig. 5. Angels Restraining the Winds, Saint-Sever Beatus, mid eleventh century. Paris, Bibliothèque nationale, MS lat 8878, fol. 119r. By permission of the Bibliothèque nationale de France.

which describe the gathering together of people from the cardinal directions and the furthest reaches of the world.[37] Later, the Lord speaks directly to the ends of the earth: "*Convertimini ad me et salvi eritis omnes fines terrae.*"[38] In his commentary on the Book of Isaiah, St. Jerome associated these prophecies with the preaching of the Apostles, and the New Testament itself uses a similar expression to refer to their mission.[39] In Acts, Christ tells the disciples that they will witness to him "*usque ad ultimum terrae.*"[40] Together, these comparisons suggest that the cartographical elements on the S. Salvatore floor are not simply a device to define the centre of the composition as mundane, but that they constitute a depiction of the ends of the earth. As such, they can be seen to have held an allusion to the extent of divine power, whether viewed in terms of the Apostolic mission or of the Messianic Kingdom to come. If this hypothesis is correct, it provides a motive for the correspondences between the pavement and the Turin Beatus map, since the function of such maps was to illustrate the mission of the Apostles.[41] Furthermore, it addresses Kitzinger's charge that Fortuna seems the autonomous mistress of the earth and lacks an opposing force.[42] The ends of the earth themselves provide such a presence, forming a conceptual as well as a physical frame that promises ultimate redemption.

While the pavement of Asti cathedral is far less conventionally cartographic than the S. Salvatore floor, it too possesses a strong spatial aspect. The mosaic, which has been dated to the second half of the twelfth century, is one of the few surviving elements of a Romanesque cathedral dedicated to the Virgin Mary.[43] It came to light during excavations in the 1980s and remains in situ in the presbytery of the present cathedral, built during the first half of the

[37] "Noli timere quoniam tecum ego sum ab oriente adducam semen tuum et ab occidente congregabo te. Dicam aquiloni da et austro noli prohibere adfer filios meos de longinquo et filias meas ab extremis terrae:" Isa. 43:5-6 Vulg.

[38] Isa. 45:22 Vulg.

[39] Jerome, *Commentariorum in Esaiam* 13.45.18/26, ed. Marc Adriaen, *S. Hieronymi Presbyteri Opera*, 1.2a, Corpus Christianorum Series Latina 73A (Turnhout: Brepols, 1963), 512-15.

[40] Acts 1:8 Vulg.

[41] Edson, *Mapping Time and Space*, 149-59.

[42] Kitzinger, "World Map," 369.

[43] Pianea, "I mosaici pavimentali," 412-13.

fourteenth century.[44] As in Turin, the episcopal complex is likely to have had palaeochristian origins. Documentary evidence indicates that the complex developed further during the ninth and tenth centuries; a new cathedral was built in the last decades of the eleventh century, and dedicated by Pope Urban II in July 1095.[45] The excavations that revealed the mosaic showed this structure to have had three aisles ending in semi-circular apses and a crypt beneath the presbytery. In the twelfth century, the crypt was abandoned and filled in, and the presbytery was divided in two by steps going up to the *sancta sanctorum*. At the same time, a mosaic pavement was laid in both parts of the presbytery.[46] Although fragments were still visible in the eastern end as late as the nineteenth century, today the mosaic survives only in the western end.

The mosaic is divided into twelve square panels, arranged in three rows of four, and includes scenes from the life of Samson and representations of King David on horseback, a cantor, and a count (Fig. 6). At the four corners of the composition are labelled personifications of the Rivers of Paradise: Phison at the top left (Fig. 7); Tigris at the top right (Fig. 8); Geon at the bottom left (Fig. 9); and Euphrates, now almost entirely lost, at the bottom right. While all the text is positioned so as to be read from the west, the figures of Geon and Euphrates are shown upside down. Further inscriptions, drawn from Genesis, associate the rivers with areas of the known world. The clearest of these—*AETHIO...*—accompanies the representation of Geon, referring to its path through Ethiopia: "*Et nomen fluvio secundo Geon ipse est qui circuit omnem terram Aethiopiae.*"[47] The Tigris scene has the inscription *VADI...RIOS*, which has been reconstructed by Alberto Crosetto to read "*vadit Assyrios,*" quoting the biblical description: "*Nomen vero fluminis tertii Tigris ipse vadit contra Assyrios.*"[48] The scene depicting Phison has the letters *EUI...AT*, which may refer to the land of "*Evilat*": "*Nomen uni Phison ipse est qui circuit omnem terram*

[44] Alberto Crosetto, "Indagini archeologiche nel medioevo astigiano. La cattedrale di S. Maria," *Quaderni della Soprintendenza Archeologica del Piemonte* 13 (1995), 255-76.

[45] Ibid., 259, 262; Paul Fridolin Kehr, *Italia Pontificia*, vol. 6, *Liguria sive Provincia Mediolanensis*, 2, *Pedemontium, Liguria Maritima* (Berlin: Apud Weidmannos, 1914), 177.

[46] Crosetto, "La cattedrale di S. Maria," 262-63.

[47] Gen. 2:13 Vulg.

[48] Gen. 2:14 Vulg.; Crosetto, "La cattedrale di S. Maria," 264.

Fig. 6. Presbytery pavement of Asti cathedral, second half of the twelfth century. By permission of the Soprintendenza Archeologica del Piemonte.

Fig. 7. Detail of the River Phison from the presbytery pavement of Asti cathedral, second half of the twelfth century.

Fig. 8. Detail of the River Tigris from the presbytery pavement of Asti cathedral, second half of the twelfth century.

Fig. 9. Detail of the River Geon from the presbytery pavement of Asti cathedral, second half of the twelfth century.

Evilat."[49] If there was once an additional label accompanying the Euphrates it has been entirely lost, but it is possible that none existed, since this river's path is not described in Genesis.

The Rivers of Paradise were popular subjects on medieval floor mosaics. They are shown as personifications holding amphorae in the cathedrals of Aosta and Novara,[50] and as streams of water at the French abbey of Cruas and the chapel of St-Nicholas in the episcopal palace at Die.[51] Records of the lost twelfth-century floor at St. Remi in Reims indicate that here, too, there were depictions of the Rivers of Paradise.[52] Finally, there is the interpretation of the green bands on the floor of Hagia Sophia ascribed to Justinian in the legendary *Narratio de S. Sophia*: "the four strips he called the Four Rivers that flow out of paradise."[53] This was part of a wider Byzantine tradition of interpreting elements in *opus sectile* pavements as rivers.[54] However, it appears that the Asti mosaic is the only example to give the biblical descriptions of their courses, suggesting that here the rivers were not intended to evoke paradise itself so much as their paths through the wider world.

If the Asti rivers are unique among surviving floors, it is also difficult to find exact parallels in other media, largely because the mosaic combines two spatial conceptions of the Rivers of Paradise. In one interpretative tradition, based on the Genesis text, they were identified with known rivers, generally the Tigris, Euphrates, Nile

---

[49] Gen. 2:11 Vulg.; Crosetto, "La cattedrale di S. Maria," 264.

[50] Renato Perinetti, "I mosaici medievali di Aosta," in Federico Guidobaldi and Andrea Paribeni (eds.), *Atti del VI Colloquio dell'AISCOM* (Ravenna: Edizioni del Girasole, 2000), 161-74; Simonetta Minguzzi, *I Mosaici Pavimentali della Cattedrale di Novara dal Tardoantico al Medioevo* (Ravenna: Edizioni del Girasole, 1995), 27-29.

[51] Barral I Altet, "Poésie et iconographie," 48, fig. 6, 52; Id., "Observations sur l'iconographie des pavements en mosaïques romans de France en relation avec le décor mural des édifices religieux," in Gabriel Biancotto et al. (eds.), *Le rôle de l'ornement dans la peinture murale du Moyen Age* (Poitiers: Université de Poitiers, Centre d'études supérieures de civilisation médiévale, 1997), 151-53.

[52] Madeline Harrison Caviness, *Sumptuous Arts at the Royal Abbeys in Reims and Braine: ornatus elegantiae, varietate stupendes* (Princeton: Princeton University Press, 1990), 38.

[53] *Anonymi Narratio de Aedificatione Templi S. Sophiae* 26, in *Scriptores Originum Constantinopolitanarum*, vol. 1, ed. Theodore Preger (Leipzig: B. G. Teubner, 1901), 102-103. English translation from Cyril Alexander Mango (ed.), *The Art of the Byzantine Empire 312-1453: Sources and Documents* (Englewood Cliffs, N. J.: Prentice-Hall, 1972), 101.

[54] A. Frolow, "Deux églises byzantines d'après des sermons peu connus de Léon VI le sage," *Études Byzantines* 3 (1945), 56-57.

and Ganges, although a Syriac tradition saw Phison as the Danube.[55] It is in this sense that their paths are indicated on *mappaemundi*.[56] In the late twelfth-century Sawley map, for example, the rivers flow from the central point of an insular paradise (Fig. 10). The Tigris and the Euphrates are also shown in Asia, immediately above and below the Tower of Babel, while the river that separates the area labelled as Ethiopia from the rest of Africa is identified as "*Gion*." Phison is not expressly labelled, but the Ganges is shown close to paradise.[57] An allegorical tradition set the Rivers of Paradise within a quadripartite world-view. By virtue of their number, the rivers were inserted early on into the tetradic cosmology inherited from Graeco-Roman thought, coming to be associated with quaternities including the cardinal directions or four regions of the earth.[58] The rivers were also linked with the Evangelists, another quaternity with a geographical dimension, since the Gospels were understood as spreading to the four parts of the world.[59] The *Liber numerorum*, an arithmological text linked with the works of Isidore of Seville, associates the number four with the Rivers of Paradise, the Evangelists and the four regions, although here the rivers are seen as flowing around the world: "*Numerus autem iste quadratus evangelistarum quatuor deputatur, qui in quatuor partes, vel in angulos mundi fusi sunt. Quanquam et quatuor sunt paradisi flumina, quae totum orbem circumfluunt.*"[60] The personification of the rivers at Asti and their disposition on the pavement, especially the posi-

[55] Alessandro Scafi, "Mapping Eden: Cartographies of the Earthly Paradise," in Denis Cosgrove (ed.), *Mappings* (London: Reaktion Books, 1999), 52; Alessandro Scafi, *Mapping Paradise: A History of Heaven on Earth* (London: The British Library, 2006), 35. Ephraim Syrus, *Commentary on Genesis* 2.6.4, trans. Edward G. Mathews Jr. and Joseph P. Amar, in Kathleen McVey (ed.), *St. Ephrem the Syrian: Selected Prose Works* (Washington: Catholic University of America Press, 1994), 100 n. 125.

[56] For detailed discussion of the cartographic representation of Paradise and its rivers see Scafi, *Mapping Paradise*, chapters 5 and 6.

[57] Cambridge, Corpus Christi College, Parker Library, MS 66, p. 2. Danielle Lecoq, "La Mappemonde d'Henri de Mayence, ou l'Image du Monde au XIIe Siècle," in Gaston Duchet-Suchaux (ed.), *Iconographie Médiévale: Image, Texte, Contexte* (Paris: Éditions du Centre National de la Recherche, 1990), 162-63, 173-75.

[58] Anna C. Esmeijer, *Divina Quaternitas. A Preliminary Study in the Method and Application of Visual Exegesis* (Amsterdam; Assen: Van Gorcum, 1978), 47-51, 59-72; Elizabeth Sears, *The Ages of Man: Medieval Interpretations of the Life Cycle* (Princeton: Princeton University Press, 1986), 1-37.

[59] Jennifer O'Reilly, "Patristic and insular traditions of the evangelists: exegesis and iconography," in A. M. Luiselli Fadda and É. Ó. Carragáin (eds.), *Le Isole Britanniche e Roma in Età Romanobarbarica* (Rome: Herder, 1998), 49-94.

[60] *Liber numerorum* 5(20) (Migne *PL* 83.183b).

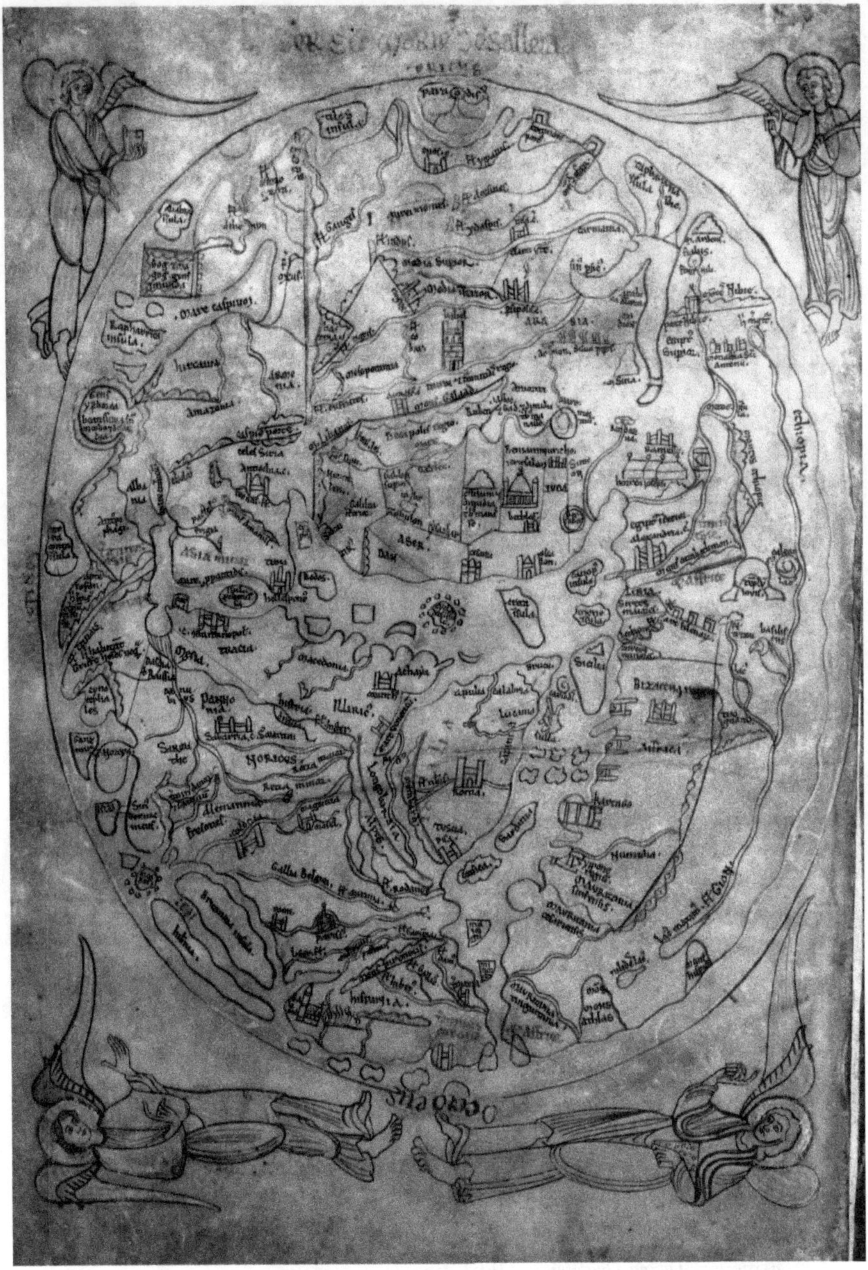

Fig. 10. Sawley Map, *Imago Mundi*, late twelfth century. Cambridge, Corpus Christi College, Parker Library, MS 66, p. 2. By permission of the Master and Fellows of Corpus Christi College, Cambridge.

tion of the lower two figures, are reminiscent of schemata featuring these and other quaternities, which often show the bottom figures upside down.[61] The rivers themselves are depicted in this way in a diagram from a late twelfth-century manuscript of the *De Laudibus S. Crucis* (Fig. 11).[62] Centred around a personification of "*Paradisus,*" the diagram shows streams of water flowing out diagonally towards labelled personifications of the rivers, with the symbols of the Evangelists and personifications of the virtues at the cardinal points. The initials A.D.A.M. placed between the Evangelists and the virtues make reference both to the cardinal directions—the initial letters of which, in Greek, spell the name ADAM—and to the seed of Adam scattered over the four regions of the world.[63]

The tetradic and topographic approaches were difficult to reconcile. No attempt was made to link particular rivers with particular directions in schemata; nor were the rivers ever personified on *mappaemundi*. However, an interest in their fourfold nature can be seen in the Burgo de Osma Beatus map of 1086, where the rivers flow from the centre of a rectangular Paradise to its corners in a diagonal cross (Fig. 12).[64] A more schematic treatment of the paths of the rivers that nevertheless exhibits concern for geographical realities can be found in the illumination preceding Genesis in the early thirteenth-century Lothian Bible (Fig. 13).[65] Here the rivers flow out of a central pool towards the four corners of the square frame to encircle small roundels that contain representations of various geographical areas, shown as ramparts from behind which peep a pair of heads. The inscriptions that identify the rivers and the regions are worn, but it is possible to make out that the bottom left-hand scene is labelled "*Perse,*" perhaps surrounded by the river Euphrates, while the bottom right-hand scene is identified as Ethiopia. The combination here of geographical specificity with a quadripartite, personified world is particularly close to the Asti mosaic. In the pavement, however, the personification and arrangement of the rivers, reminiscent

[61] See Esmeijer, *Divina Quaternitas*, 183, figs. 1a-b, 3; 184, fig. 6; 200, figs. 56a,c; 201, fig. 57b; 211, fig. 84b; Sears, *Ages of Man*, fig. 12.

[62] Munich, Bayerische Staatsbibliothek, Clm. 14159, fol. 5v. Esmeijer, *Divina Quaternitas*, 65-67, 201, fig. 57b.

[63] Esmeijer, *Divina Quaternitas*, 60-61, 65-67.

[64] Burgo de Osma, Cathedral, Cod. 1, fols. 34v-35r. Williams, *The Illustrated Beatus*, vol. 4, 17-25, fig. 5.

[65] New York, Pierpont Morgan Library, MS. M. 791, fol. 4v.

Fig. 11. Paradise quaternities, *De Laudibus S. Crucis,* 1170-85. Munich, Bayerische Staatsbibliothek, Clm. 14159, fol. 5v. By permission of the Bayerische Staatsbibliothek.

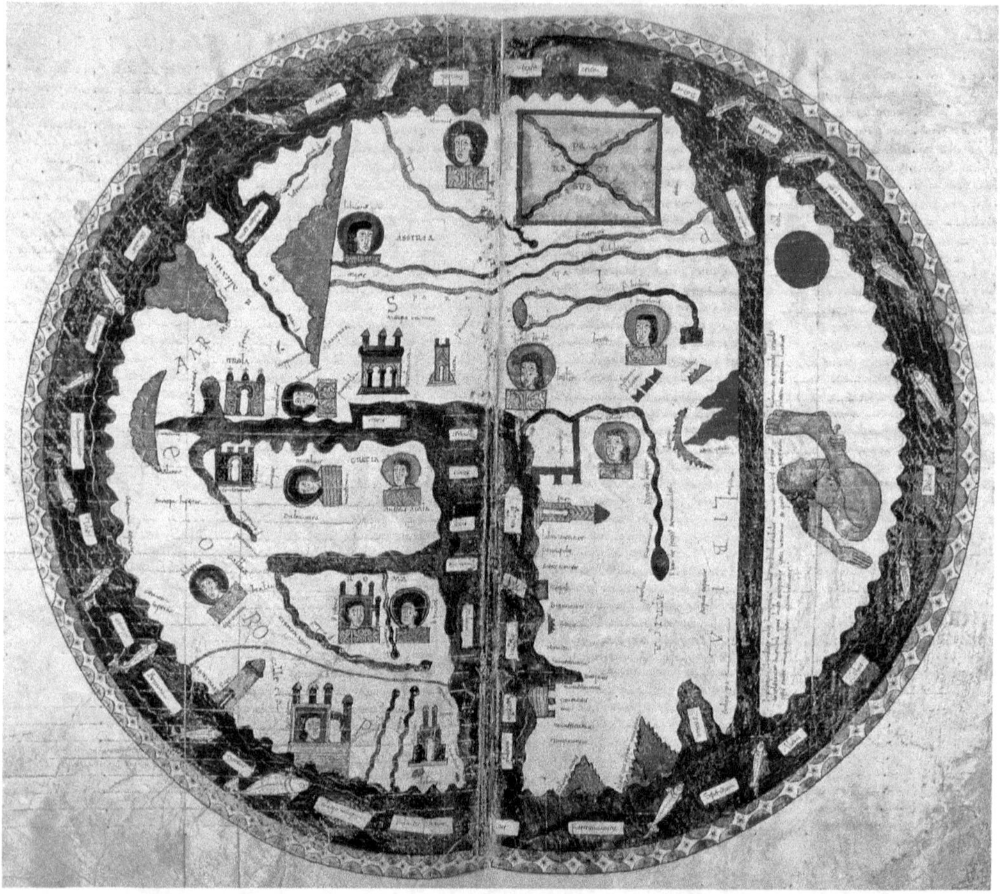

Fig. 12. World map, *Commentary on the Apocalypse* by Beatus of Liébana, 1086. Burgo de Osma, Archivo Capitular, Cod. 1, fols 34v-35r. By permission of the Archivo Capitular de Osma-Soria.

of diagrams featuring multiple quaternities, served not only to map the paths of the rivers but also to evoke the spreading of the Gospels throughout the world.

The relationships of both the S. Salvatore and Asti pavements to *mappaemundi* and other geographical depictions are complex. The fact that there are no perfect matches for the mosaic compositions suggests that their geographical elements are not the results of simple copying, but rather of the thoughtful and deliberate choice of particular elements to suit their context. What, therefore, were the impli-

Fig. 13. Trinity and Creation Scenes, Lothian Bible, c. 1220. New York, Pierpont Morgan Library, MS M 791, fol. 4 v. By permission of the Pierpont Morgan Library, New York.

cations of evoking the entire earth, whether through its confines or its corners, on the presbytery floor? In the first place such an evocation can be related to the way in which the church building itself embodied the world. This relationship was discussed by Kitzinger with reference to the idea of the church as cosmos, with the floor representing the earth just as the dome or ceiling represented heaven.[66] However, this idea best corresponds to the vertical hierarchy of a centrally-planned Byzantine church. More important for the medieval West was the issue of orientation, both as regards the building as a whole and for various participants in the liturgy. In common with the majority of *mappaemundi*, the S. Salvatore mosaic has east at the top of the composition, at least when seen from the steps into the presbytery. In the case of the pavement, this arrangement also corresponds to the internal logic of the church building and reinforces it.[67] Both the east-west and the north-south axes of a church were imbued with significance, and liturgical directions were commonly expressed in terms of the cardinal points. Praepositinus of Cremona, for example, writing in the late twelfth century on the division of infant catechumens between the left- and right-hand sides of the church during pre-baptismal scrutinies, assigned the male children to the south and the female to the north: "*masculi ad austrum statuuntur, femine ab aquilone.*"[68] On occasion, directions expressed in this way might even be given an allegorical interpretation that drew on conceptions of a particular region. One example concerns the reading of the Gospel on, or towards, the northern side of the church. As discussed by Joseph Jungmann, this practice had its origins in practical considerations, and was only later given an allegorical meaning.[69] The first instance of such an explanation comes in the *Expositio Missae* of Remigius of Auxerre, which characterises north as the region of the devil, paraphrasing Isaiah Chapter Fourteen: "*O*

[66] Kitzinger, "World Map and Fortune's Wheel," 371-72.

[67] Ibid., 372.

[68] Praepositinus of Cremona, *Tractatus de officiis* 1.150, ed. James A. Corbett (Notre Dame; London: University of Notre Dame Press, 1969), 89.

[69] Joseph A. Jungmann, *The Mass of the Roman Rite: Its Origins and Development*, trans. Francis A. Brunner (Blackrock: Four Courts Press, 1986), 1.413-19. See also Barbara Maurmann, *Die Himmelsrichtungen im Weltbild des Mittelalters: Hildegard von Bingen, Honorius Augustodunensis und andere Autoren* (Munich: Wilhelm Fink Verlag, 1976), 184-85.

Lucifer, qui dicebas in corde tuo: Sedebo in lateribus aquilonis."[70] In the same context, Ivo of Chartres associated the north with paganism and the coldness of unbelief, against which the Gospel should be preached.[71] A similar explanation is found in the *Gemma Animae* of Honorius Augustodunensis: "*Per aquilonem enim infidelis populus denotatur, cui Evangelium praedicatur, ut ad Christum convertatur.*"[72] Given the presence of the north-western islands on the S. Salvatore floor and the way in which their remote and unusual nature is emphasised in the inscriptions, this liturgical characterisation of the north as a region in need of evangelisation is particularly telling.

Other rationales reside in the particular qualities of the floor surface. One of the most obvious and overlooked properties of pavement decoration is that it is necessarily viewed from above. Indeed larger compositions can only be appreciated from an elevated position, such as a pulpit or galleries if the church possessed them. Today, floor mosaics are generally seen reproduced in photographs taken from specially-built vantage points inaccessible to contemporaries. The significance of viewpoint for geographical imagery is suggested in an article by Patrick Gautier Dalché that likens the view of the earth in *mappaemundi* to the divine vision of the world sometimes vouchsafed to the pious in dreams.[73] According to Gregory the Great, such visions, in which the whole world was seen as if from a great height, were possible because divine light enlarged and raised the soul, enabling it to comprehend the limitations and insignificance of all creation.[74] Although the pavements do not provide the wealth of detail given in the visions, they do evoke the world in its entirety. Furthermore, in both Turin and Asti the mosaics were laid as part of works to the presbytery, which sought to raise, enclose and articu-

[70] Remigius of Auxerre, *Expositio Missae*, in Pseudo-Alcuin, *De Divinis Officiis Liber* (Migne *PL* 101.1250d-1251a). The reference is to Isa. 14:13. For the attribution to Remigius, see Ludwig Eisenhofer, *Handbuch der katholischen Liturgik*, vol. 1 (Freiburg im Breisgau: Herder, 1932), 121.

[71] Ivo of Chartres, *Sermo V, Sive opusculum de convenientia veteris et novi sacrificii* (Migne *PL* 162.550a-b).

[72] Honorius Augustodunensis, *Gemma Animae* 1.22 (Migne PL 172.551c).

[73] Patrick Gautier Dalché, "De la glose à la contemplation: Place et fonction de la carte dans les manuscrits du haut moyen âge," in *Testo e immagine nell'alto medioevo*, Settimane di studio del centro italiano di studi sull'alto medioevo 41 (Spoleto: Centro italiano di studi sull'alto medioevo, 1994), 752-64.

[74] Gregory the Great, *Dialogues* 2.35, ed. Adalbert de Vogüé and trans. Paul Antin (Paris: Les Éditions du Cerf, 1979), 2.236-43.

late this especially sacred area of the church. Considering also that the mosaics would be viewed by a primarily or exclusively clerical audience, it seems reasonable to identify the geographical elements as subjects for their contemplation that evoked this visionary perspective while reserving the full picture for God. Such a notion both accords with Kitzinger's interpretation of the S. Salvatore mosaic as representing earthly vanities, and supports the idea that this bounded and finite earth was understood to lie within a divinely ordered universe.

Liturgical sources also testify to the role of the pavement in the definition of sacred space within the church. Especially relevant to the iconography of the S. Salvatore and Asti mosaics is the rite of church consecration, which involved inscribing the Greek and Latin alphabets in a diagonal cross of sand or ashes scattered on the pavement, as shown in a diagram in an eleventh-century pontifical from Arezzo (Fig. 14).[75] A form of the *abecedarium* is first documented in the "*Ordo quomodo ecclesia debeat dedicari,*" found in pontificals dating back to the first quarter of the ninth century.[76] The ritual was included in two dedication *ordines* in the Romano-Germanic Pontifical and, by the late twelfth century, was normal practice in both the Roman and Ambrosian rites.[77] The bishop "wrote" the alphabets on the pavement with his *cambutta* or staff, moving from the left-hand eastern corner of the church towards the right-hand western corner, and then from the right-hand eastern corner towards the left-hand western corner. An interpretative tradition drew on tetradic thought to give these actions symbolic significance of a markedly geographical nature. In a commentary included in the Romano-Germanic Pontifi-

[75] Oxford, Bodleian Library, MS Canonici Liturgical 359, fol. 2r. I discuss the ceremony in relation to floor mosaics in Claudia Angelelli (ed.), "Mosaici pavimentali medievali nell'Italia settentrionale e i loro rapporti con la liturgia," *Atti del X Colloquio dell'AISCOM* (Tivoli: Edizioni Scripta Manent, 2005), 505-509.

[76] *Ordo Romanus XLI,* in *Les Ordines Romani du Haut Moyen Age*, vol. 4, *Les Textes (Ordines XXXV-XLIX*), ed. Michel Andrieu (Louvain: Université Catholique, 1956), 339-49, esp. 340-41.

[77] *Pontificale Romano-Germanicum*, 33.8, 40.25-26, ed. Cyrille Vogel & Reinhard Elze, *Le Pontifical Romano-Germanique du Dixième Siècle*, vol. 1 (Rome: Biblioteca Apostolica Vaticana, 1963), 83, 135-36. For an example of the rite in an eleventh-century Ambrosian manuscript see: *Ordo Ambrosianus ad Consecrandam Ecclesiam et Altaria*, in *Antiche Reliquie Liturgiche Ambrosiane e Romane, con un excursus sui frammenti dogmatici Ariani del Mai*, ed. Giovanni Mercati (Rome: Tipografia Vaticana, 1902), 21-27.

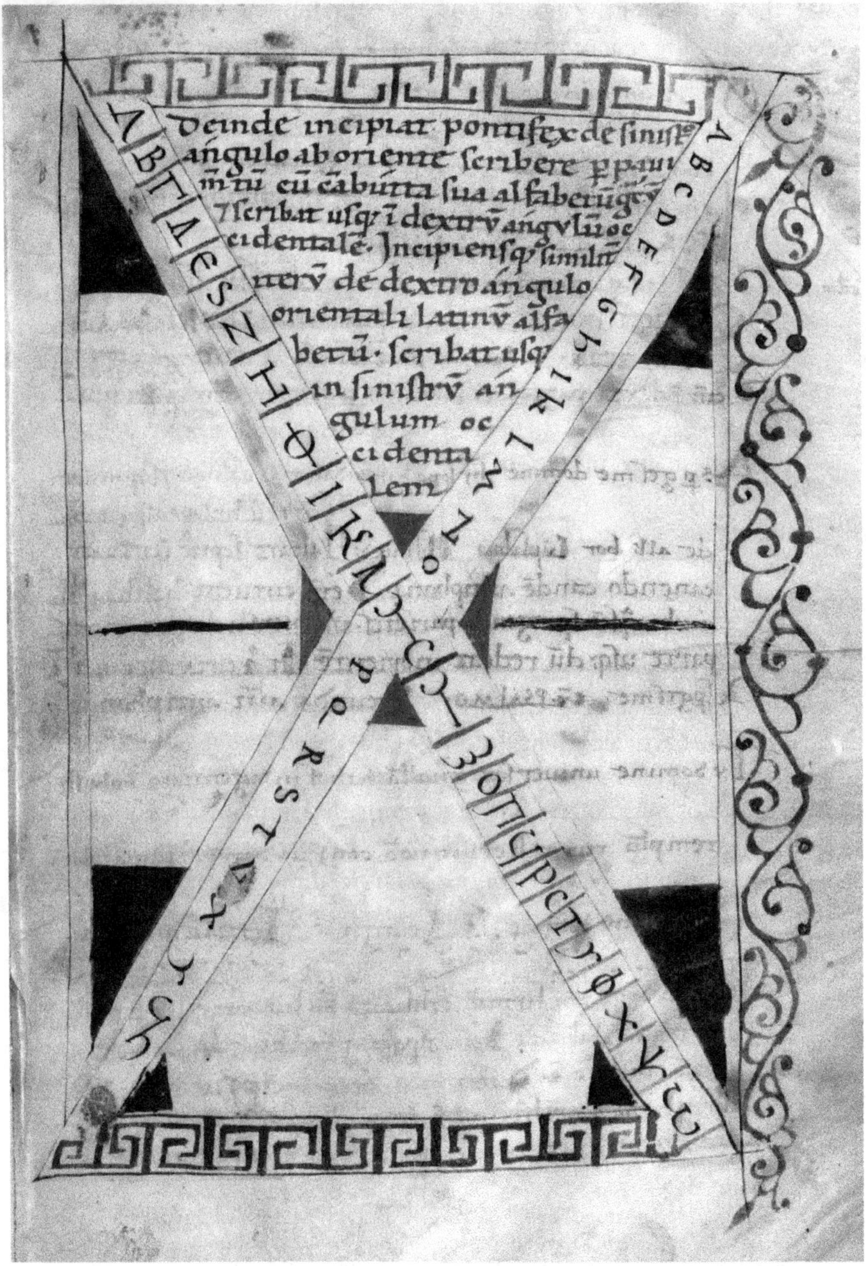

Fig. 14. Diagram of the Alphabet Cross, pontifical from Arezzo, eleventh century. Oxford, Bodleian Library, MS Canonici Liturgical 359, fol. 2r. By permission of the Bodleian Library, University of Oxford.

cal, the corners of the church represent the four regions of the world reached by the teaching of the Church—"*quattuor anguli basilice quattuor designant plagas mundi ad quas pervenit doctrina ecclesiastica*"—while the alphabets represent the preaching of the Gospels.[78] The letters are written in the form of a cross to represent Christ's death, with the left-hand eastern corner standing for his birth among the Jews and the right-hand western corner for his Ascension. The alphabet written from the right-hand eastern corner towards the west signifies the salvation of Israel. This symbolism was adopted and elaborated by a succession of later commentators.[79] For Ademar of Chabannes, for example, the ash was shaped into the form of a cross because through Christ Christianity is spread to the four parts of the world: "*quia cristianitas in Christo et per Christum dilatata est per qua(t)tuor partes orbis.*"[80] These interpretations are of relevance for both the Piedmontese pavements. In the case of the S. Salvatore floor, the emphasis on the corners of the church and their identification with the regions of the world finds a close parallel in the orientation of the mosaic and its depiction of the winds and the ends of the earth. In common with the allegorical interpretations of reading the Gospel towards the north, the concept of the diagonal cross as representative of the spreading of Christianity throughout the world reinforces the hypothesis that the earth's corners and edges possessed associations of evangelisation. Given the common association of the Rivers of Paradise with the Gospels, correspondences with the rite can also be found in the Asti mosaic. The location of the rivers at the corners of the composition, and the inscriptions relating them to various parts of the world, echo both the form and meaning of the cross of ash and letters.

In conclusion, it must be understood that the two floor mosaics examined here do not directly reproduce maps in other media. Even so, the S. Salvatore pavement shares elements with both *mappaemundi*

[78] *Pontificale Romano-Germanicum*, 35.15-18 (1.97-99); Brian Repsher, *The Rite of Church Dedication in the Early Medieval Era* (Lewiston, N.Y.; Lampeter: The Edwin Mellen Press, 1998), 176-77.

[79] For discussion, note Xavier Barral I Altet, "Quelques observations sur les mosaïques de pavement, l'architecture, la liturgie et la symbolique de l'édifice religieux médiéval," *Hortus Artium Medievalium* 9 (2003), 255-60.

[80] Victor Mortet (ed.), *Recueil de textes relatifs à l'histoire de l'architecture et à la condition des architectes en France au moyen age, XIe—XIIe siècles* (Paris: Alphonse Picard et Fils, 1911), 282.

and representations of the Apocalypse. The Asti mosaic has features in common with schemata of the quaternities and diagrammatic illustrations of Genesis. However, bearing in mind Harley and Woodward's definition of maps, the spatial sense displayed by the mosaics suggests that both are engaged in a process of mapping. In a literal sense, the first maps the edges and corners of the earth, while the second plots the course of the Rivers of Paradise with reference to the four regions of the world. Symbolically, the mosaics can also be seen to refer to the boundaries of divine rule and the spread of salvation. This view is supported by their position on the floor of the presbytery, where they would be seen from above as in a divinely inspired vision, as well as by the similar themes found in allegorical interpretations of liturgical directions. By depicting the earth—whether through its confines or its corners—in the holiest area of the church building, the mosaics evoked a spiritual empire that stretched to the ends of the earth.

MAPS IN CONTEXT: ISIDORE, OROSIUS, AND THE MEDIEVAL IMAGE OF THE WORLD

Evelyn Edson

For the past century scholars of medieval cartography have been proposing various systems of classification for the variety of world maps that have been preserved. The earliest approach to a system was made by Théophile Simar, who in 1912 sorted medieval maps by their sources: those derived from the Roman view, limited to the inhabited world or *orbis terrarum;* those derived from the Greek heritage, imagining the earth as part of the cosmos; and those which combined the two.[1] In 1926 Michael Andrews, after examining 600 *mappaemundi*, proposed a system which classified maps by form: ecumenical maps, which showed the inhabited world as a circle; hemispheric maps, in which the ecumene appeared in the northern half of the circle; and an intermediate group, which featured a southern continent, such as the maps associated with the late eighth century *Commentary on the Apocalypse* of Beatus of Liebana.[2] Andrews believed that the first category was evidence of belief in a flat earth, while the second two categories revealed remnants of the ideas of Greek scientists, that the earth was a globe. Marcel Destombes, in his catalogue (1964), added to Andrews's classification system a fourth category, Category D, maps of the ecumene that included many names and some geographical configurations.[3] Although this group included several important early maps, most of the examples (76 out of 95) were post 1300.

[1] Théophile Simar, "La géographie de l'Afrique Centrale dans l'antiquité et au Moyen-Age," *Revue Congolaise* 3 (1912-13), 1-23, 81-102, 145-69, 225-52, 289-310, 440-41.

[2] Michael C. Andrews, "The Study and Classification of Medieval Mappae Mundi," *Archaeologia* 75 (1925-26), 61-76.

[3] Marcel Destombes (ed.), *Mappaemondes A.D. 1200-1500. Catalogue préparé par la Commission des cartes anciennes de l'Union géographique internationale* (Amsterdam: N. Israel, 1964), 3-23.

In the first volume of the *History of Cartography* (1987), David Woodward rearranged the categories once again, his main contribution being the creation of a separate class of "transitional maps," those of the fourteenth and fifteenth centuries that incorporated features of late medieval sea charts.[4] His proposed classification included (1) tripartite world maps, (2) zonal maps, (3) quadripartite maps, that is, those with a fourth continent, and (4) transitional maps. He took Destombes' fourth category (D) and put the earlier of these maps in with the T-O maps of Group 1, dividing this group into schematic and non-schematic types. Of the first three categories, which predate 1300, Woodward wisely noted that they were not mutually exclusive. Someone drawing a circular map of the inhabited world did not necessarily believe that it showed the entire globe, and we have not only maps of different types in the same work, but also maps which attempted to combine both views, such as an ecumenical map placed in the northern hemisphere of a zonal map.[5] The biggest problem with Woodward's classification is that the tripartite division becomes an unwieldy amalgam of a majority of examples, many having little in common with the others. For example, this classification groups together the maps of Cosmas Indicopleustes (sixth century) and Ranulf Higden (fourteenth century).

In 1993 Patrick Gautier Dalché wrote a critique of the Andrews/Destombes system, observing that one result of their work was to regard map forms as though they represented different views of the world, when actually they were derived from the same reality.[6] Instead of sorting maps by form, he chose to classify them by their function within the manuscript. He laid out a three-part division. First are schematic diagrams or theoretical maps, integrated into the manuscript page and directly related to the text. This classification

[4] David Woodward, "Medieval Mappaemundi," in *The History of Cartography* vol. 1 (Chicago: University of Chicago Press, 1987), 294-99.

[5] Examples would be the two maps in the British Library (Cotton MS, Tiberius B. V. 1, mid-eleventh century), one of which is the famous rectangular world map, and the other a zonal map. The Ripoll map in the Vatican Library (MS Reg. Lat. 123, fols. 143v.-144r., 1055) has an ecumenical map drawn in the northern half of a hemispheric map.

[6] Patrick Gautier Dalché, "De la glose à la contemplation: Place et fonction de la carte dans les manuscrits du Haut Moyen Âge," *Testo e Immagine nell'Alto Medioevo, Settimane di Studio del Centro Italiano di Studi sull'Alto Medioevo* (Spoleto: Presso la Sede del Centro, 1994), XLI, 700-704. He does not discuss Woodward's system in full, because he is dealing principally with maps before Woodward's "transitional" era.

would include both T-O maps and zonal maps, such as those illustrating the work of Macrobius, *Commentary on the Dream of Scipio*, when they were designed to illustrate a textual description.[7] Second are maps that were more or less autonomous within the manuscript—at the end, at the beginning, on a separate page. An example would be the well-known ninth-century world map of St. Gall,[8] which shows the world surmounted by Christ on a cross. The book contains Isidore's *Etymologies*, but the illustration is on the fly-leaf and probably has nothing to do with the text. The third category was reserved for maps rich in content and large in size which led the reader to contemplate them as a sort of spiritual exercise.[9] In this group would be the great *mappaemundi* of the High Middle Ages.

Gautier Dalché also points out that Destombes's catalogue of medieval maps, on which we all rely, is incomplete, noting that he himself had a list of four hundred pre-1200 manuscripts containing maps, compared to Destombes's 283,[10] and doubtless the number has continued to grow since then. In addition, a number of the maps are wrongly dated and attributed. Destombes himself imagined his work as provisional, hoping that there would be "new discoveries" leading to "agreeable surprises" in the field.[11]

Considering these learned proposals, one might inquire about the function of the classification systems, aside from a pure intellectual exercise. An ideal system would illuminate questions of transmission, from Greco-Roman antiquity, and from era to era and place to place in the medieval world. It might also help us learn more about how people thought about maps, what they were used for, how they were made, and how they got into various manuscripts. For example, studying maps in the context of computus manuscripts helped to show that the diagrammatic quality of some of these, such as the early twelfth-century world map in the St. John's College, Oxford,

---

[7] Macrobius, who lived in the fifth century, wrote a handbook of astronomy and geography which was very popular in the Middle Ages. In his text, he describes a number of diagrams, including a world map on the zonal model. See his *Commentary on the Dream of Scipio*, trans. William Harris Stahl (New York: Columbia University Press, 1952 and 1990).

[8] Cod. 237, Stiftsbibliothek, St. Gall.

[9] ". . . les cartes plus complexes de la terre conduisant ceux qui les contemplaient, par leur taille et par la richesse de leur contenu, à un veritable exercice spirituel:" Gautier Dalché, "De la glose à la contemplation," 704.

[10] Ibid., 702. Gautier Dalché is working on a revision of Destombes's catalogue.

[11] Destombes, *Mappaemondes A.D. 1200-1500*, 20.

manuscript, were of a piece with the other diagrams which inhabit that well-illustrated volume. This study also made clear the intimate link between the calculation of time and the representation of space in the Middle Ages.[12] A mistaken system, on the other hand, could lead us astray. For example, the distinction between a flat earth and a spherical one was the premise of earlier classification systems, from Simar to Destombes, and yet scholars now agree that almost no medieval thinker or mapmaker believed in the former.

Rather than attempting yet another system of classification, this paper concentrates on a smaller subtopic, the question of Isidorian and Orosian map models. Woodward classified a number of early medieval world maps in his first category of "Tripartite Maps" as either "Isidorian" (schematic) or "Orosian" (non-schematic). The works of Isidore and Orosius, two great authors of late antiquity, were widely available in the Middle Ages and provided the basis of numerous geographical texts as well as maps. To begin with Isidore: he was not only Bishop of Seville in the early seventh century, but the energetic compiler of many books and treatises, which were often copied and highly valued in the Middle Ages. The ones of concern here are the *De Natura Rerum*, which was a sort of Dummies' Guide to the Universe, and the *Origins* or *Etymologies*, an encyclopedic survey of all knowledge, guided by the theme of the derivation of names. The world map in many copies of the *Etymologies* can be found at the opening of chapter 3, "Asia," of Book XIV, "*De Terra et Partibus*," the beginning of Isidore's extensive survey of geography, which goes on to cover Europe, Africa, the islands, promontories, and mountains, ending with what lies beneath the earth.[13] The map in many manuscripts is a simple T-O bearing the names of the three continents and acting almost as an illuminated initial (Fig. 1). Some artists embroidered further, adding the names of several winds, the cardinal directions, the sons of Noah, or labeling the boundaries between the continents. As early as the eighth century there appeared a slightly more elaborate map, which departed from the T-O format by adding the Sea of Azov (*Paludes Meotides*) in an elbow shape

[12] MS 17, St. John's College, Oxford (1110).

[13] Isidore also covered geographical topics in Books IX ("*De Linguis, Gentibus, Regnis, Militia, Civibus, Affinitatibus*") on peoples, XIII ("*De Mundo et Partibus*") on seas, gulfs, lakes, and rivers, and XV ("*De Aedificiis et Agris*") on cities. *Isidori Hispalensis Episcopi Etymologiarum sive Originum Libri XX*, W. M. Lindsay, ed. (Oxford: Clarendon Press, 1911), 2 vols.

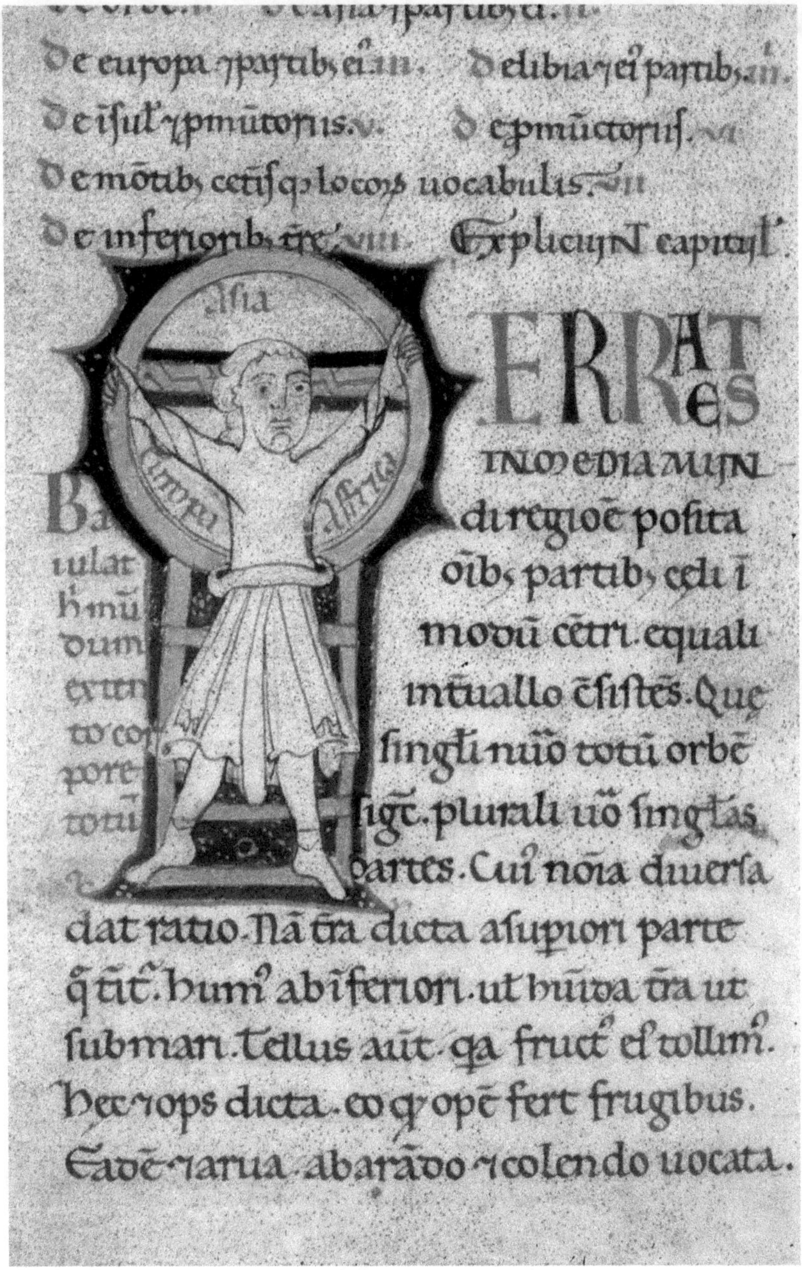

Fig. 1. T-O with human figure. Isidore, *Etymologies*. Florence, B M-L, Conv. Sopp. 319, f. 99. Thirteenth century.

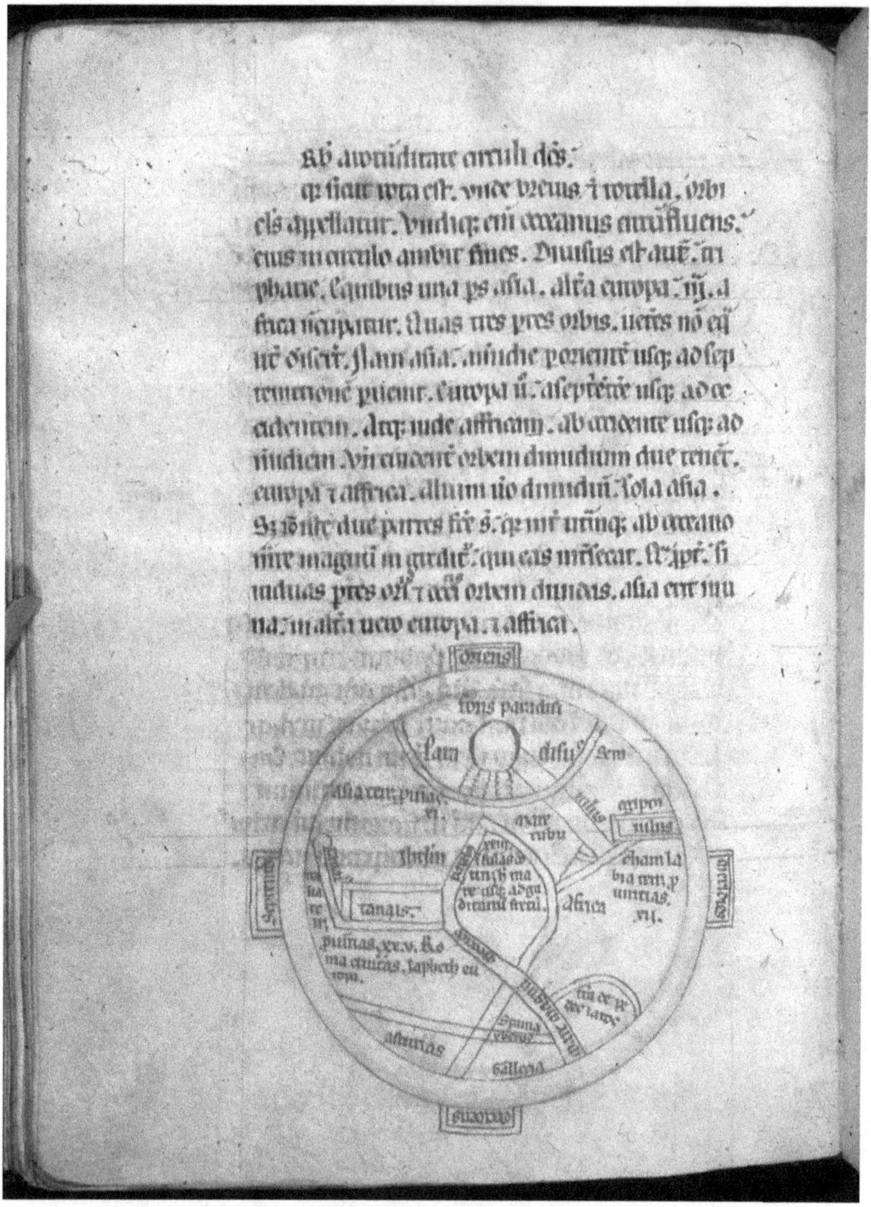

Fig. 2. "Elbow" T-O, Isidore, *Etymologies*. Florence, B M-L, Plut. 27, sin. 8, f. 64v. Thirteenth century.

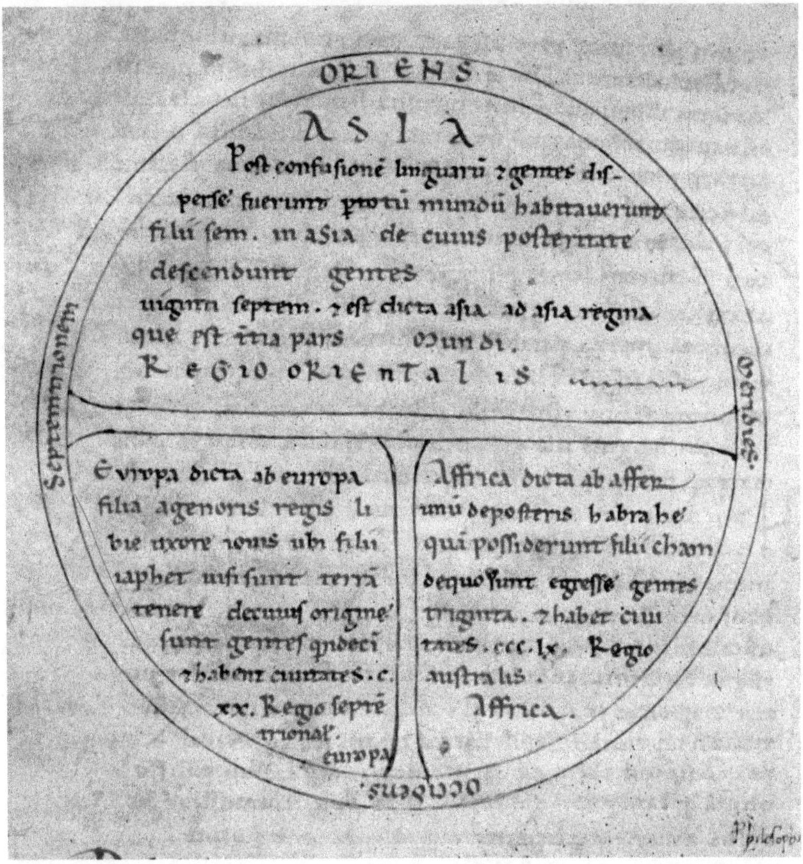

Fig. 3. T-O, "Post confusionem linguarum," Isidore, *De Natura Rerum*. Florence, Biblioteca Medicaea-Laurenziana, Plut. 29.39, f. 19v. Thirteenth century.

(Fig. 2). Whether a map was part of Isidore's original manuscript, again, is unknown. A map is integrated into the page in most cases, but sometimes it is added in the margin, perhaps by a reader who felt the lack of one. In any case, to call this simple diagram Isidorian is misleading, because the T-O description and its diagram appear in many other contexts, such as the center of a cosmic diagram in astronomical and computus manuscripts, not to mention the histories of Sallust and Lucan. This little diagram probably pre-dates Isidore by some centuries.

The *De Natura Rerum* of Isidore was nicknamed *Liber Rotarum* because of the collection of circular diagrams that accompanied it.

A T-O map sometimes appears at the end of the last chapter, in which Isidore describes the division of the earth into three continents. This diagram frequently contains a text that describes the division of the earth among the descendants of Noah after the fall of the Tower of Babel (Fig. 3). The diagram does not appear in the earliest manuscripts, and its position at the end leads us to believe it was a later addition, as was apparently the final chapter itself.[14] Unlike the other diagrams in *De Natura Rerum*, which are heralded with a *"haec est figura,"* there is no reference to it in the text of the brief chapter. It is therefore questionable whether the map is the work of Isidore or of an early copyist.

As for the more complex maps that appear in Isidore's manuscripts, each of these has a story of its own. There was great excitement when an elaborate eighth-century map was found in the Vatican Library, in the course of researches sponsored by Prince Youssouf Kamal in the 1920s (Fig. 4). Richard Uhden suggested that this map might be a direct descendant of Isidore's original construction, but this theory became untenable when an examination of the entire manuscript revealed that, while containing some excerpts of Isidore's works, it did not include the geographical chapters of the *Etymologies* and actually formed part of a computus collection.[15] Another interesting map is found in an early twelfth-century copy of Isidore's *Etymologies*, now in the Bavarian State Library in Munich. It takes up most of the page and appears at the beginning of Book XIV, rather than at the opening of chapter 3 on Asia, as is usual (Fig. 5). This colorful and detailed world map is so different from the simple diagrammatic style of almost all Isidore maps that Wood-

---

[14] Jacques Fontaine, *Isidore de Séville, Traité de la Nature* (Bordeaux: Féret, 1960), 79-80. However, Wesley Stevens argues that the chapter and diagram were added by Isidore himself in his second recension; see "The Figure of the Earth in Isidore's 'De Natura Rerum'," *Isis* 71 (1980), 272-73.

[15] MS Vat. lat. 6018, Biblioteca Apostolica Vaticana, Vatican City , fols. 63v.-64. The first study of the map was by Richard Uhden, "Die Weltkarte des Isidorus von Sevilla," *Mnemosyne. Bibliotheca Classica Batava* 3 (1936), 1-28. On the map in its computus context, see Evelyn Edson, *Mapping Time and Space: How Medieval Mapmakers Viewed Their World* (London: British Library, 1997), 61-62; and Leonid Chekin, "Easter Tables and the Pseudo-Isidorean Vatican Map," *Imago Mundi* 51 (1999), 13-23. A brilliant analysis of the contents of the manuscript may be found in Alessandro Scafi, *Mapping Paradise: A History of Heaven on Earth* (Chicago: University of Chicago Press, 2006), 95-104.

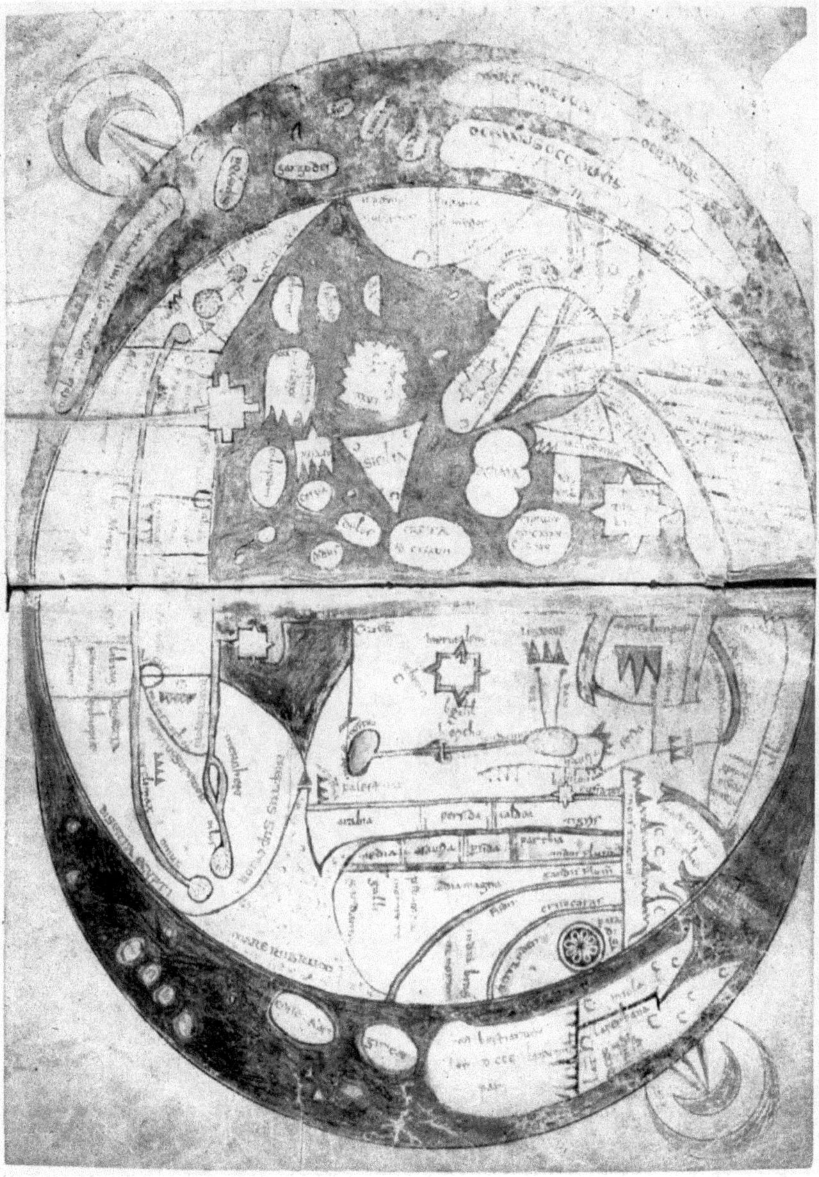

Fig. 4. World maps in a miscellany. Biblioteca Apostolica Vaticana, Vat. lat. 6018, fols. 63v.-64. Eighth century.

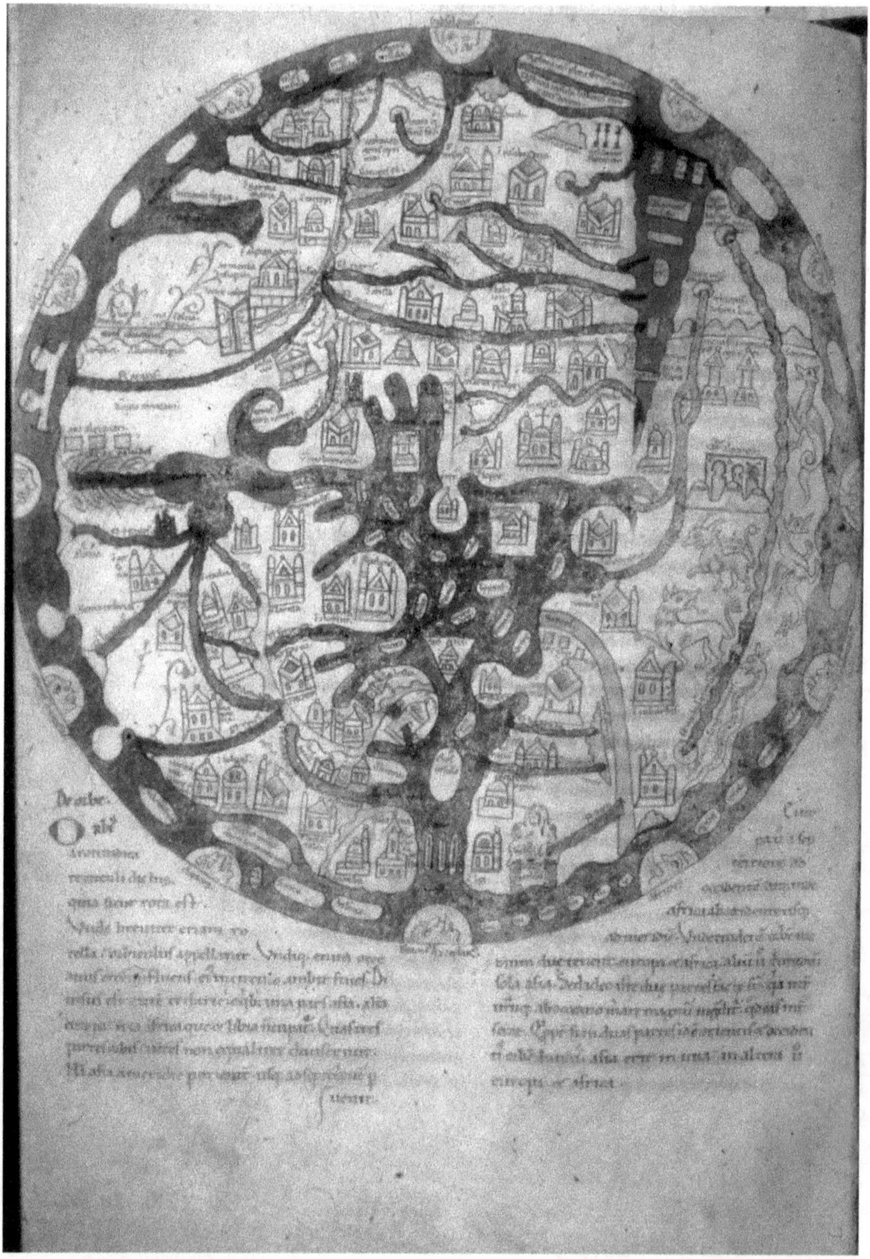

Fig. 5. "Munich Isidore." Bayerische Staatsbibliothek, CLM 10058, f. 154v. Twelfth century.

ward, in despair, called it an "Orosian-Isidorian" map.[16] Its presence here in a manuscript of Isidore clearly has little to do with Isidore, and reflects the state of cartography in the twelfth century, not the seventh.

Paulus Orosius, who flourished in the early fifth century of our era, was born in Spain and traveled to Africa and the Holy Land, where he met Augustine and Jerome. His book, *Historia Adversum Paganos Libri VII (Seven Books of History Against the Pagans)*, was a world history with a didactic purpose.[17] He opened it with a chapter on geography, which was often copied separately. Orosius describes himself as looking at the world, "as if from a watchtower," beginning with its division into three continents, and then going into further details about the provinces, rivers, mountains, and other physical features to be found. His text is rich in names for bodies of water, such as the gulfs of the Mediterranean Sea, but he names few cities. We do not know whether he was looking at a map when he wrote, though that seems likely,[18] nor whether he made a map or had one made to accompany his work. Woodward warned us that there was no strong map tradition associated with manuscripts of the text, and that this made it difficult to identify Orosian influence. Nevertheless he himself went on to describe as "Orosian" several important medieval world maps, such as the Albi and Sawley maps. His example has been widely followed.

In Destombes's catalogue there are exactly four manuscripts of Orosius' work which contain maps, out of more than 200. These four are the late eighth-century Albi manuscript,[19] a ninth-century

[16] MS CLM 10058, Bayerische Staatsbibliothek, Munich, fol. 154v. See Patrick Gautier Dalché, *La "Descriptio Mappe Mundi" de Hugues de Saint Victor* (Paris: Études Augustiniennes, 1988) for a study of this map. Woodward, "Medieval Mappaemundi," 348-49.

[17] Paulus Orosius, *Seven Books of History Against the Pagans,* Roy J. Defferari, trans. (Washington, D. C.: Catholic University Press, 1964). There is no modern critical edition.

[18] This is the opinion of Yves Janvier, based on Orosius's use of "dynamic language:" *La Géographie d'Orose* (Paris: Societé d'Edition "Les Belles Lettres," 1982), 165-68.

[19] MS 29, Bibliothèque Rochegude, Albi, f. 487; Destombes, *Mappemondes,* 22.1.

manuscript in St. Gall,[20] a thirteenth-century manuscript in Paris,[21] and a fifteenth-century manuscript in Tours.[22] The Albi map is the most detailed, and it has added interest as one of our oldest world maps (Fig. 6). This map appears in a manuscript containing a collection of excerpts, which might have served as a reference book for a priest or a teacher. It heads a section devoted to geography, and is immediately followed by a list of winds and seas, then Orosius' geographical chapter, the names of the provinces by Polemius Sylvius, the *Notitia Galliarum*, and the *De Nominibus Gallicis*. Other texts are Biblical commentaries, notes on grammar and language, and some historical texts, such as the six ages of the world. The form is roughly rectangular and appears to have been drawn freehand. The continents are not named, nor does the map use the T-O structure. Fifty geographical names appear, of which forty-nine can be found in Isidore's *Etymologies* and forty-one in Orosius' chapter. This does not mean too much, as all the names were common currency and appear in many works. For example, Jerusalem—a place actually visited by Orosius and of course vital to his history of the world—is not on the map. The map displays several other features missing from his work. One is the configuration of the rivers of Paradise in the east (although Paradise itself is not named) and the presence of Mt. Sinai. The Tanais river, which Orosius mentions several times as the boundary between Europe and Asia, is not here, nor are the islands of Taprobana and Hibernia (Ireland). Spain is not a triangle, as Orosius describes it, and the positions of Corsica and Sardinia are reversed. Thus the map should probably be considered as an addition to the volume's geographical materials rather than as an illustration specially intended for the chapter by Orosius.

As for the other examples, the ninth-century St. Gall manuscript has sketched in its margins three maps that were added in the mid-

[20] Cod. 621, Stiftsbibliothek, St. Gall, p. 35; Destombes, *Mappemondes*, 22.4.

[21] MS lat. 17543, Bibliothèque Nationale, Paris, fol. 2; Destombes, *Mappemondes*, 49.22.

[22] MS 973, Bibliothèque Municipale, Tours, fol. 2; Destombes, *Mappemondes*, 51.30. See also J.M. Bately and D.J.A. Ross, "A Check List of Manuscripts of Orosius, 'Historiarum Adversum Paganos Libri Septem'," *Scriptorium* 15.2 (1961), 329-33; D. J. A. Ross, "Illustrated Manuscripts of Orosius," *Scriptorium* IX (1955), 35-56.

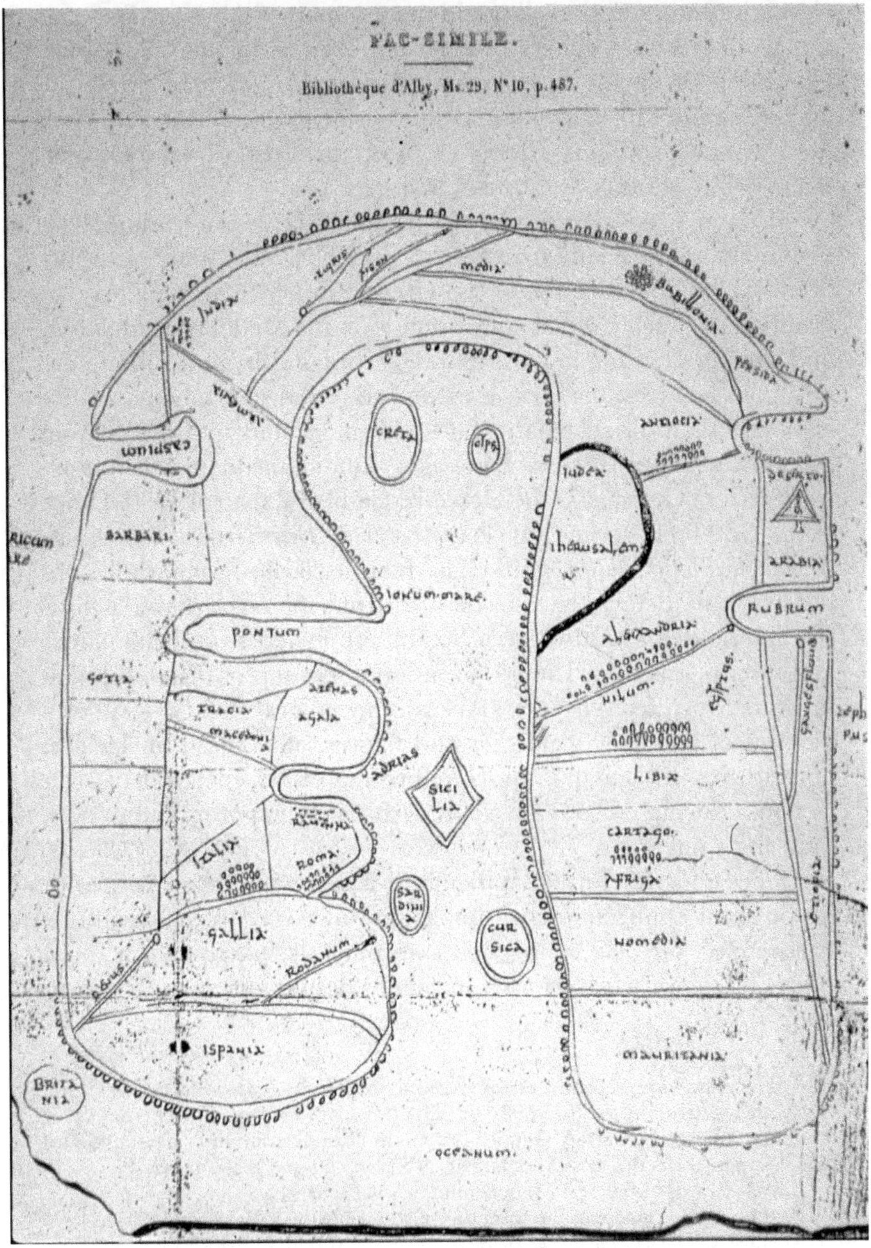

Fig. 6. World map. Albi, Bibliothèque Municipale, MS 29. Eighth century.

eleventh century.[23] A small world map, a map of the Near East and a map of Italy are glosses rather than integral to the manuscript. The Paris Orosius is illustrated with a simple T-O map, while the Tours manuscript is a pictorial version of the same format. With these as our possibilities, therefore, it is fair to describe the Orosian manuscript tradition as virtually map-free.

If there is no map model in any manuscript of Orosius, then, one might ask what an Orosian map would be. Woodward describes "Orosian" maps as closer to the Greco-Roman than the Christian tradition, characterized by an emphasis on the Mediterranean basin, and having "undulating coastlines."[24] Judging by the subsequent examples he gives, the term is employed to describe any map more elaborate than the schematic T-Os which accompany Book XIV in so many copies of Isidore's *Etymologies*. Among the maps that Woodward terms "Orosian" is the Hereford Cathedral map of 1300. Down in the right hand corner it is clearly labeled "*Descripcio Orosii de Ornesta Mundi sicut interius ostenditur.*"[25] The map itself, however, while containing some details that are derived from the *Seven Books*, such as the lighthouse in northwestern Spain, also includes over 200 names and inscriptions from Isidore. Paradise at the top of the world with the four rivers, references to monstrous races and exotic animals and pictures of both, mythical sites and figures, all come from Isidore, not Orosius. In actuality the Hereford map is an encyclopedic collection, drawing on so many sources that it is inappropriate to tie it to a single one.[26]

Since the Hereford Mappamundi is a free-standing work with no physical connection to a book, the only way to determine its Isidorian vs. Orosian connection is through its textual and pictorial content. Isidore was very rich in names despite the succinctness of

[23] The maps are reproduced by Konrad Miller in *Mappaemundi: die ältesten Weltkarten* (Stuttgart: J. Roth, 1895-98), VI, *Rekonstruierte Karten*, 62. The dating of the maps is noted by Patrick Gautier Dalché in "Les diagrammes topographiques dans les manuscrits des classiques latins," *Bibliologia* 20 (n.d.): 300 n. 41.

[24] Woodward, "Medieval Mappaemundi," 341 and 347.

[25] "Orosius's account of the 'ornesta' of the world, as shown within:" see Paul D. A. Harvey, *Mappa Mundi: The Hereford World Map* (London: British Library, 1996) for a translation of the marginal texts of the map.

[26] A complete text of all the inscriptions on the map and information about the source for each can be found in Scott D. Westrem, *The Hereford Map: A Transcription and Translation of the Legends with Commentary* (Turnhout: Brepols, 2001).

the T-O maps often found in his works, and his text included many more cities than that of Orosius, as well as references to mythology and to both secular and sacred history. Around 1900 Konrad Miller, in the course of his great work on the *mappaemundi*, attempted to draw maps based on various early geographical works, including those of Orosius and Isidore.[27] At first glance his reconstructed maps look almost identical (Plates VIII and IX). Then we notice the vestigial "fourth continent" in the south, which is the greatest difference in form between the two. This feature is derived from Isidore, who wrote a few enigmatic lines on the subject: *"Extra tres autem partes orbis quarta pars trans Oceanum interior est in meridie, quae solis ardore incognita nobis est; in cuius finibus Antipodes fabulose inhabitare produntur."*[28] This fourth continent appears on a number of medieval maps, most notably those illustrating Beatus of Liebana's *Commentary on the Apocalypse.* It may be a reflection of the zonal map, which often shows an equatorial ocean and a southern continent beyond it, although John Williams stresses the point that the ocean is described by Isidore as "interior," that is, part of the ecumene, and therefore possible to cross, though maybe not easily. Several Beatus maps show either a Sciopod, a one-legged man, or have an inscription describing him. According to Isidore, the Sciopod lived in Libya, not in a continent below the equator.[29] The other differences between Miller's two reconstructions are mostly names, with Isidore having many more of them. Isidore also adds several islands—Chryse and Argire, the gold and silver islands in the Far East—and says that there are five islands west of Africa instead of Orosius' three. Otherwise the two reconstructed maps are much the same. If anything, the Isidore map, far from being a simple T-O, is more detailed than the Orosius map. As for "undulating coastlines," they are not mentioned in either book.

---

[27] Miller, *Rekonstruierte Karten*, Tafs. II and III.

[28] "Outside the three parts of the world is a fourth part across the interior Ocean to the south, which is unknown to us due to the heat of the sun, and in this region the Antipodes dwell, according to the story:" Isidore, *Etymologies*, XIV.iii.6.

[29] John Williams, "Isidore, Orosius, and the Beatus Map," *Imago Mundi* 49 (1997), 7-32, especially p. 18. Isidore describes the Sciopod in Book XI. See also Serafín Moralejo, "Las Islas del Sol. Sobre el Mapamundi del Beato del Burgo de Osma (1086)," *A Imagem do Mundo na Idade Média, Actas do Colóquio Internacional, 1989* (Lisbon: Ministério da Educaçao, Instituto de Cultura e Língua Portuguesa, 1992), esp. 43-44.

Another way to distinguish Isidorian from Orosian maps would be by characteristic words, phrases or place names. Orosius names Calearsus, Parethonius, Dagusa, Mossylon Emporium, Mevania, and the altars of Alexander, none of which are used by Isidore. Isidore, on the other hand, goes into more detail in some regions, such as Greece, where he lists Arcadia, Thebes, Marathon and Mount Parnassus; in Italy he lists Baiae, Lake Avernus, Umbria, Etruria and Campania. Isidore also has some memorable phrases which appear on maps, either literally or graphically, such as his description of Jerusalem as the *"umbilicus regionis totius."*[30] Orosius gives more information about the two possible variants of the course of the Nile, one placing its source near the Red Sea in the east, and the other putting it by the Atlas Mountains in the west. In the latter case the river was supposed to disappear beneath the sands, re-emerging in the waters of a central African lake and flowing east to Ethiopia, then finally north to the delta. This configuration appears on many medieval world maps, the Hereford Cathedral map among them. Isidore simply says that the Nile flows through Egypt, making the land rich by its flooding, that it is used for commerce, and that the mouth is at Canopus.[31]

John Williams—the leading authority on the illuminated manuscripts of Beatus' *Commentary on the Apocalypse*—turned his attention to the maps in a 1997 article in *Imago Mundi*.[32] After a thorough consideration of the problem—the inclusion of a fourth continent, the tradition of maps in manuscripts of Isidore's work, various views of the Antipodes held by early Christian thinkers—he concluded that the original Beatus map was based on an Orosian model. Despite the presence of a southern landmass, he argued that it could not represent a continent in the southern hemisphere, for the whole purpose of Beatus' map was to show the missions of the apostles which would encompass the entire world. Even though no apostle is shown on the southern continent, it would undermine Beatus' thesis if a part of the world were inaccessible. Williams found it significant that Beatus in his commentary drew heavily on sources from North Africa, where Orosius was living when he completed his book in 418.

[30] *Etymologies*, XIV, iii.21.

[31] Ibid., XIV.iii.28 and XIII.xxi.6.

[32] See note 29 above. Williams' definitive study is *The Illuminated Beatus* (London: Harvey Miller, 1994-2003), 5 vols.

Fig. 7. Beatus, *Commentary*. Morgan Library, MS 644, fols. 334v.-34. 926.

Williams finally formed the opinion that the Beatus maps are most similar to the "Orosian" map at Albi, which was roughly contemporary with the writing of the *Commentary*. Hence, he concluded that the Beatus maps are Orosian rather than Isidorian (Fig. 7).

With all due respect, the time has come to dispense with these two categories completely. As we have seen, there is no clear manuscript tradition which associates a particular type of map exclusively with either work; T-O maps are too common, and Orosius' work has virtually no maps at all. As for the content—place-names and geographical forms—Isidore himself drew upon Orosius' work, and the two authors used many of the same sources, such as Sallust, Pliny and Solinus. Both give a description of the three continents and their boundaries as a framework for a world map. If one or the other includes more details on a subject—Orosius on the Nile, or Isidore on foreign animals or gems—these are not really contradictory. Instead, one supplements the other, and this is how they were used by medieval geographical writers and cartographers.

Medieval mapmakers took whatever they had at hand, including in even the most impoverished monastic library the works of Orosius and Isidore, as well as other intermediate sources such as Rabanus Maurus and Honorius Augustodunensis. Maps were also copied from one manuscript to another, even if the texts were different. So an allegedly Isidorian T-O can appear in any geographical or historical work, or indeed anywhere that a world diagram seemed called for; meantime a so-called Orosian map might be placed where more detail was thought to be appropriate, or where the mapmaker had the space or energy to create a large, more elaborate world picture. All medieval maps, from the seventh century onwards, could fairly be described as Isidorian-Orosian.

MEDIEVAL MAPS IN A RENAISSANCE CONTEXT: GREGORIO DATI AND THE TEACHING OF GEOGRAPHY IN FIFTEENTH-CENTURY FLORENCE

Raymond Clemens

Gregorio Dati, a silk merchant who rose to Florence's highest civil office, wrote his *Sfera* (The Spheres) sometime before his death in 1435, in part as a textbook of world geography directed at children of the Florentine merchant class, but clearly envisioning other audiences as well. Unique in both language and layout, this first geography primer written in the vernacular employed a complex illustration cycle that drew on several medieval maps: *mappaemundi* (both realistic and schematic), maps of the Holy Land, and portolan charts. Each of these maps found a new context in Dati's work, especially the portolan charts, which were fragmented both to fit them into the margins of the work and to arrange them in a linear fashion to accompany his text. The maps in the *Sfera* served three interrelated purposes: education, entertainment, and commerce, demonstrating the complex relationship between knowledge and power expressed and transmitted through verbal and pictorial mapping in the fifteenth century. In addition, Dati's *Sfera* reveals that those in the merchant class were not only patrons and consumers of humanist culture, but also producers, adapting existing models to new uses and creating new forms for their own vernacular production and consumption.

Gregorio Dati was neither a typical merchant nor a typical humanist. Rather, he was a member of a group of humanist merchants who viewed the humanities not simply as path for personal advancement, but as means for expressing the virtues of civic humanism that combined the study of letters with service to the state. Dati's best known work is his *History of Florence*, characterized by Hans Baron as the first modern (as opposed to medieval) work of history.[1] In it, the history of Florence is recounted through the eyes of one of its own citizens. The *Sfera*, on the other hand, is a description of the world

[1] Hans Baron, *The Crisis of the Early Italian Renaissance* (Princeton, 1966), 167-188.

outside of Florence seen from a global perspective in which neither Florence nor even Italy are represented. It demonstrates how Florentines visualized the exotic and distant lands that most would never travel to except in their imaginations.[2]

The *Sfera* is divided into four, equal-length books, beginning with the heavens and ending with the earth. The first book treats cosmography and astronomy, the second cosmographical phenomena such as eclipses and phases of the moon, the third compasses, timekeeping, and navigation, and the fourth maps out the Middle East and southern and southeastern Mediterranean. Each book is divided into eight equal-length stanzas of rhymed lines, a common form for educational works produced during the Renaissance.[3] The marginal illustrations facilitated the understanding of the text and served as mnemonic aids and, happily, may also have saved the *Sfera* from destruction when its content was made obsolete by later discoveries.[4] The manuscripts that survive, many with extensive illumination, may represent only a small portion of the manuscripts originally composed because, in addition to obsolescence, books intended for the classroom often saw heavy use that eventually destroyed them.

Dati's first two books contain information available from a wide range of sources, but the source he most probably used was the *Tractatus de sphaera* (before 1220) by John of Holywood (often called by the Italian *Sacrobosco*), either in the original or through one of its many commentators. Sacrobosco's *Sphaera* was intended for use as

---

[2] Traveling for intellectual curiosity and religious devotion was a distinguishing feature of the Renaissance, but lack of money and the dangers inherent in travel made imaginary journeys more appealing for all but the most adventurous. Although presented as a guidebook, Petrarch's *Itinerarium ad sepulchrum domini nostri Yhesu Christi* took the reader to a place Petrarch had never been, allowing the reader to take an imaginary journey with him. On Petrarch's love and fear of travel, see *Petrarch's Guide to the Holy Land. Itinerary to the Sepulcher of Our Lord Jesus Christ. Facsimile edition of Cremona, Biblioteca Statale, Deposito Libreria Civica, manuscript BB.1.2.5. With an introductory Essay, Translation, and Notes by Theodore J. Cachey Jr.* (South Bend: University of Notre Dame Press, 2002) 22 and 46-49 n. 93.

[3] Paul F. Gehl, *A Moral Art: Grammar, Society, and Culture in Trecento Florence* (Ithaca: Cornell University Press, 1993), 80; Mary Carruthers, *The Book of Memory: A Study of Memory in Medieval Culture* (Cambridge: Cambridge University Press, 1990), 162-70; Paul Grendler, *Schooling in Renaissance Italy: Literacy and Learning, 1300-1600* (Baltimore: Johns Hopkins University Press, 1989), 196-97.

[4] See Sandra Hindman, Michael Camille, Nina Rowe and Rowan Watson, *Manuscript Illumination in the Modern Age: Recovery and Reconstruction* (Evanston: Mary and Leigh Block Museum of Art, Northwestern University, 2001).

an introductory university textbook, and it describes many of the same phenomena that Dati treats in his first two books.[5] Dati's illustrations are similar to those found in illuminated copies of Sacrobosco's work, but if he relied on Sacrobosco, he significantly adapted the text by casting it in a much simpler format, translating it into Italian, and arranging it in verse.[6] Dati's version was designed for classroom instruction of young children in the vernacular, whereas Sacrobosco's original was part of the university curriculum, designed originally for those who read Latin.[7] The *Sfera* also placed the earth in its cosmological context, explained solar and lunar eclipses with precise diagrams, showed the workings of a compass, and described the two most common schemas of the earth's surface: Macrobius's Zonal map (tilted on its side so that, like all of Dati's maps, East is at the top) and Isidore of Seville's T-O map (Fig. 1 and Fig. 2). Beneath the Isidorian T-O map Dati included a simple *mappamundi*, which varies in detail from manuscript to manuscript, but can often be quite complex. One can see from this illustration that the shape of Europe and Africa can be made out, as can the Mediterranean, the Nile, the Black and Caspian Seas, and the Indian Ocean.

When turning to the specifics of the earth's geography, however, Dati did something completely original that is not found in any exemplar. He included in the margins of his book fragments of portolan charts, the detailed maps of the coastline used by sailors in the Mediterranean (Fig. 3), which he arranged from east to west (as in the earliest collections of portolan charts), from the Fortunate Islands in the Atlantic to the Black Sea (Plates X and XI).[8] The fourth book

[5] Lynn Thorndike, *The Sphere of Sacrobosco and Its Commentators* (Chicago: University of Chicago Press, 1949). Sacrobosco's work was, in turn, based on ancient, medieval and Arabic authors. "[Sacrobosco] welded together Macrobius and Ptolemy and frosted it over with Alfraganus, and his book stayed in style for five centuries." Thorndike, *The Sphere of Sacrobosco...*, 21.

[6] A Dominican in Florence, Ugo de Castello, completed a commentary (no longer extant) on Sacrobosco in 1337. Ibid., 36.

[7] It was required for several degree programs, including those at the major universities in Paris, Vienna, Oxford, and Bologna. Ibid., 42-43.

[8] For the mental structure of maps in atlases, see James R. Akerman, "From Books with Maps to Books as Maps: The Editor in the Creation of the Atlas Idea," in *Editing Early and Historical Atlases*, ed. Joan Winearls (Toronto: University of Toronto Press, 1995), 1-5. However, Akerman is careful to note that we do not have an atlas proper until Ortelius' 1570 *Theatrum Orbis Terrarum* (see 3-48).

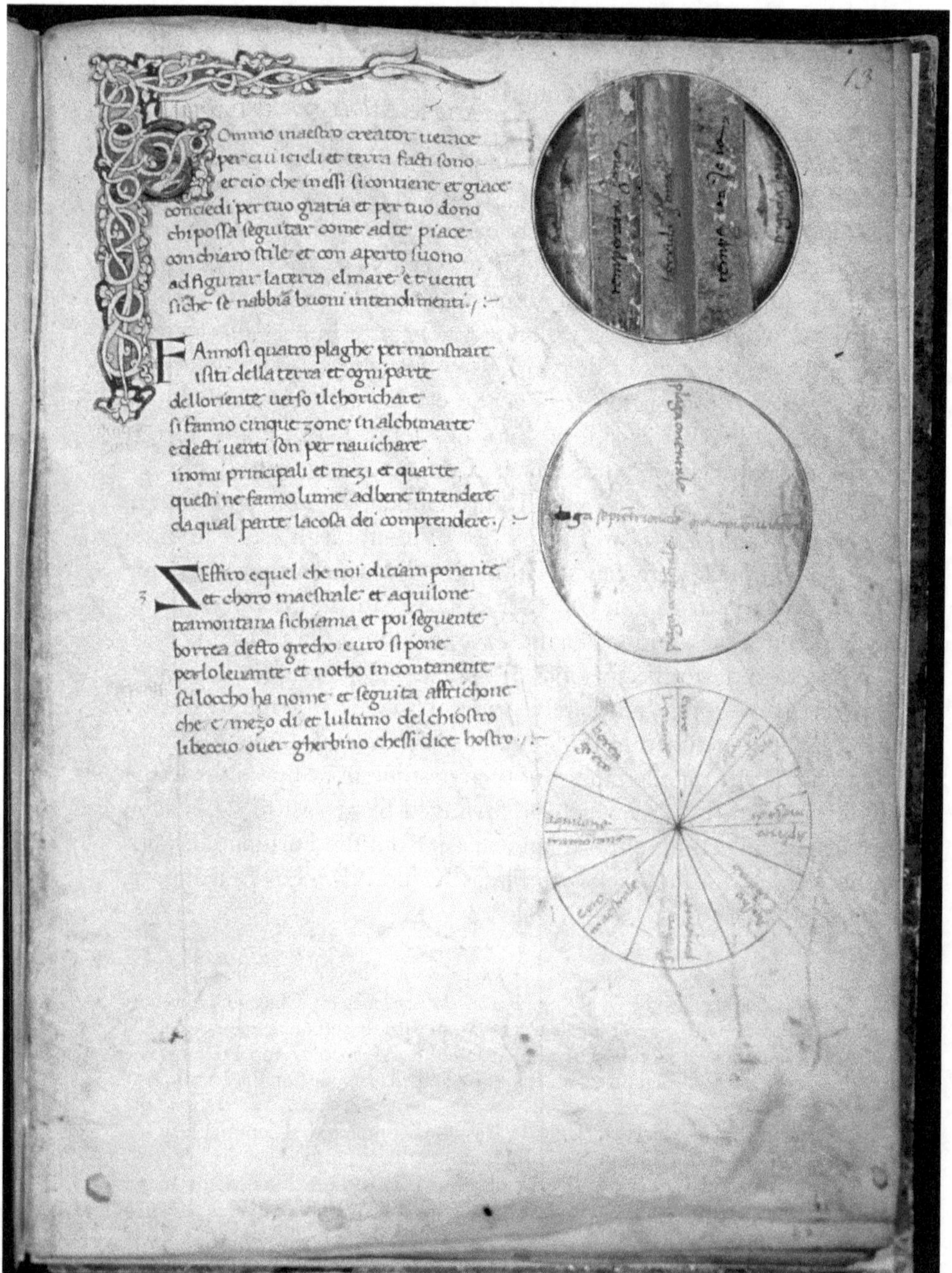

Fig. 1. Gregorio Dati, *Sfera.* Newberry Ayer MS Map 1, f. 13r.

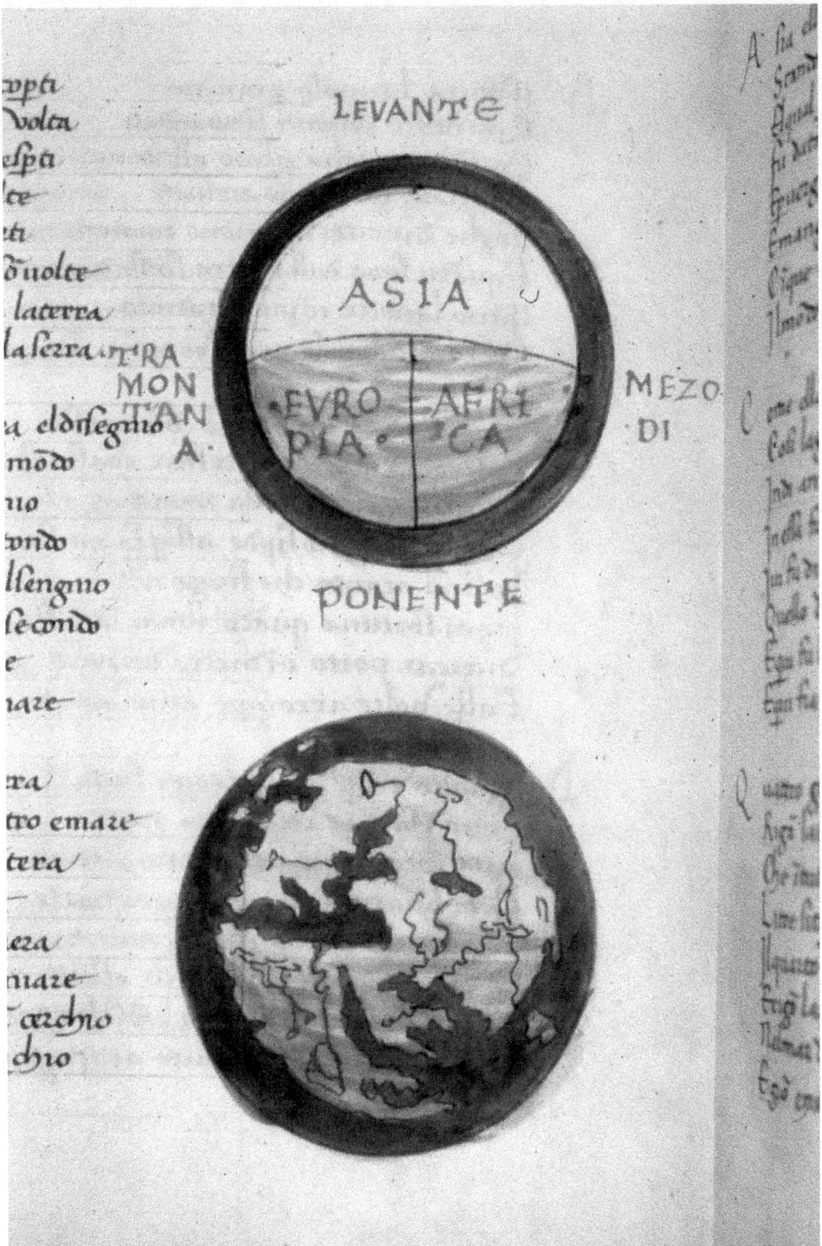

Fig. 2. Gregorio Dati, *Sfera*. Laurenziana, Mediceo-Palatino MS 88, f. 14v.

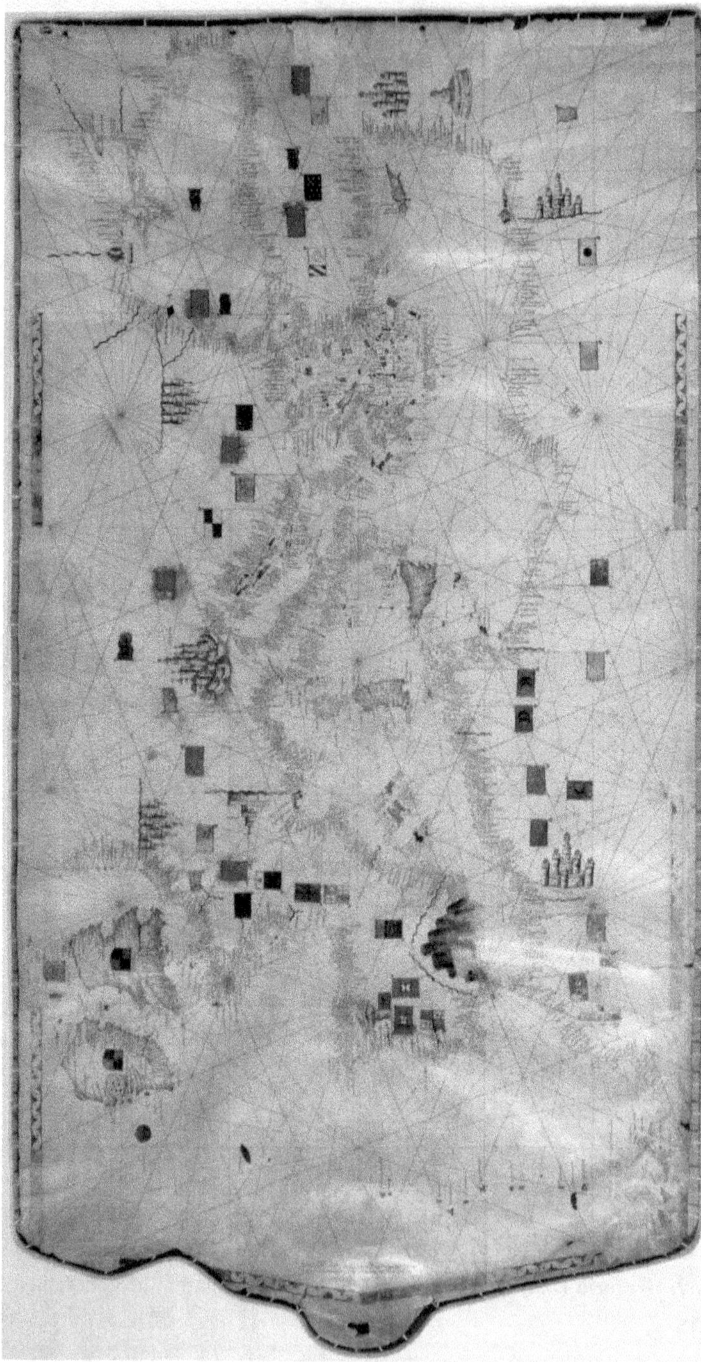

Fig. 3. Portolan Chart by Petrus Roselli, 1461. Newberry Library, Ayer MS, Map 3.

of Dati's *Sfera* may be the first European proto-atlas since Ptolemy to have a consistent orientation, with East always at the top.[9]

The history of the development of these graphic portolans is vexing.[10] The earliest surviving portolan is the *Carte Pisane*, dated from the end of the thirteenth century, but it is unlikely that this was the first portolan, in large measure because it is already fully formed, with detailed landforms and coasts accurately depicting the location of important port cities and indicating submerged dangers to sailors around the Mediterranean.[11] In the early part of the twentieth century, it was the contention of the Scandinavian explorer and historian of cartography, Nils A. E. Nordenskiöld, that Dati's maps were based on now-lost sketch maps and skipper's charts, and that these same lost maps formed the basis for the earliest portolans; but this theory has been thoroughly put to rest by Roberto Almagià.[12] Instead, Dati seems to have used at least two sources: a graphic portolan and a verbal portolan. Graphic portolans provided outlines of the coast, while verbal portolans provided distances between the major port cities; distances between ports are not normally included in graphic portolans. The verbal portolans may also have provided Dati with some of the text in Book IV, which incorporates in verse much of the same information including descriptions of towns and their relationship to geographic features such as rivers, mountains and lakes.[13] The maps rarely align perfectly with the text. One often has to turn the page to see the map that is described in the text. The maps also present information in greater detail than that found in the text. They include a greater number of ports and the distances between

[9] Tony Campbell, "Portolan Charts from the Late Thirteenth Century to 1500," in *The History of Cartography*, vol. 1, 378.

[10] The best study is still Tony Campbell's "Portolan Charts from the Late Thirteenth Century to 1500," 371-463.

[11] Paris, BnF, Cartes et Plans, Rés. Ge B 1118. The coasts of northern Europe and England and Ireland were no more than vague outlines, but the conventions utilized and the detail of the lands surrounding the Mediterranean suggest that this is not the fledging work of a novice mapmaker.

[12] Nils A. E. Nordenskiöld, "Dei disegni marginali negli antichi manoscritti della *Sfera* del Dati," *Bibliofilia* 3 (1901-2): 49-55; Roberto Almagià, *Monumenta cartographica Vaticana*, vol. 1 (Rome, 1944), 128-29.

[13] For editions of several early verbal Italian portolans, see Konrad Kretschmer, *Die italienischen Portolane des Mittelalters: Ein Beitrag zur Geschichte der Kartographie und Nautik* (Berlin, 1909), 233-552.

them, and the islands are colored or have distinctive markings to indicate possession, a trait common in portolan charts.

Like Marino Sanudo, Dati mixes several types of maps in his book.[14] The two regional maps that are not based on portolan charts are both of the Middle East, an area of special religious and economic interest to his readers. The maps are a digression within the narrative framework of the atlas, both in form and in function.[15] The first and larger of the two appears to be based on the Ptolemaic maps of Asia (Plate XII and Fig. 4). The realism of the coastline gives way to the development of inland features with the coast serving merely to anchor the geographic space. In the manuscripts that survive, the map almost always spreads across two facing folios, a difficult feat because pages are normally illustrated before they are bound together, making it difficult to align the two halves.[16] It is possible Dati had access to Ptolemaic maps through Jacobus de Angelus, who was translating Ptolemy's *Geographia* from Greek to Latin in Florence; he completed this work in 1406/7, almost thirty years before Dati's death. While that manuscript is not thought to have had maps, the earliest Byzantine Ptolemy manuscripts with maps date to the thirteenth century, so it is possible that Jacobus de Angelus's exemplar had maps, or that some were obtained after interest grew in the work

[14] Marino Sanudo's arrangement of Pietro Vesconte's maps in his *Liber secretorum fidelium crucis super Terrae Sanctae recuperatione et conservatione* (1306-21), which encouraged political leaders to retake the Holy Land, may represent a similar collection of maps arranged according to a recognizable geographic space, although his usage is not consistent. In BL Add. MS 27376*, dated 1321, Sanudo uses portolan charts, arranged from west to east, until he reaches the Middle East, where, like Dati, he has more detailed maps of the region; he then continues with portolan maps of Greece and the Black Sea. However, he follows those with a world map (with portolan influence) and a map of the Holy Land, breaking up the geographic order established by the arrangement of the earlier maps. See Akerman, "From Books with Maps to Books as Maps," 16.

[15] For the narrative structure of atlases, see Denis Wood, "Pleasure in the Idea: The Atlas as Narrative Form," in *Atlases for Schools: Design Principles and Curriculum Perspectives*, ed. R.J.B. Carswell, G.J.A. de Leeuw, and N.M. Waters, *Cartographica* monograph no. 36 *Cartographica* 24 (Toronto, 1987), 24-45. Although Wood's central topic is modern school atlases, his observations on the nature and structure of atlases holds for earlier periods too.

[16] My observation of *Sferas* in the Newberry Library, British Library, and in several Florentine libraries indicates that the normal order of composition was followed, with the illustrations continuing into the gutter of the quire.

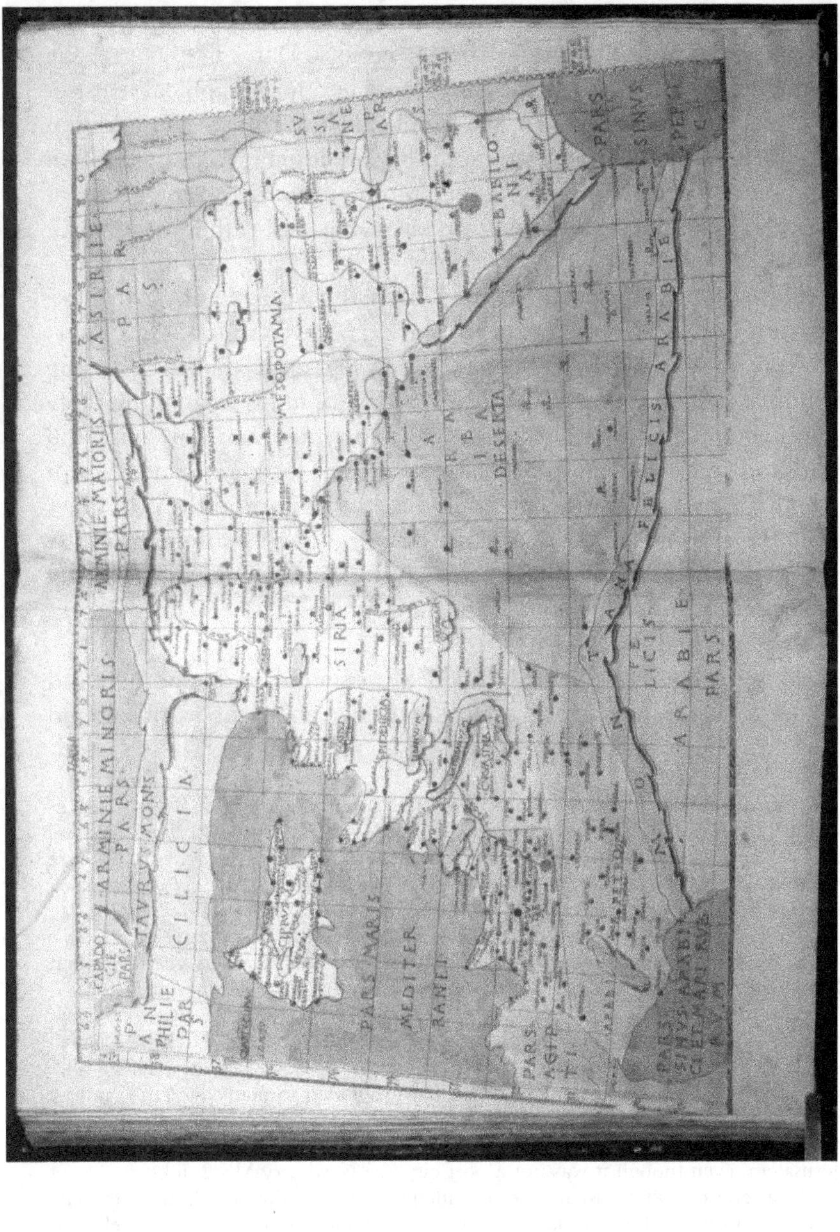

Fig. 4. Ptolemy, *Cosmographia* (Bologna, 1477). Bodleian Library Auct. K. 1. 17, t. 18.

once it was translated into a more accessible language.[17] Pirrus de Noha's world map, tenatively dated to 1414, shows clear Ptolemaic features decades before Dati's *Sfera* (Fig. 5).[18] In contrast to his portolan-based maps, Dati's Middle Eastern map, like Ptolemy's, has no port cities and no distances between cities. Dati's map covers the same region and many of the same cities and topographical features as Ptolemy's map, but with two significant departures: the addition of several elements from Biblical history, including Noah's ark, often represented as a ship in the Armenian mountains; and the tilting of the map to conform to the pattern of having East rather than North at the top.[19]

The second non-portolan map is of the Holy Land, and here Dati's model seems to have been Marino Sanudo's map or an intermediate copy based on it (Plate XIII and Fig. 6). Rather than adding features, as Dati did with the Ptolemaic map, in his map of the Holy Land he reduced the information to a bare minimum, removing the grid and many topographic features so that what remained was a schematic map. Onto this he grafted the coastline and the port cities from the graphic portolan, and the distances between them from the verbal portolan. These maps satisfied the reader's need for placing the events of Biblical history in a spatial context, a need only imperfectly addressed in display portolans.[20]

These maps may have fulfilled an essential and new demand among Florentines for geographic knowledge about the world around

[17] Rome, BAV, *Urbinas Graecus* 82. In the Byzantine world there is a tenth-century witness to a Ptolemy with maps and Maximos Planoudes (1260-1310) had a Greek copy which he executed with his own maps based on Ptolemy's tables. See George Tolias, *The Greek Portolan Charts 15th-17th centuries. A Contribution to the Mediterranean cartography of the modern period* (Athens: Olkos, 1999), 17-18, esp. n. 2.

[18] Rome, BAV, Archivio di San Pietro, H.31, fol. 8r. Pirrus copied Ptolemy's *Cosmographia* for cardinal Orsini (d. 1438), but is otherwise unknown as a cartographer. See David Woodward, "Medieval *Mappaemundi*," in *The History of Cartography*, vol. 1, 358; Marcel Destombes, ed., *Mappaemondes A.D. 1200-1500. Catalogue préparé par la Commission des cartes anciennes de l'Union géographique internationale* (Amsterdam: N. Israel, 1964), 187-88 (51:34).

[19] Biblical elements not in Ptolemy's text are added to many Ptolemaic maps. Noah's ark, for example, is represented in several printed Ptolemaic maps, such as the *Geographia* (Münster, 1540), *tertia tabula Asiae*. Portolan charts often represent Jerusalem, even though it was not a port city (see Newberry MS 2, fol. 6r).

[20] Several display portolans (i.e. not intended for use shipboard), have the city of Jerusalem marked in very large letters and may include other places of religious interest.

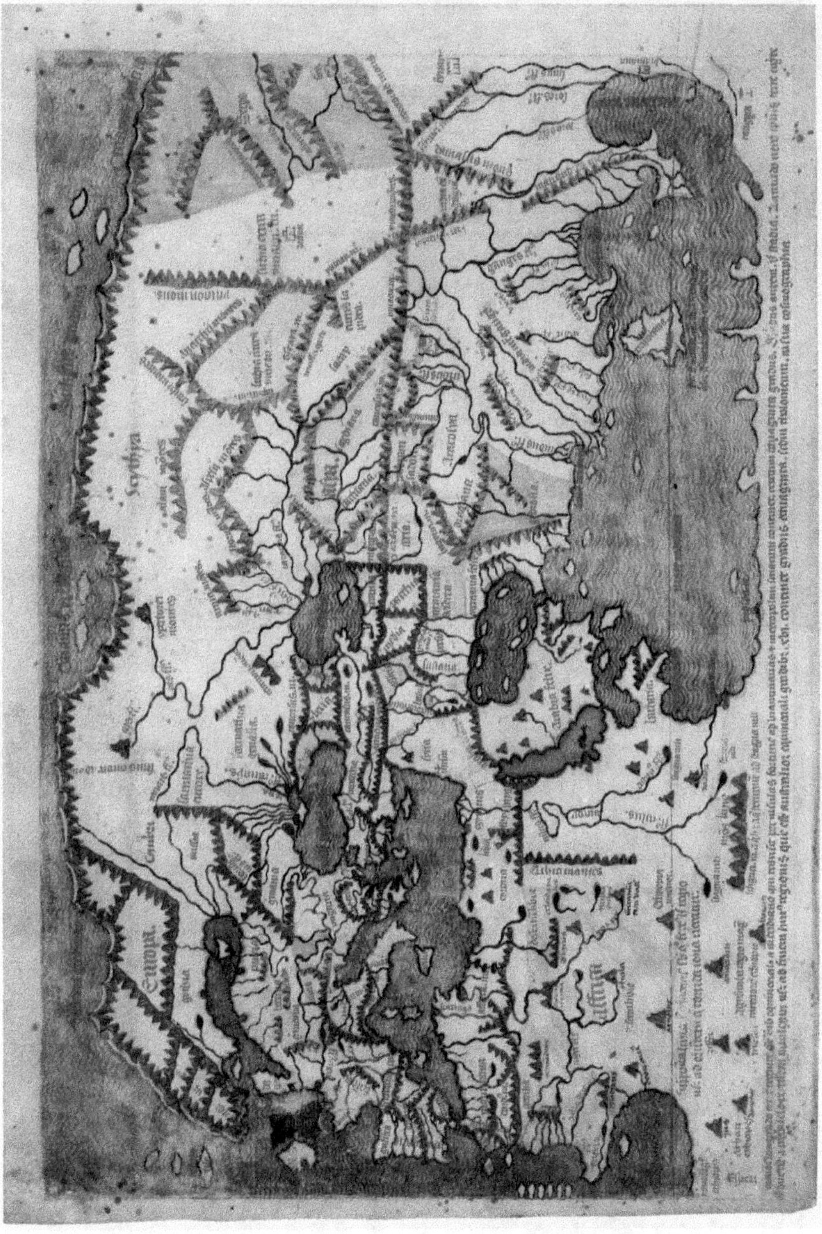

Fig. 5. Pirrus de Noha, world map in Pomponius Mela, *De cosmographia*. Biblioteca Apostolica Vaticana, Archivio di San Pietro, H.31, f. 8r.

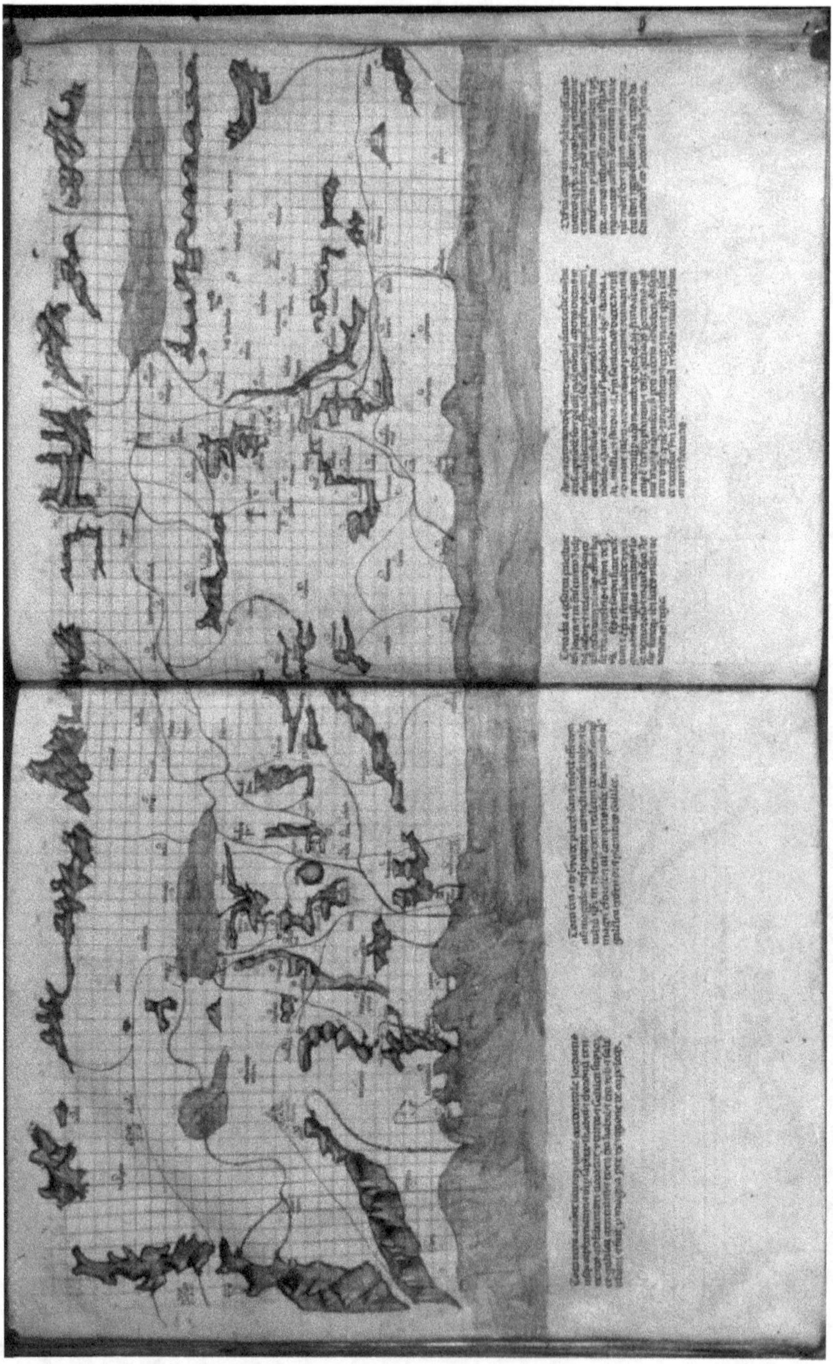

Fig. 6. Pietro Vesconte, map of Palestine in Marino Sanudo, *Secreta fidelium crucis*. Bodleian MS Tanner 190, f. 205v-206r.

them during a time of immense social and economic change and expansion. To meet the changing requirements of educating Florentine children, there were two distinct schools that had very different curricula and served very different purposes and people. The most studied in contemporary scholarship are the Latin schools, which focused on classical literature and the development of morality through the *studia humanitatis*.[21] The Latin schools were in part a product of the humanist revolution in education and were designed for, and attended by, the highest levels of Italian society. Often a preliminary to a university education or a career in letters, the curriculum sought to develop the mind and character of the student through the study of grammar, rhetoric, history, poetry, and moral philosophy. The vernacular schools, on the other hand, had an entirely different focus: to prepare students to survive and thrive in the world of business. The vernacular schools taught students the skills of reading and writing through simple religious tracts and secular tales of adventure. Much of the curriculum was overtly moralizing, including books such as the *Fior di Virtù*, a collection of moral tales illustrating the virtues and vices. But the type of ethics that the vernacular schools sought to instill were different from those found in the Latin schools, where the curriculum was based on pre-Christian classics. The *Fior* and similar Christian works presented a clear sense of wrong and right, and provided models of how the reader should behave as a Christian.[22] Students might also have read Christian hagiography, much of it written in *ottava rima* (eight rhymed lines, like Dati's *Sfera*), an extremely popular form for vernacular didactic literature perhaps because its simple cadence facilitated memorization. From there the student might have proceeded to vernacular epics and romances, which were often in verse. Obviously these later

---

[21] See Paul Gehl, *A Moral Art*; also id., "Latin Readers in Fourteenth-Century Florence: Schoolkids and Their Books," *Scrittura e civiltà* 13 (1989), 387-440; Paul Grendler, *Schooling in Renaissance Italy*, esp. Part 2, pp. 111-271. Bibliographies in the books by Gehl and Grendler treat this topic in depth.

[22] Paul Grendler has written at length about the use of the *Fior di Virtù* in the Venetian vernacular schools. See *Schooling*, pp. 278-280, and "What Zuanne Read in School: Vernacular Texts in Sixteenth-Century Venetian Schools," *Sixteenth Century Journal* 13 (1982), 44-47. Several of the miscellanies that contain Dati's *Sfera* also contain the *Fiore di Virtù*, which is either the same work or a close cousin of the *Fior*. Florence BNC Magl. VII 845; Magl. XXI, 169; and Panciatichiano 66 (Bertolini, *Censimento*, 507, 531; 573).

works were not meant to instill virtue. Instead they captivated their readers with heroes who overcame divine and man-made obstacles, and where true love (usually adulterous) conquered all.

The other component of the vernacular curriculum was *abbaco*, which provided schooling in the type of practical mathematics needed by merchants: how to compute interest, convert currencies, determine and reconcile weights and measures, and divide shares of profit and loss. Vernacular *abbaco* books taught these skills largely through complicated word problems patterned on situations that merchants might encounter regularly, such as determining the total debt when money had been borrowed from different people at different rates of interest, or computing an inheritance when the deceased had several children by several wives, each of whom had a different dowry. The *abbaco* books also contained tables of multiplication, tables of weights and measures, lists of tariffs of various sorts, and methods for computing using a counting board or one's fingers. There was, in short, a vast variety of material necessary to conduct day-to-day business in the Mediterranean world. The manuscript *abbaco* books for the most part were composed not by learned men, but by men literate in the vernacular who taught *abbaco* and by the merchants who used it.[23] Interestingly, the manuscript *abbaco* books that survive today were probably not textbooks, although once printed, they may have been used for instruction. Rather, they were designed to be reference works for merchants who had already studied and for teachers who would adapt the word problems for their classes.[24] Unlike vernacular literature, which required a text in the hands of a student if he or she was to learn to read, *abbaco* seems to have been geared towards memorization. Figuring was most likely done on wax tablets or similar materials that could be easily erased and reused. The manuscript *abbaco* served to refresh the mind for instruction or employment.

Similarly, Dati's *Sfera* was a hybrid of literature that entertained and instructed. It was the sort of practical information that the merchant would need to know to conduct business in the world outside

[23] Warren Van Egmond, *Practical Mathematics in the Italian Renaissance: a catalog of Italian abbacus manuscripts and printed books to 1600* (Florence: Istituto e museo di storia della scienza, 1980), 14.

[24] This is the conclusion of Van Egmond, who has cataloged the manuscript and print *abbaco* books in Italy. See *Practical Mathematics...*, 30-31.

of Florence. The diagrams provide information about the distances between ports, how a compass works, and what forces govern the physical world of which he was a part. The text that accompanied the illustrations could be easily memorized, providing a verbal map of the same region as the portolan charts; both text and image situate the essential historical and holy places of Judaism, Christianity, and Islam, the primary religious groups with which the Florentine merchant had contact.

This explanation may account for the simply illustrated copies, but it does not explain the many lavishly illuminated copies that survive. In fact the illuminations in the *Sfera* and similar cartographic works rarely functioned just to explain the text; rather, they possessed an aesthetic quality that conveyed a distinct type of knowledge, one whose vocabulary was composed of images rather than words. In that sense Dati's *Sfera* was like Christoforo Buondelmonti's *Liber insularum arcipelagi*, often called by its genre *Isolario* (Fig. 7). Composed in the same place around the same time or a little earlier (c. 1420), Buondelmonti's *Isolario*, like Dati's *Sfera*, is an oddity, one that seems to have had no direct forbearers. Richly illuminated, it is an accounting of many of the islands in the Eastern Mediterranean, most of which Buondelmonti claimed to have visited and drawn from sight or memory as he traveled in search of Greek texts for his patron, Cardinal Orsini.[25] His textual description of the history and culture of the islands is equally rich, odd facts mixed with traditional, mostly fictional, histories of the islands and their peoples. His islands, like Dati's coastlines, were almost certainly taken from portolan charts. They have the same level of detail and employ many of the same cartographic conventions to depict ports, landforms, and other features. So while his images were accurate, his text, which is in an easily readable Latin, is mostly fictional, made up of fantastic tales often lifted directly from earlier authors. While some medieval authors added maps at the beginning or end of their work, such a system did not suit Buondelmonti's purpose because he wanted to

[25] Buondelmonti's *Isolario* was translated into Greek in the sixteenth century. See the critical edition, with French translation, Émile Legrand, ed., *Version Grecque du 'Liber Insularum Archipelagi' c. 1420 par un anonyme; Publiée d'après le manuscrit du Sérail, avec une traduction en Français; Suivie de la 'Descriptio Insule Candie' et de la 'Descriptio Cretae' et précédée d'une préface historique et critique* (Paris: E. Leroux, 1897). No Greek manuscript survives with illustrations. George Tolias, *The Greek Portolan Charts*, 22.

Fig. 7. Map of Crete. Cristoforo Buondelmonti, *Liber insularum arcipelagi*, Bodleian, MS Canon. Misc. 280, f. 14v.

show details, such as boats, ruins, forests, and monasteries. The scale of a portolan chart sewn into the end of his book would not have allowed him to do that. Instead, he surrounded each island with text, both above and below, but in general the text and the image rarely compete for the same space, with the possible exception of the title page. Even here, however, where the waters flow to the end of the page, effectively without frame, the space between text and image is respected, unlike in Dati's illustrations where seas routinely flow into the text (see Plate XII).

As in Dati's *Sfera*, the islands are not represented to scale.[26] However, this inconsistency is a less serious breach than Dati's, perhaps because the islands, as clearly self-contained units, do not need to have the same relationship to each other as the seas and fractured coastlines in Dati's *Sfera.* More germane is the fact that Buondelmonti's *Isolario* was not primarily intended to be a didactic or even an instructive work. It was meant, instead, to engage and entertain its readers, especially the author's patron and dedicatee Cardinal Orsini. Its easy Latin would have made it possible for most contemporaries fluent in Italian to read it, and in many ways the book could be enjoyed without the text, because the images themselves tell their own story, much like a medieval block-book. The illustrations in both books served several purposes: they allowed the reader to travel in their imaginations to exotic places across the Mediterranean, to explore the ancient and Biblical history of the region, and even to satisfy their curiosity about the physical world around them.

There are several reasons Florentines might have desired books about Mediterranean geography, even if, traditionally, it was not part of a school curriculum. The years in which Dati and Buondelmonti lived and wrote saw the Florentines nearly crushed by the despotic Visconti of Milan. Dati and others saw the war in much broader terms than city-state against city-state. Rather, it was a clash of two governing systems: the tyrants of Milan against the Republican Florentines.[27] While such focus may have stirred up Florentines to a better defense of their city, there were additional issues tied its

[26] This detail seems not to have troubled historians in the same way that Dati's maps have, although both draw on portolan charts and both intentionally misrepresent scale.

[27] The full title of Dati's *Istoria* is, in English, "A History of the long and most important Italian war which took place in our day between the Tyrant of Lombardy

unique geography which made Florence vulnerable to attack, both militarily and (perhaps more important) economically. Although Florence was a leader in European trade, it had no direct access to the Mediterranean. The Arno, which flowed through Pisa to the Tyrrhenian Sea, was not deep enough to enable ships to sail directly to the city. Instead, most of the trade went overland to Pisa or, less often because it was more distant and therefore more costly, to Venice. During the conflict with Milan, the Visconti had cut Florence off from the Pisan ports, strangling its trade. After the war, the Florentines were anxious to establish and maintain an active merchant shipping fleet. In addition to harnessing the Pisan ports, the Florentines established galleys that sailed around the Mediterranean and the Atlantic as far as England, from 1422 until 1478.[28] The Florentine economy was founded primarily on the production of luxury goods, and so relied on merchants to bring the necessary raw materials to the city and to export and sell the finished goods abroad. Florentine merchants needed information about the islands and ports throughout the Mediterranean, and both Buondelmonti's and Dati's books may have furnished the basic outlines of the Mediterranean in an aesthetically pleasing manner. Because they were not navigating the ships, the level of detail did not need to be precise, but a general familiarity with the Mediterranean would have been extremely useful for merchants shipping goods overseas. Dati himself traveled to Spain on several occasions, and he may have encountered portolan charts on board ship. There are indications that some portolan charts were used by merchants rather than sailors. One surviving collection of charts, for example, has tariffs on goods for the port of Alexandria on it.[29]

Looking at the book as an expression of Florentine mercantile and political power is problematic, however. If the *Sfera* was intended to solidify and further Florentine domination of sea-based trade routes, presumably it would have gone full circle around the Mediterranean and up the western coast of Europe at least as far as England or even

---

and the magnificent Commune of Florence," in Hans Baron, *The Crisis of the Early Italian Renaissance*, 168.

[28] Michael E. Mallett, *The Florentine Galleys in the Fifteenth Century* (Oxford: Clarendon Press, 1967).

[29] Campbell, "Portolan Charts from the Late Thirteenth Century to 1500," 444.

to the Hanseatic ports of Hamburg and Luebeck. Florence's most important trading partners were England and Flanders, both of which were markets for Dati's silk as well as many other Florentine luxury goods.[30] The easiest answer is to remove any authorial intentionality and conclude that Dati died before completing the work, but that he would have continued the *Sfera* past book four had he lived. Clearly, for most of the fifteenth century, Florentines felt that the *Sfera* was useful in its "incomplete" state. There are several possible explanations. First, Dati may have assumed that his audience knew the geography of Italy and the northwestern Mediterranean so well that there was little need for instruction or memorization. Florentines had been concerned with Italian geography for decades because of the continuing rivalry with Venice, the wars with the Visconti of Milan, conflicts with the papal states, and particularly the need to keep the port of Pisa open for trade with the rest of the known world. Because of its history of warfare and trade, educated Florentines must have had a detailed knowledge of the place of Florence and its relationship to other cities on the Italian peninsula.

Second, Dati might have seen a need for knowledge about areas where Florentines typically did not trade, areas that were ripe for the expansion of Florentine markets if the Florentines felt they could compete militarily with the dominant Venetians and Turks and escape the plundering of the Barbary corsairs, as they attempted in the sixteenth century.[31] Third, the book may have been intended to provide information about 'the other', against which Florentines might define themselves, specifically the lands dominated by Moslems, but also by eastern orthodox Christianity. In fact, the maps directly trace the Islamic empire from northern Africa to Constantinople, and from Constantinople to the cities surrounding the Black Sea. No regions in which Roman Catholicism was practiced are depicted in the *Sfera*'s fourth book. This religious curiosity might explain the centrality of the maps devoted to the Holy Land that

[30] As Richard Goldthwaite notes in *Wealth and the Demand for Art in Italy, 1300-1600* (Baltimore: Johns Hopkins University, 1993), Florence was in a unique position among European cities because it produced most of the luxury goods that were sought by other regions, and so imported very little, creating a cash surplus.

[31] Luigi Monga, *The Journal of Aurelio Scetti: A Florentine Galley Slave at Lepanto (1565-1577)* (Tempe: Arizona Center for Medieval and Renaissance Studies, 2004), 19-22.

depict Jewish, Christian, and Islamic historical places, as well as the inclusion of a map that closely resembles the map of Palestine by Marino Senudo to guide crusaders to the Holy Land. Finally, if Dati's *Sfera* functioned in the same way as Buondelmonti's *Isolario* does, to explore from the comfort of one's study places and events distant from the everyday world, then it would serve its purpose well by focusing on those areas less central in Florentine consciousness.

While the uses to which Florentines put Dati's *Sfera,* and the reasons why his text does not extend fully around the Mediterranean, are still open questions, the fact that it was highly valued is evident in the large number of surviving copies and the great expense involved in illuminating a good proportion of them.[32] Dati's *Sfera* gives a unique insight into how Florentines might have viewed their world. It also demonstrates that the line between humanists writing in Latin and merchants writing in the vernacular was a fine one at best, one that may have existed more in the minds of the twentieth-century interpreters of Florentine civic humanism than in Florence itself.

---

[32] In her "Censimento dei manoscritti della Sfera del Dati," *Annali della Scuola normale superiore di Pisa: Classe di Lettere e Filosofia*, 3rd series, Lucia Bertolini has cataloged 77 manuscripts (41 of them with an illustration cycle) in the Biblioteca Laurenziana," Biblioteca Riccardiana, Biblioteca Nazionale Centrale di Firenze and the Biblioteca dell'Archivio di Stato di Firenze: "Manoscritti della biblioteca Laurenziana," 12:2 (1982), 666-705; "Manoscritti della Biblioteca Riccardiana," 15:3 (1985), 889-940; and "I manoscritti della Biblioteca Nazionale Centrale e dell'Archivio di Stato di Firenze," 18:2 (1988), 417-588. Filiberto Segatto, "Un'immagine quattrocentesca del mondo: la *Sfera* del Dati," *Memorie, Accademia nazionale dei Lincei: Classe di scienze morali, storiche e filologiche*, 8th series, 27 (1983), 169-70, lists another eleven in Italian archives, but omits to specify which ones are illustrated. There are twelve in the Vatican, of which ten are illustrated (Roberto Almagià, *Monumenta cartographica vaticana,* 118-29). Another dozen or so are scattered in European and American archives: see Marcel Destombes, ed., *Mappaemondes A.D. 1200-1500.*

CARTES ET CHRONIQUES: MAPPING AND HISTORY IN LATE MEDIEVAL FRANCE

Camille Serchuk

French medieval cartographic studies have progressed little since the first volume of Harley and Woodward's *History of Cartography* was published in 1987. In fact, Paul Harvey, in his treatment of regional mapping in Europe (chapter 20), essentially revisited the discoveries of François de Dainville that had first been published in 1970. Harvey's overview of the European tradition led him to emphasize points of cartographic continuity across national boundaries, and of course, French medieval cartography did not develop purely in isolation. In particular, maps in French historical manuscripts share many features with maps elsewhere that illustrate historical works from other traditions. There were notable points of convergence with world maps, particularly in the way that time and space are compressed, presenting a setting for events described in the text. French maps in luxury manuscripts also share features of more modest vernacular maps. Like the images of rather crude manufacture produced to resolve territorial disputes, they emphasize boundaries and notable landmarks. Indeed, where French maps in historical manuscripts depart from the broader medieval tradition is in their synthesis of these two distinct models of mapping—the universal and the local, the compendious and the contentious. This novel combination of features is, it seems, the result of a particular conjunction of circumstances of production, notably the cross-fertilization of function and genre made possible by the makers of these maps, who were themselves artists fluent in many pictorial modes. The introduction of regional and national maps into French historical manuscripts appears to have been the contribution of artists evidently familiar with the simplified components of the local maps designed to serve as legal documents, as well as with the practice of introducing historical texts with world maps, and also with the distinctly complex conventions of such maps. This intersection of legal practice and book culture ultimately illuminates a variety of themes central to the history of cartography, including the relationship of map-making to the manuscript tradition, the invention of maps and manipulation of precedents, and the connection of maps to texts.

A brief analysis of some maps that appear in French historical manuscripts serves to illustrate the importance of these versatile map-makers, and also allows for a reassessment of *The History of Cartography's* contribution to the study of French mapping in the Middle Ages. Although a wide variety of factors contributed to the presence of maps in such manuscripts, most important was the role of the scribe/editor known as the *remanieur*. Such individuals copied, edited, expanded and revised historical texts. Though little is known of their identity, they are believed to have ranged from clerks to minor nobles.[1] *Remanieurs* shaped historical texts by bringing together a variety of sources in a more or less unified fashion. Competing accounts might be reconciled, or by contrast might be allowed to co-exist without resolution. The practice of *remaniement* allowed for the preservation of particular conventions and sources, as well as the updating of historical texts to include, for example, the reigns of additional kings and also the incorporation of supplemental information. As creators of historical texts, *remanieurs* were thus more than copyists, and some of their compilations and formats had enduring legacies.

The introduction of maps to French historical manuscripts is likely to have been a result of the intervention of these dynamic editors. Familiar with a wide variety of texts and sources, the *remanieurs* transformed the practice of introducing historical works with geographical material, either textual or cartographic; they went so far as to add new maps to their works. While still respecting the tradition of a geographical introduction to an historical text established by Orosius' *Historiae adversus paganos*, *remanieurs* evidently dispensed for the most

[1] For discussion of the *remanieurs* and their role in shaping the historical text *A tous nobles* (see further below), note Marigold Anne Norbye, "The King's blood: royal genealogies, dynastic rivalries and historical culture in the Hundred Years War. A case study of *A tous nobles qui aiment beaux faits et bonnes histoires*" (PhD diss., University College London, 2004), 61-84. For the practice of historical compilation relating to unofficial histories of France, see the discussion (with bibliography) by Sanford C. Zale, "Unofficial Histories of France in the late Middle Ages" (PhD diss., The Ohio State University, 1994), 39-41. For the role of *remanieurs* in shaping literary texts, see Sylvia Huot, "Authors, Scribes, Remanieurs: A Note on the Textual History of the *Romance of the Rose*," in Kevin Brownlee and Sylvia Huot, *Rethinking the Romance of the Rose. Text, Image, Reception* (Philadelphia: University of Pennsylvania Press, 1992), 203-33.

part with much textual excursus, and instead inserted a map into a chronicle to perform the introduction's typical function. Although very few of the maps considered here actually appear at the opening of a volume, as the Orosian tradition would seem to expect, the legacy of the geographical excursus in the *Historiae adversus paganos* may well have instigated the introduction of maps to French vernacular historical manuscripts; some of these cite Orosius as a source. In a similar fashion, *remanieurs* may have borrowed from the related tradition of introducing a universal chronicle with a world map; in historical works of narrower scope, too, they may have conceived of comparable cartographic illustrations. Those *remanieurs* in particular who were clerks and notaries may have facilitated the introduction of maps into historical manuscripts, not least because in the latter capacity they would have participated in the preparation of maps for legal proceedings. In the French tradition, the scribal and documentary dimensions of notarial practice may thus have had an instrumental role in the introduction of maps into vernacular history books.

The importance of Orosius to history writing in the Middle Ages has been well examined, although his specific importance to the French tradition still requires further study.[2] Chronicles may have been fertile ground for a variety of forms of experimentation, but they also belonged to an historical tradition that originated with Orosius, where historical narratives began with geographical descriptions. Orosius' text survives in over 245 copies, of which one-third date from the fourteenth to sixteenth centuries.[3] Only four of these include maps. Another four are elaborately illustrated in other ways.[4]

[2] For a discussion of the text itself, see Yves Janvier, *La Géographie d'Orose* (Paris: Les Belles Lettres, 1982). For the legacy of Orosius, see Bernard Guenée, *Histoire et culture historique dans l'Occident médiéval* (Paris: Aubier-Montaigne, 1980), 166-72.

[3] J.M. Bately and D.J.A. Ross, "A Check List of Manuscripts of Orosius 'Historiarum Adversus Paganos Libri Septem'," *Scriptorium* 15.2 (1961), 329-33; http://www.tertullian.org/rpearse/manuscripts/orosius_history.htm

[4] D. J. A. Ross, "Illustrated Manuscripts of Orosius," *Scriptorium* 9 (1955), 35-56. Ross notes that there are several other copies of the *Historiae adversus paganos* with marginal drawings, but he considers only four to have complete programs of illumination. The Orosian maps are described in Marcel Destombes, ed. *Mappaemondes, AD 1200-1500: Catalogue préparé par la Commission des cartes anciennes de l'Union géographique internationale* (Amsterdam: N. Israel, 1964), 46, 172, 186. For a more recent assessment of the Orosius manuscripts, see Evelyn Edson, *Mapping Time*

Chronicle writers often borrowed from Orosius' model of introducing an historical text with a geographical introduction, so the influence of the *Historiae adversus paganos* was frequently indirect. Works like the *Grandes Chroniques de France* include a geographical description in the reign of Clodio (c. 395-447), a semi-legendary king of the Franks who was the son of Pharamond and the predecessor of Meroveus. The description thus appeared in the text just before the beginning of the Merovingian dynasty, and significantly linked the emergence of the French people and their king to the territory where they lived. No illustrations of this passage survive; possibly it was never illustrated.[5] The author of the abridged French chronicle called *A tous nobles* states in his prologue that he drew on Orosius, although the text lacks a geographical component. The statement may reflect the prestige of the Late Antique text, or simply demonstrate that it was conventional to pay homage to it.[6] Jean Mansel's *Fleur des Histoires*, an historical digest also written around 1450, frequently cites Orosius.[7] But while this enthusiasm for him may have contributed to the interest in introducing geographical features into historical texts, it cannot have been the only factor. The enthusiasm may, however, explain why maps appear to feature in historical manuscripts more than in other genres in the French tradition. The pervasiveness of this medieval practice remains difficult to gauge because of the poor survival rate of manuscripts, but in light of the existing evidence it is possible to speculate that maps were considered suitable illustrations for texts of an historical nature, even if they were only rarely included.

---

and Space. How Medieval Mapmakers Viewed their World (London: The British Library, 1997), 31-35.

[5] None of the surviving manuscripts described by Hedeman appears to illustrate this passage. See Anne D. Hedeman, *The Royal Image. Illustrations of the Grandes Chroniques de France 1274-1422* (Berkeley: University of California Press, 1991), 187-268.

[6] On the historiography of the geographical introduction, see A. H. Merrills, *History and Geography in Late Antiquity* (Cambridge: Cambridge University Press, 2005); for a thorough examination of *A tous nobles*, see M. A. Norbye, "The King's blood...."

[7] He also clearly relied on Jean Corbichon's translation of Bartholomeus Anglicus. See Lisa Deam, "Mapping the Past: the *Fleur des Histoires* (Brussels Bibliothèque Royale MS 9231-9232) in the Context of Fifteenth-Century Burgundian Historiography" (PhD diss., University of Chicago, 2001).

The correlation between geography and history in the form of world maps in universal chronicles has been well documented and thoughtfully studied.[8] Such chronicles, usually written in Latin, begin with the Creation, and illustrate a view of the world with Paradise at the top, itself the beginning of the story.[9] The world maps include a large number of the places described. They serve not only as frontispieces to the volumes, but also as concise and compendious illustrations of the entire text. In them, time and space are conflated so that all of the settings for the subsequent events are shown simultaneously. These *mappaemundi* unify and link the events, but also minimize the time that elapsed between them.

Despite the fact that the practice of illustrating universal chronicles with world maps appears to have been fairly widespread in late medieval Europe, there are very few surviving French examples, particularly in vernacular manuscripts. Elsewhere, the introduction of *mappaemundi* into chronicle texts highlights their theological dimension as well as their historical one.[10] The presence of such a map in one French historical chronicle invites speculation about how its meaning and function might have changed in such a context. This *mappamundi* appears on the final folio in the earliest surviving copy of the *Grandes Chroniques de France* (Fig. 1). Probably produced around 1274 for Philip the Bold, the work used both images and text to glorify French kingship and dynastic continuity.[11] Anne D. Hedeman, who has examined the manuscript closely, believes the map to be the only work of one of the four artists who contributed to this volume. Other scholars question whether the map formed part of

[8] Patrick Gautier Dalché, "L'espace de l'histoire: le role de la géographie dans les chroniques universelles," in Jean-Philippe Genet (ed.), *L'Historiographie médiévale en Europe* (Paris, CNRS: 1991), 287-300; David Woodward, "Reality, Symbolism, Time and Space in the Medieval World Map," *Annals of the Association of American Geographers* 75 (1985), 510-21; Anna Dorothee von den Brinken, "Mappa mundi und Chronographia. Studien zur imago mundi des abendländischen Mittelalters," *Deutsches Archiv für Erforschung des Mittelalters* 24 (1968), 118-86.

[9] Alessandro Scafi, "Mapping Eden: Cartographies of the Earthly Paradise," in D. Cosgrove (ed.), *Mappings* (London: Reaktion, 1999), 50-70; id., *Mapping Paradise. A History of Heaven on Earth* (Chicago: University of Chicago Press, 2006).

[10] Evelyn Edson, *Mapping Time and Space. How Medieval Mapmakers Viewed their World* (London: The British Library, 1997), 145-63.

[11] Bibliothèque Ste. Geneviève MS 782, f. 374. For the book's treatment of these themes and the circumstances of its production (albeit without a discussion of the map), cf. Hedeman, *The Royal Image,* 11-29; 257-58.

Fig. 1. Paris, Bibliothèque Ste Geneviève, Ms 782, f. 374. Mappamundi.

the initial program to decorate the manuscript.[12] The now-ochre-coloured paint added to some of the labels, islands, and landmasses on the map does indeed appear to be a later addition, since it also highlights the signature of King Charles V (reigned 1360-1380), which must have been added in his lifetime.

Intriguingly introduced at the end of the text rather than at the beginning, this map appears alone on the page and has no clear connection to the text. Many of the miniatures from the manuscript seem to have been copied from other sources; the same can almost certainly be said of the map too.[13] Paris apart, the cities illustrated on the map are primarily Biblical sites (including Jerusalem, Nazareth, Babylon, Nineveh), or ancient historical sites (including Troy, Athens, Rome), as well as sites of particular importance to the recurrent discussion of a Crusade (including Damascus, Alexandria, Mecca, Constantinople, Antioch). Paradise appears at the top, and Hell at the bottom. The removal of this map from one context to another evidently called for little, if any, modification. The French already traced their history back to Troy and believed themselves the Chosen People, so sacred and ancient geography were appropriate settings for their prehistory. Even so, the placement of the map at the end rather than the beginning of the volume challenges this reading. As the concluding image, it looks more to the future than to the foundation of France and its history.

Because it directly follows the last element of the text, a life of St. Louis, and because, extraordinarily, it includes the city of Mecca among the sites pictured, the map may propose a Christian triumph led by the French, or perhaps serve to highlight the role of France as the heir, protector and transmitter of the history narrated within the text. In either case, the map collapses time and space by simultaneously illustrating sites that provide the setting for events from both the past and the present. In this respect it is quite similar to other world maps in historical chronicles, and yet the location of the map as a coda along with the curious choice of sites included suggest that it is deploying a familiar formula in a new way. It is presumably an ideal view of the world as the French would like to see it. This quality in turn connects it more to maps of a smaller, regional scale

[12] http://liberfloridus.cines.fr/textes/biblio_fr.html

[13] Hedeman, *The Royal Image*, 12.

than to other world maps; it not only collapses epochs of historical time, but also comprises a modern view of the world. Its inclusion of Mecca marks a departure from earlier maps and introduces the presence of Islam,[14] which seems necessary and appropriate, given the map's location at the end of the volume, following the Life of St. Louis. Like many world maps, this one places Jerusalem at the center. While that choice surely followed both cartographic and theological convention, it also served to highlight the site at the center of the conflict of the Crusades. These centered upon control over Jerusalem, and indeed of the world, and the map's placement as the culmination of the narrative underscores the point. Rather than heralding a claim to empire, the image affirms the role of the French as the defenders of the faith and the "Most Christian" of nations.[15] Thus this map of the world relies upon the pictorial conventions of *mappaemundi*, but it also functions like some smaller, regional maps. It stakes a claim to a territory under dispute.

To a large extent, the world map in the manuscript of the *Grandes Chroniques de France*, on the one hand, and the small, informal regional maps more commonly used for the purposes of settling legal disputes, on the other, represent two ends of the cartographic spectrum in late medieval France. The former focuses on a large body of territory—the whole world—and forms part of a luxury manuscript made for a royal patron. The latter represents a smaller area and is usually almost sketch-like in appearance and rarely detailed, serving as an administrative document for a wide variety of patrons. Examples of this latter group can be found across Europe in fact; just like the *mappaemundi*, they are not a distinctly French phenomenon.

French regional maps were first documented in an article by Dainville in 1970.[16] His aims here were many. He sought to counter

[14] Pascal Arnaud, "Les villes des cartographes: vignettes urbaines et reseaux urbains dans les mappemondes de l'Occident médiéval," *Mélanges de l'Ecole française de Rome. Antiquité* 96 (1984), 537-602.

[15] Hedeman, *The Royal Image,* 11-29, argues that the program of illumination is decidedly not imperial. For the French as "Most Christian," see J. R. Strayer, "France: The Holy Land, the Chosen People, and the Most Christian King," in *Medieval Statecraft and the Perspectives of History* (Princeton: Princeton University Press, 1971), 300-14.

[16] F. de Dainville, "Cartes et contestations au XVe siècle," *Imago Mundi* 24 (1970), 99-121, reprinted in *La Cartographie reflet de l'histoire* (Geneva: Slatkine, 1986), 177-99.

assumptions—widely held at that time—about the absence of a robust tradition of mapmaking in France in the Middle Ages. His research had revealed the opposite. While thereby bringing French medieval mapmaking out of obscurity, he also sought to elucidate the particular context of these maps in French legal practice in the fourteenth century. Befitting their status as legal documents, they were often quite schematic, and often monochrome. Some, such as a map of the region of the Gapençais, are quite crude. They are simple sketches that evince little if any visual elaboration on the part of the draftsman (Fig. 2). They usually focus on monuments, boundaries and waterways. Some include topographical features such as hills, while others show trees and fields. Some depend upon a combination of ground plan and bird's-eye view, and use multiple orientations. One example, a map that shows three towns in Burgundy,[17] relies on the same spatial convention, but—perhaps in deference to a distinguished patron, in this case the Duke of Burgundy—is colored and more detailed.

The *mappamundi* in the *Grandes Chroniques* manuscript is a rare example of the convergence of luxury and simplicity. Maps in high status manuscripts in the French medieval tradition are relatively uncommon. There are only a few such examples from this period, and they are clearly copied from known sources.[18] There are, however, a larger number of maps in the middle of the spectrum, in manuscripts of cruder production; these do not have clearly identified precedents, and instead appear to be the unique and spontaneous inventions of their designers. These maps raise important questions about their origins and their relationship to other modes of mapping. Conceivably, the makers of these unusual maps were artists, professional or amateur, who produced both legal documents and illuminated manuscripts. They were proficient in the relatively

[17] For an illustration of this map (Dijon, Archives départementales de la Côte-d'Or, B 263A), see *The History of Cartography*, vol. 1, plate 36.

[18] Two other examples are maps of Rome. During the Avignonese Papacy, Rome itself may have been viewed as disputed territory. One of these maps (which are clearly copied from an Italian model) appears in the *Tres Riches Heures* made by the Limbourg Brothers for Jean, Duke of Berry around 1410 (Musée Condé ms. 65, f.); the other is in a manuscript of Sallust's *Catilinarius* made by the Orosius Master around 1418 (Private Collection). For both maps, see Millard Meiss, *French Painting in the Time of Jean of Berry. The Limbourg Brothers and Their Contemporaries I* (New York: George Braziller, 1974), 209-14.

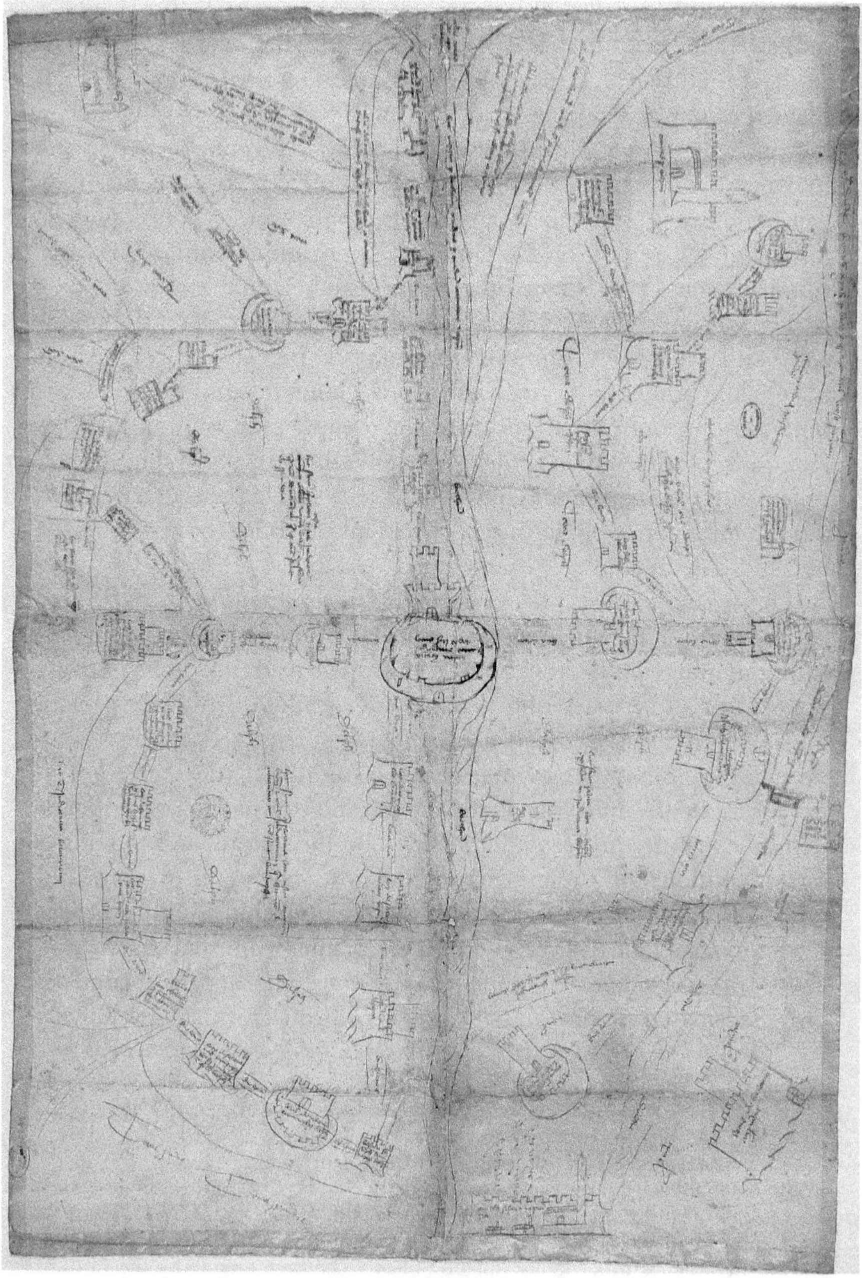

Fig. 2. Grenoble, Archives de l'Isère, B 3751. Plan de la judicature delphinale de Gapençais.

different pictorial languages of both, thus enabling them to introduce images shaped by a legal and administrative context to manuscripts outside that milieu. This explanation may also elucidate some of the more practical reasons why—as Paul Harvey observes in the *History of Cartography*—in France there is "no consistent style that would suggest a regional tradition of mapmaking."[19]

This hypothesis may prove impossible to confirm at an archival level. Even so, an account in Dainville's article relates how artists came to be connected to maps and mapmaking, and offers some insight into late medieval artistic practice. With this foundation, it becomes possible to speculate about the fundamental role played by versatile artists in introducing maps into illuminated manuscripts in the French tradition. Dainville describes documents, today in Grenoble, that relate how in 1423 the Dauphin Charles VII charged the notary Antoine Actuhier to have a map made of the Counties of the Valentinois and the Diois, both in the region of the Drome, in order to allow a more exact accounting of their chateaux, villages and lands, and thus to clarify the legacy of the late Count Louis XII of Poitiers.[20] The surviving documents—which include an accounting of Actuhier's expenses, but do not include the map—explain how he sought the services of an artist in Romans named Jean d'Ecosse, who insisted upon seeing the territory with his own eyes.[21] Actuhier evidently deemed this step unnecessary, since he tried to find another artist to take the job.[22] But he ultimately acceded to the artist's request and made arrangements for him to draw the sites *en plein air*, while traveling on horseback in the company of a guide named Pierre de Sarragosse and of Actuhier himself. Apart from these documents, nothing is known of the career of Jean d'Ecosse, and to judge by his fee at least, he was not one of the luminaries of his day. He was willing to set aside the important work that he had in town in order to accept one *franc* a day, although he initially asked for two *écus*, to travel the countryside during time of war.[23]

[19] *The History of Cartography*, vol. 1, 488.

[20] F. de Dainville, "Cartes et contestations au XVe siècle," 180. The documents are in the Archives de l'Isère, B 3495 and B 3505.

[21] *Requisito dicto pictore quod faceret dictam mapam mundi dixit quod vellet videre loca principalia et situationem patrie aliter non se intromicteret.*

[22] *...et quia non referiebam alium pictorem...*

[23] Op. cit. n. 20 above.

Despite Actuhier's failed efforts to find another artist for the job, Jean d'Ecosse cannot have been the only painter ever called upon to make a map for documentary purposes. From what is known of artistic practice in the late Middle Ages, it is fair to assume that Jean d'Ecosse picked up work like the map and then returned to his atelier to work on a manuscript, or a shield, or another type of object. Documentary evidence from the period confirms that artists performed a great variety of tasks.[24] This versatility may have stimulated cross-fertilization across genres, notably the appearance of images that normally belonged to legal practice in manuscripts unrelated to the law.

One example of such cross-fertilization is a map of the city of Constantinople made for a mutilated and now fragmentary chronicle of the years 1449 to 1453. The codex started life as a roll, but was a larger manuscript before, at some point, it was cut into leaves (Fig. 3).[25] The map carries the rubric, "Here follows the image of the beautiful city of Constantinople,"[26] and appears in the narrative with the description of the siege and capture of that city by Mehmet II on May 29, 1453. This image is the only surviving illustration in the manuscript, and occurs on folio 21. Some components of the map are quite crude. The tents, the trenches with the figures (labeled *mines*), the cannon and the ships all appear to be rather hastily drawn.[27] Other elements are more sophisticated. The architecture is all shown with a consistent scale, and the fortifications display some effort at modeling. Hagia Sophia, here labeled *Ste. Sophie*, is drawn

[24] The evidence for this diversity of artistic practice comes primarily from both French and Italian texts and documents. See in particular Vasari's account of the life of Giotto, where he describes a story written by Franco Sacchetti, in which Giotto rejects a request to paint a shield, apparently deeming such a task to be beneath him; Giorgio Vasari, *The Lives of the Artists*, trans. Julia Conaway Bondanella and Peter Bondanella (Oxford: Oxford University Press, 1991), 34-35. Documents also show that in 1461 the French painter Jean Fouquet was instructed by the leaders of Tours to design scaffolds and scenery ("*échafauds et mysteres*") for the royal entry of the new king Louis XI into the city; Nicole Reynaud, *Jean Fouquet* (Paris: Réunion des Musées Nationaux, 1981), 4.

[25] Paris, Bibliothèque nationale de France, Ms Fr 6487, f. 21.

[26] "*Sensuit la pourtraiture de la belle cité de Costantinoble.*"

[27] Richard Unger has observed that the ships also appear to belong to a period earlier than the text, which must have been written after 1453. His observation would support the assumption that the ships were copied from another image. See further R. Unger, *The Ship in the Medieval Economy 600-1600* (Montreal: McGill-Queen's University Press, 1980), 144-46.

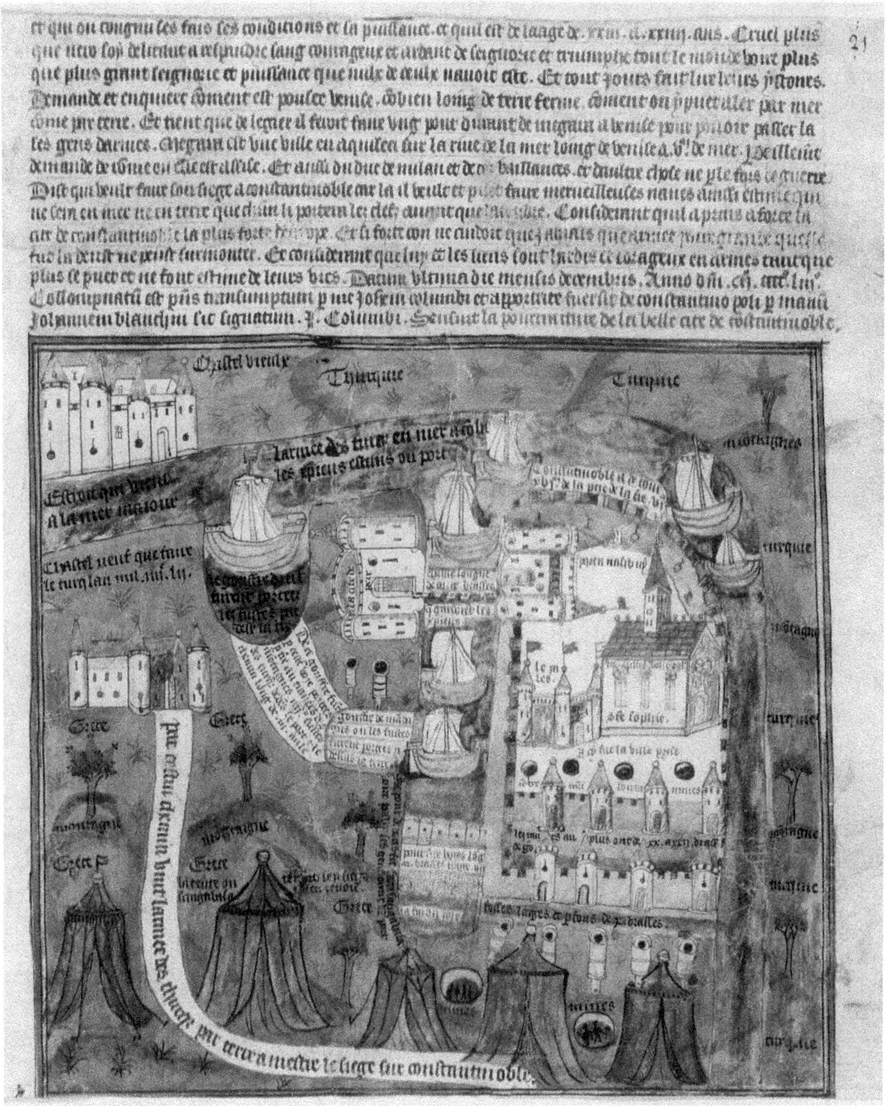

Fig. 3. Paris, Bibliothèque nationale de France Ms Fr 6487, f. 21 Map of Constantinople.

in rudimentary perspective. Although the identity of the miniaturist is unknown, it is possible to speculate that he had much experience representing some of these forms and very little with others. It is worth noting that maps made for the purposes of dispute resolution usually included buildings and landscape, but rarely, if ever, showed figures, armaments or tents. Moreover, while this map is primarily oriented to be read with the text, some of its features—the city of Pera, for example, and certain of the labels—would require that the page be turned for correct viewing. Multiple orientations are a common feature of medieval maps, but were only very rarely used in illuminations in the fifteenth century. These factors suggest that the maker of this map may have been conversant with a variety of images, and that he drew on all of them to produce this one.

In accordance, too, with the conventions of maps made for legal proceedings, the map is quite scrupulously labeled. *Grece* and *Turquie* are identified repeatedly, as are the mountains, which appear at regular intervals, more decorative than documentary. These territorial boundaries and topographical features add interest to the scene, but are also elements that would be identified on a map made in the service of dispute resolution. Other labels on the map are more descriptive, and fall into one of two categories. Some describe strategic features of the city, often in meticulous detail. These include a variety of measurements, including the circumference of the city, the width of its walls and moat, and the length of the chain bridge that traversed the Golden Horn and protected the city: "The chain of 1200 *brasses* [the length of two arms spread wide, thought to be about 1.85m] that protected the Christians."[28] A second set of labels recounts the progress of the siege: "By this route came the army of the Turks by land to lay siege to Constantinople."[29] It is an image that narrates and laments the defeat of the city and simultaneously prepares for its liberation. Like the *mappamundi* in the *Grandes Chroniques* manuscript, this map of Constantinople relies on those conventions of regional mapping that emphasized features useful or necessary to clarify and resolve territorial disputes. The map also compresses

---

[28] "*Chaine longue de m. cc. brasses que gardoient les xpiens.*"

[29] "*Par cestui chemain vint larmee des thurcqs par terre a mectre le siege sur constantinoble.*"

time, by including elements of both past and present in the manner of world maps in universal chronicles.

The temporal compression of the map of Constantinople becomes more obvious when it is compared to a contemporary image of the siege that does not take the form of a map, but instead shows the scene in a bird's-eye view (Fig. 4).[30] The distinction, though subtle, is an important one, since while images that conform to cartographic conventions are quite rare in late medieval French manuscripts, bird's-eye views are fairly common.[31] This particular bird's-eye view of the siege features as an illustration in a traveler's account called *Le Voyage d'Outremer* written in the 1430's by one Bertrandon de La Broquière. The author traveled, sometimes disguised in native costume, overland from Damascus to Dijon under the sponsorship of Philip the Good, Duke of Burgundy, for whom this deluxe copy of his narrative was made.[32] Philip may have commissioned the journey and the narrative as a means to gather intelligence in support of a crusade to liberate Jerusalem. His plans were thwarted by the fall of Constantinople, pictured in the manuscript. Yet despite this context of intrigue and espionage, the image of the siege is an historical rather than a strategic one. Though many of its features are labeled, they are overshadowed by the army in the foreground and the flotilla of ships in the background. If the image was intended to rouse the viewer to action, it was only by means of the outrage that such a meticulous representation of past events might inspire. In other words, it does not compress the past and the future in the same way as the map does when it simultaneously recounts the siege and enumerates the barricades intended to withstand cannon fire. The image only captures the action of the past. The difference between the two representations of the siege of Constantinople stems from both their forms and their functions. The bird's-eye view in the luxury manuscript provides a suitable setting for the tragic sequence of the siege;

---

[30] Paris, Bibliothèque nationale de France, Ms Fr 9087, f. 207.

[31] The calendar in the *Tres Riches Heures* of the Duke of Berry provides many examples (Musée Condé ms. 65, f.).

[32] Bertrandon de La Broquière, *The Voyage d'outremer*, ed. and trans. Galen R. Kline (New York: Peter Lang, 1988). For recent bibliography, see Rima Devereaux, "Reconstructing Byzantine Constantinople: Intercession and Illumination at the Court of Philippe le Bon," *French Studies* 59/3 (2005), 297-310.

Fig. 4. Paris, Bibliothèque nationale de France Ms Fr 9087, f. 207. View of the siege of Constantinople.

the map, by contrast, transcends the temporal boundaries of the siege to project forward to the future reconquest of the city.

This map is all the more complicated and problematic because it is so difficult to link to any specific precedent in the French manuscript tradition. It does not employ the conventions of world maps at all. Instead, it looks more like the map (n. 17 above) showing three towns in Burgundy drawn to resolve a boundary dispute between Duke Philip the Good of Burgundy and King Charles VII of France. That map is in fact somewhat more elegantly drawn, but its trees, buildings, and the water, too, have much in common with the map of Constantinople. Because the features of the latter map are much more frequently consistent with local mapping conventions than with narrative images in illuminated manuscripts, it may be that it was produced by an artist who worked in both genres. In the absence of a pictorial precedent to illustrate the many facets of the city under siege, he might have considered another genre of imagery which was expressly designed to document territory under dispute.

Such cross-fertilization across pictorial types and contexts seems all the more likely because of the crudeness of the chronicle image. It evinces the work of a talented amateur rather than an established and well-respected professional illuminator such as the artist of the Constantinopolitan view, who had plenty of luxury manuscript commissions to keep him busy— he was quite prolific—and so probably did not accept much other work.[33] Journeyman illustrators like Jean d'Ecosse were often affiliated with the workshops of those *remanieurs* (whether private or commercial), who as copyists and editors of manuscripts often sent out manuscripts to be illuminated by artists—including rather mediocre ones of Jean d'Ecosse's caliber—or might draw the images themselves. The *remanieurs* were sometimes the secretaries or notaries of local administrators or aristocrats, exactly the same individuals called upon to draw up maps, as the documents cited by Dainville demonstrate. *Remanieurs* were especially important to the editions of chronicles, since such texts required regular updating and often integrated a wide variety of diverse sources. Although no specific individuals can be identified, it seems fair to suggest that

[33] This miniature is usually attributed to the workshop of Jean Tavernier. See Georges Dogaer, *Flemish miniature painting in the 15th and 16th centuries*, trans. Anna E.C. Simoni *et al.* (Amsterdam: B.M. Israël, 1987).

remanieurs may have played a pivotal role in the introduction of maps in French manuscripts, since they were charged with the responsibilities of both mapmaking and manuscript production.

The role of *remanieurs* as editors of texts is particularly important to the development of regional and national maps in manuscripts because many such maps, produced without pictorial prototypes, are also deeply connected to their textual sources. The map of Constantinople has considerable correlation with the text that describes the siege in the chronicle; the possibility arises, therefore, that the two components may have been the work of the same individual, whether that person was an author or copyist of the text.

The strong reliance on text, the crude draftsmanship, and the unprecedented appearance of a map in a text that exists in sixty-five other copies, is likewise a feature of a map of France in an historical chronicle almost certainly produced by a *remanieur* (Plate XIV).[34] Produced in the late 1450s, this map of France appears in a copy of a text known as *A tous nobles,* a chronicle and genealogy composed around 1400. *A tous nobles*, as mentioned previously, explicitly claims Orosius as one of its sources. This copy includes no additional textual features that link it to the earlier work, but it is the only known manuscript of the surviving copies of the text to include a map.[35] Although there are some small families of manuscripts, for the most part each of the copies of *A tous nobles* is unique, including its images (that is, if it is illustrated at all). Marigold Anne Norbye has made a convincing argument that this wide variety of formats is the hallmark of a text heavily edited and shaped by its *remanieurs*.[36]

This map of France includes some of the same features seen in both the *mappamundi* in the *Grandes Chroniques* manuscript and the map of Constantinople. It is thoroughly labeled, illustrates architecture and topography, uses multiple orientations, and compresses time and space, suggesting that even though the images are entirely unprecedented, its designer evidently had some familiarity with these distinctive dimensions of the several varieties of cartographic illustration. To some extent, certain conventions of this French map may have been borrowed from the *mappaemundi.* Its unusual orientation,

[34] Paris, Bibliothèque nationale de France, Ms Fr 4991, f. 5v.

[35] For a full discussion of this map, see my "Picturing France in the Fifteenth Century: The Map in BnF MS Fr 4991," *Imago Mundi*, 58/2 (2006), 133-49.

[36] Norbye, "The King's blood," 67 ff.

with southeast at the top, may point to the representation of France in T-O type world maps as a possible source of inspiration, since France is so oriented in these images. Like the image of the world in many *mappaemundi*, sites are arranged around a central focus. In the map of France, Paris takes the place of Jerusalem. Like the world map in the *Grandes Chroniques*, however, this map both sets the stage for the historical events described in the text and also stakes a claim for disputed territory. The map appears in the text at the reign of Marcomir, who is credited there with changing the name of his kingdom from Gaul to France; so his reign literally marks the beginning of the history of France. It marks it figuratively, too, since it is here that the genealogy of the kings of France begins. A rubric on the genealogical tree (shown by a green line) that runs through the map announces "Here begins the tree of the kings of France."[37] But the map does not illustrate the France of Marcomir's day. Rather, it shows the present, compressing time and space. Another rubric, unfortunately fragmentary and abraded, explains: "This image contains all the kingdom of France which is now described within, and relates how in the time of the first name of France the kingdom was not so large, but later was larger and then smaller…because of certain actions…within these chronicles."[38] France was once bigger, and was then reduced. Its contours in the map include sites both lost and regained over time, showing this expansion and contraction.

In terms of content, this map of France, like the map of Constantinople, is also very closely related to the text that it illustrates. Because the scribal hand of the labels is quite close to that of the text, it seems probable that one individual produced both the text and its images. The style of the illuminations is fairly crude although, as with the map of Constantinople, there is a marked distinction between the representation of buildings and landscape features which are in some cases quite complex, and that of the figures which are quite crude. This distinction may point to a designer who had some

[37] "*ycy commence l'arbre des rois de france.*"

[38] "*Ceste figure contient tous le royaume de France, qui est a present excript d-s et relait savoir combien que au temps de premier nom de France le royaume nestoit pas si grant mais aprez il a este plus grant et depuis reduit…Par certains fais… aprez en ces presentes croniques.*" Paris, Bibliothèque nationale de France, Ms Fr 4991, f. 5v. For further discussion of this rubric, see Serchuk, "Picturing France in the Fifteenth Century," 143.

experience making maps. Indeed, the modest style of the manuscript exemplifies the tradition of works produced by *remanieurs*.

Other features of this map of France also relate it to the milieu of legal arbitration, and particularly to the representation of disputed territory.[39] Like legal maps, it isolates the territory in question from its broader context; the rest of Europe, for example, is dispensed with entirely. That omission narrows the focus of the image to the object of the claim, which was France itself. Claimants here were the kings of England and France, and the theater of their dispute was the Hundred Years War. This map, produced at the very end of the conflict, puts the French crown's claim into pictorial form, much as would be done before a court of French law. Paul Harvey observed that such a practice was not customary in England, so it is conceivable that the French sought the upper hand by relying on local custom and legal procedure, just as they relied on the French laws governing royal succession to counter English hereditary claims. Thus this extraordinary image may well have been conceived by a designer with experience of using pictorial evidence to resolve territorial claims, perhaps an artist or even a *remanieur* who worked in both a legal and a chronicle context.

In Paul Harvey's view, the small number of known maps makes it difficult to draw "general conclusions about local maps in medieval France."[40] One explanation for this dearth is that many different artists, both amateur and professional, were called upon to draw these maps, and the lack of a uniform style may reveal as much about the diversity of artistic practice across regional boundaries as it does about regional mapmaking. Expanding on the work begun by Dainville, it seems that the link between mapping and disputed territory may extend beyond a strictly legal milieu. While French mapping of the Middle Ages is still largely *terra incognita*, the role of journeyman artists and *remanieurs* was certainly critical to the introduction of maps into historical manuscripts. Many of these still remain to be discovered.

---

[39] Serchuk, "Picturing France in the Fifteenth Century," 138.
[40] *The History of Cartography*, vol. 1, 488.

BIBLIOGRAPHY

Adler, Marcus N., ed. and trans. *The Itinerary of Benjamin of Tudela*. London: Frowde, 1907.

Akerman, James R. "From Books with Maps to Books as Maps: The Editor in the Creation of the Atlas Idea," in *Editing Early and Historical Atlases. Papers Given at the Twenty-Ninth Annual Conference on Editorial Problems, University of Toronto 5-6 November 1993*, ed. Joan Winearls, 3-48. Toronto: University of Toronto Press, 1995.

Akerman, James R., ed. *Cartographies of Travel and Navigation*. Chicago: University of Chicago Press, 2006.

Albu, Emily. "Imperial Geography and the Medieval Peutinger Map," *Imago Mundi* 57 (2005), 136-48.

Alexander, J. J. G. "The Illustrated Manuscripts of the Notitia Dignitatum," in *Aspects of the* Notitia Dignitatum*: Papers presented to the conference in Oxford December 13 to 15, 1974*, ed. R. Goldburn and P. Bartholomew, 11-49. Oxford: British Archaeological Reports, 1976.

Alexandre, Monique. "Entre ciel et terre: les premiers débats sur le site du Paradis (Gen. 2,8-15 et ses réceptions)," in *Peuples et pays mythiques*, ed. B. Deforge, 187-224. Paris: Belles Lettres, 1988.

Almagià, Roberto. *Monumenta cartographica vaticana iussu Pii Xii P.M. consilio et opera procuratorum Bibliothecae Apostolicae Vaticanae*. Citta del Vaticano: Biblioteca Apostolica Vaticana, 1944.

Althoff, Gerd. *Otto III*. trans. Phyllis G. Jestice. University Park: Pennsylvania State University Press, 2003. German edition Darmstadt: Wissenschaftliche Buchgesellschaft, 1996.

Anastos, Milton V. "The Alexandrian Origin of the Christian Topography of Cosmas Indicopleustes," *Dumbarton Oaks Papers* 3 (1946), 73-80.

Anderson, A.R. *Alexander's Gate, Gog and Magog, and the Inclosed Nations*. Monographs of the Medieval Academy of America, 5. Cambridge, Massachusetts: The Medieval Academy of America, 1932.

Andrews, Michael C. "The Study and Classification of Medieval Mappae Mundi," *Archaeologia* 75 (1925-26), 61-76.

Andrieu, Michel, ed. *Les Ordines Romani du Haut Moyen Age*. Vol. 4, *Les Textes (Ordines XXXV-XLIX*). Spicilegium Sacrum Lovaniense 28. Louvain: Université Catholique, 1956.

Angerer, Carl, and Alexander Göschl. *Peutingeriana tabula itineraria in Bibliotheca Palatina Vindobonensi asservata nunc primum arte photographica expressa*. Vienna, 1888.

Anonymi Narratio de Aedificatione Templi S. Sophiae. In *Scriptores Originum Constantinopolitanarum*, vol. 1, ed. Theodore Preger, 74-133. Leipzig: B. G. Teubner, 1901.

Arnaud, Pascal. "L'affaire Mettius Pompusianus, ou le crime de cartographie," *Mélanges de l'École française de Rome: Antiquité* 95 (1983), 677-99.

Arnaud, Pascal. "Les villes des cartographes: vignettes urbaines et reseaux urbains dans les mappemondes de l'Occident medieval," *Mélanges de l'Ecole française de Rome. Antiquité* 96 (1984), 537-602.

Arnaud, Pascal. "L'origine, la date de rédaction et la diffusion de l'archétype de la table de Peutinger," *Bulletin de la Société nationale des antiquaires de France* (1988), 302-20.

Arnaud, Pascal. "*Plurima orbis imago*. Lectures conventionnelles des cartes au Moyen Age," *Médiévales* 18 (1990), 33-51.

Arnaud, Pascal. "L'*Itinéraire d'Antonin*: un témoin de la littérature itinéraire du Bas-Empire," *Geographia Antiqua* 2 (1993), 33-47.

Arnaud-Lindet, M.-P., ed. L. Ampélius, *Aide-mémoire*. Paris: Budé, 1993.

Asche, Ulrike. *Roms Weltherrschaftsidee und Aussenpolitik in der Spätantike im Spiegel der Panegyrici Latini*. Bonn: Habelt, 1983.

G. Aujac, ed. *Optima Hereditas. Sapienza giuridica romana e conoscenza dell'ecumene*. Milan: Scheiwiller, 1992.

Bagnall, Roger S., and J.A. Sheridan. "Greek and Latin Documents from 'Abu Sha'ar, 1992-1993," *Bulletin of the American Society of Papyrologists* 31 (1994), 109-20.

Baron, Hans. *The Crisis of the Early Italian Renaissance*. Princeton: Princeton University Press, 1966.

Barral I Altet, Xavier. "Poésie et iconographie: un pavement du XIIe siècle décrit par Baudri de Bourgueil," in *Studies on art and archaeology in honour of Ernst Kitzinger on his seventy-fifth birthday*, ed. William Tronzo, Irving Lavin and Hans Belting, *Dumbarton Oaks Papers* 41 (1987), 41-54.

Barral I Altet, Xavier. "Observations sur l'iconographie des pavements en mosaïques romans de France en relation avec le décor mural des édifices religieux," in *Le rôle de l'ornement dans la peinture murale du Moyen Age, Actes du Colloque international tenu à Saint-Lizier du 1er au 4 juin 1995*, Civilisation Médiévale 4, ed. Gabriel Biancotto, Robert Favreau, and Piotr Skubiszewski, 149-55. Poitiers: Université de Poitiers, Centre d'études supérieures de civilisation médiévale, 1997.

Barral I Altet, Xavier. "Quelques observations sur les mosaïques de pavement, l'architecture, la liturgie et la symbolique de l'édifice religieux médiéval," *Hortus Artium Medievalium* 9 (2003), 255-60.

Baudri of Bourgueil. *Adelae comitissae*, in Baudri de Bourgueil, *Poèmes*, ed. Jean-Yves Tilliette, vol. 2. Paris: Belles Lettres, 2002.

Baumgärtner, Ingrid. "Die Welt im kartographischen Blick: Zur Veränderbarkeit mittelalterlicher Weltkarten am Beispiel der Beatustradition vom 10. bis 13. Jahrhundert," in *Der weite Blick des Historikers: Einsichten in Kultur-, Landes-, und Stadtgeschichte. Peter Johanek zum 65. Geburtstag*, ed. Wilfried Entrecht, Angelika Lampen, Franz-Joseph Post and Mechthild Siekmann, 527-49. Cologne, Weimar and Vienna: Böhlau, 2002.

Beard, Mary, *et al. Literacy in the Roman World*, ed. John H. Humphrey, *Journal of Roman Archaeology Suppl.* 3, 1991.

Beatus. *Sancti Beati a Liebana Commentarius in Apocalypsin*, ed. E. Romero-Pose. Rome: Typis Officinae Polygraphicae, 1985.

Beazley, Charles R. *The Dawn of the Modern Geography: A History of Exploration and Geographical Science from the Conversion of the Roman Empire to AD 900*, 3 vols., London: J. Murray, 1897-1906.

Beit-Arié, Malachi. *Catalogue of the Hebrew Manuscripts of the Bodleian Library*, ed. R.A. May. Oxford: Clarendon Press, 1994.

Bejaoui, F. "Iles et villes de la Méditerranée sur une mosaïque d'Ammaedara," *Comptes Rendus des séances de l'Académie des Inscriptions et belles-lettres* (1997), 827-60.

Berger, H. *Die geographischen Fragmente des Eratosthenes*. Leipzig: Teubner, 1880.

Berggren, J. Lennart, and Alexander Jones. *Ptolemy's* Geography: *an Annotated Translation of the Theoretical Chapters*. Princeton: Princeton University Press, 2000.

Berliner, Abraham, ed. *Rashi on the Pentateuch*. Berlin, 1866.

Berschin, Walter. *Eremus und Insula: St. Gallen und die Reichenau im Mittelalter—Modell einer lateinischen Literaturlandschaft*. Wiesbaden: Reichert Verlag, 1987.

Bertolini, Lucia. "Censimento dei manoscritti della Sfera del Dati," *Annali della Scuola normale superiore di Pisa: Classe di Lettere e Filosofia*, 3rd series: "Manoscritti della biblioteca Laurenziana," 12:2 (1982), 666-705; "Manoscritti della Biblioteca Riccardiana," 15:3 (1985), 889-940; "I manoscritti della Biblioteca Nazionale Centrale e dell'Archivio di Stato di Firenze," 18:2 (1988), 417-588.

Bieler, Ludwig. "The Text Tradition of Dicuil's *Liber de mensura orbis terrae*," *Proceedings of the Royal Irish Academy* 64 C 1 (1965), 1-31.

Birkholz, Daniel. *The King's Two Maps: Cartography and Culture in Thirteenth-Century England*. Studies in Medieval History and Culture, 22. New York: Routledge, 2004.

Borst, A. *Der Turmbau von Babel. Geschichte der Meinungen über Ursprung und Vielfalt der Sprachen und Völker*. Stuttgart: Hiersemann, 1957.

Bouloux, Nathalie. "Les usages de la géographie à la cour des Plantagenêts dans la seconde moitié du XIIe siècle," *Médiévales* 24 (1993), 131-48.

Braude, B. "The sons of Noah and the construction of ethnic and geographical identities in the medieval and early modern periods," *The William and Mary Quarterly* 54 (1997), 103-42.

Brincken, Anna-Dorothee von den. "Mappa mundi und Chronographia. Studien zur imago mundi des abendländischen Mittelalters," *Deutsches Archiv für Erforschung des Mittelalters* 24 (1968), 118-86.

Brincken, Anna-Dorothee von den. "'...ut describeretur universus orbis.' Zur Universalkartographie des Mittelalters," *Miscellanea Mediaevalia* 7 (1970), 249-78.

Brincken, Anna-Dorothee von den. "Weltbild der lateinischen Universalhistoriker und -kartographen," in *Popoli e paesi nella cultura altomedievale*, I, Settimane di studio del Centro italiano di studi sull' alto medioevo, 29. Spoleto, 1983.

Brincken, Anna-Dorothee von den. *Fines Terrae: Die Enden der Erde und der vierte Kontinent auf mittelalterlichen Weltkarten*. Hannover: Hahnsche Buchhandlung, 1992.

Brodersen, Kai. *Terra Cognita: Studien zur römischen Raumerfassung*. Hildesheim: Olms, 1995; ed. 2, 2003.

Brodersen, Kai. "Mapping the Ancient World," *Ad Familiares: Journal of the Friends of Classics* 17 (Autumn, 1999), 2-4.

Brodersen, Kai. "The presentation of geographical knowledge for travel and transport in the Roman world. *Itineraria non tantum adnotata sed etiam picta*," in *Travel and Geography in the Roman Empire*, C. Adams, R. Laurence, eds., 7-21. London and New York: Routledge, 2001.

Brodersen, Kai. "Mapping (in) the ancient world," *Journal of Roman Studies* 94 (2004), 183-90.

Broquière, Bertrandon de La. *The Voyage d'outremer*. Ed. and trans. Galen R. Kline. New York, Peter Lang: 1988.

Brubaker, Leslie. "The Relationship between Text and Image in the Byzantine MSS of Cosmas Indicopleustes," *Byzantinische Zeitschrift* 70 (1977), 42-57.

Brubaker, Leslie. "The Tabernacle Miniatures of the Byzantine Octateuchs," *Actes du XVe Congrès International d'Etudes Byzantines* II: *Art et Archéologie*, 73-92. Athens: Association Internationale des Etudes Byzantines, 1981.

Bühl, Gudrun. *Constantinopolis und Roma: Stadtpersonifikationen der Spätantike*. Kilchberg and Zurich: Akanthus, 1995.

Burstein, Stanley M., ed. Agatharchides, *On the Erythraean Sea*. London: Hakluyt Society, 1989, 2nd series no. 172.

Calzolari, M. *Introduzione allo Studio della Rete Stradale dell'Italia Romana: L'* Itinerarium Antonini, *Atti della Accademia Nazionale dei Lincei* (Classe di Scienze Morali, Storiche e Filologiche) *Memorie* Ser. IX vol. VII fasc. 4 (1996), 369-520.

Campbell, B., ed. *The Writings of the Roman Land Surveyors: Introduction, Text, Translation and Commentary*. London: Society for the Promotion of Roman Studies, 2000.

Carettoni, G. et al., eds. *La Pianta Marmorea di Roma Antica. Forma Urbis Romae*. Rome: Arti grafiche M. Danesi, 1960.

Carey, Sorcha. *Pliny's Catalogue of Culture: Art and Empire in the* Natural History. Oxford: Oxford University Press, 2003.

Carruthers, Mary J. *The Book of Memory: A Study of Memory in Medieval Culture*, Cambridge Studies in Medieval Literature, 10. Cambridge: Cambridge University Press, 1990.

Caruso G., and R. Volpe. "Preesistenze e Persistenze delle Terme di Traiano," in *Romanization and the City: Creation, Transformations, and Failures*, ed. E. Fentress, 42-56, *Journal of Roman Archaeology Suppl.* 38, 2000.

Casson, Lionel, ed. *The Periplus Maris Erythraei*. Princeton: Princeton University Press, 1989.

Castagnoli, F. "Un Nuovo Documento per la Topografia di Roma Antica," *Studi Romani* 33 (1985), 205-11.

Caviness, Madeline Harrison. *Sumptuous Arts at the Royal Abbeys in Reims and Braine: ornatus elegantiae, varietate stupendes*. Princeton: Princeton University Press, 1990.

Chaplin, Tracey D., and Robin J. H. Clark, Alison McKay, and Sabina Pugh. "Raman spectroscopic analysis of selected astronomical and cartographic folios from the early 13th-century Islamic *Book of Curiosities of the Sciences and Marvels for the Eyes*," *Journal of Raman Spectroscopy* 37 (2006), 865-77.

Chekin, Leonid. "Easter Tables and the Pseudo-Isidorean Vatican Map," *Imago Mundi* 51 (1999), 13-23.

Ciancio Rossetto, P. "Circus Maximus," in *Lexicon Topographicum Urbis Romae* I, ed. Eva Margareta Steinby, 272-77. Rome: Quasar, 1993.

Ciancio Rossetto, P. "Rinvenimenti e restauri al portico d'Ottavia e in piazza delle Cinque Scole," *BullCom* 97 (1996), 267-79.

Clark, Ch. U. *The text tradition of Ammianus Marcellinus*. New Haven: The Author, 1904.

Coarelli, F. "Pax, Templum," in *Lexicon Topographicum Urbis Romae* IV, ed. Eva Margareta Steinby, 67-70. Rome: Quasar, 1999.

Conticello De' Spagnolis, M. *Il Tempio dei Dioscuri nel Circo Flaminio*. Rome: De Luca, 1985.

Cosmas Indicopleustes. *The Christian Topography of Cosmas, an Egyptian Monk*. Ed. and trans. J. W. McCrindle. London: Hakluyt Society, 1897.

Courcelle, P. "La vision cosmique de saint Benoît," *Revue des études augustiniennes* 13 (1967), 97-117.

Cressedi, G. "Contributo per la datazione della Forma Urbis," *Bullettino della Commissione Archeologica Comunale di Roma* 73 (1949-50), 91-95.

Crone, Gerald R. *The world map by Richard of Haldingham in Hereford Cathedral, circa A.D. 1285*. London: Royal Geographical Society, 1954.

Crone, Gerald R. "New light on the Hereford map," *The Geographical Journal* 131 (1965), 447-58.

Crosetto, Alberto. "Indagini archeologiche nel medioevo astigiano. La cattedrale di S. Maria," *Quaderni della Soprintendenza Archeologica del Piemonte* 13 (1995), 255-76.

Cuntz, Otto, ed. *Imperatoris Antonini Augusti Itineraria Provinciarum et Maritimum*. Leipzig: B. G. Teubner, 1929.

Dainville, François de. "Cartes et contestations au XVe siecle," *Imago Mundi* 24 (1970), 99-121.

Daniélou, Jean. "Terre et Paradis chez les Pères de l'Église," *Eranos Jahrbuch* 22 (1954), 433-72.

Daunicht, Hubert. *Der Osten nach der Erdkarte al-Ḫuwārizmīs. Beitrage zur Historischen Geographie und Geschichte Asiens. Bd. I: Rekonstruktion der Karte, Interpretation der Karte: Südasien.* Bonner Orientalistische Studien, Neue Serie, 19. 4 vols. Bonn: Selbstverlag des Orientalischen Seminars der Universität, 1968.

Deam, Lisa. "Mapping the Past: the Fleur des Histoires (Brussels Bibliothèque Royale MS 9231-9232) in the Context of Fifteenth-Century Burgundian Historiography," PhD diss., University of Chicago, 2001.

Delano-Smith, Catherine. "Geography or Christianity ? Maps of the Holy Land before AD 1000," *Journal of Theological Studies* 42 (1991), 143-52.

Delano-Smith, Catherine, and Mayer I. Gruber. "Rashi's Legacy: Maps of the Holy Land," *The Map Collector* 59 (1992), 30-35.

Delforge, Th. "Songe de Scipion et vision de saint Benoît," *Revue bénédictine* 69 (1959), 351-54.

Deliyannis, Deborah. "Charlemagne's Silver Tables: The Ideology of an Imperial Capital," *Early Medieval Europe* 12 (2003), 159-77.

de Rosa, Antonio, and Mauro Perani, eds. *Giovanni-Ovadiah da Oppido, proselito, viaggiatore e musicista dell'età normanna. Atti del convegno internazionale Oppido Lucano, 28-30 marzo 2004.* Florence: Giuntina, 2005.

Destombes, Marcel, ed. *Mappaemondes A.D. 1200-1500. Catalogue préparé par la Commission des cartes anciennes de l'Union géographique internationale.* Amsterdam: N. Israel, 1964.

Detlefsen, D. *Ursprung, Einrichtung und Bedeutung der Erdkarte Agrippas.* Berlin: Weidmann, 1906.

Devereaux, Rima. "Reconstructing Byzantine Constantinople: Intercession and Illumination at the Court of Philippe le Bon," *French Studies* 59/3 (2005), 297-310.

Dicuil. *Dicuili liber de mensura orbis terrae.* Ed. J. J. Tierney and Ludwig Bieler. Dublin: Dublin Institute for Advanced Studies, 1967.

Dilke, Oswald A.W. *The Roman Land Surveyors: an Introduction to the Agrimensores.* Newton Abbot: David and Charles, 1971.

Dilke, Oswald A.W. *Greek and Roman Maps.* London: Thames & Hudson, 1985.

Dillemann, L. *La Cosmographie du Ravennate.* Collection Latomus, vol. 235. Brussels: Latomus, Revue d'Etudes Latines, 1997.

Dinzelbacher, P. "Voli celesti e contemplazione del mondo nelle letteratura estaticovizionaria del medioevo," in *Cieli e terre nei secoli XI-XII. Orizzonti, percezioni, rapporti,* Atti della tredicesima settimana internazionale di studio, Mendola, 22-26 agosto 1995, 215-33. Milan, 1998.

Dodwell, C. R. "Reichenau Reconsidered: A Reassessment of the Place of Reichenau in Ottonian Art," *Warburg Institute Surveys, II.* The Warburg Institute, University of London, 1965.

Dogaer, Georges. *Flemish miniature painting in the 15th and 16th centuries.* Trans. Anna E.C. Simoni et al. Amsterdam: B.M. Israël, 1987.

Donkin, Lucy E. G. "*Ornata decenter.* Figurative Ecclesiastical Floor Mosaics in Northern Italy, 1030-1213." Ph.D. diss., University of London, 2005.

Donkin, Lucy E. G. "Mosaici pavimentali medievali nell' Italia settentrionale e i loro rapporti con la liturgia," in *Atti del X Colloquio dell' Associazione Italiana per lo Studio e la Conservazione del Mosaico, Lecce, 18-21 febbraio 2004,* ed. Claudia Angelelli, 503-14. Tivoli: Edizioni Scripta Manent, 2005.

Dormeier, Heinrich. "Die Renovatio Imperii Romanorum und die "Außenpolitik" Ottos III. und seiner Berater," in *Polen und Deutschland vor 1000 Jahren: die Berliner Tagung über den "Akt von Gnesen,"* ed. Benjamin Scheller, 163-91. Berlin: Akad.-Verlag, 2001.

Dueck, Daniela. "The Augustan Concept of 'an Empire without Limits'," in *Creating and Representing Sacred Spaces,* ed. Michael Dickhardt and Vera Dorofeeva-Lichtmann, *Göttinger Beiträge zur Asienforschung* 2-3, 211-27. Göttingen: Peust und Gutschmidt, 2003.

Dümmler, Ernst. "Gedichte aus dem elften Jahrhundert," *Neues Archiv* 1 (1876), 175-85.

Duft, Johannes. *The Abbey Library of Saint Gall,* trans. James C. King and Petrus W. Tax. St. Gallen: Verlag am Klosterhof, 1985.

Edney, Matthew H. "David Alfred Woodward (1942-2004)," *Imago Mundi* 57 (2005), 75-83.

Edney, Matthew H. *The Origins and Development of J.B. Harley's Cartographic Theories. Cartographica* 40,1-2 [2005] = Monograph 54. Toronto: University of Toronto Press.

Edson, Evelyn. "The Oldest World Maps: Classical Sources of Three VIII Century Mappaemundi," *The Ancient World* 24 (1993),169-84.

Edson, Evelyn. "World maps and Easter tables: medieval maps in context," *Imago Mundi* 48 (1996), 25-42.

Edson, Evelyn. *Mapping Time and Space: How Medieval Mapmakers Viewed Their World.* London: British Library, 1997.

Edson, Evelyn, and Emilie Savage-Smith, *Medieval Views of the Cosmos.* Oxford: Bodleian Library, 2004.

Eisenhofer, Ludwig. *Handbuch der katholischen Liturgik.* 2 vols. Freiburg im Breisgau: Herder, 1932-1933.

Elweskiöld, B. "Cosmas Indicopleustes and John Philoponus. A Christian controversy on the structure of the world in sixth-century Alexandria." PhD diss., Lund University, 2005.

Encyclopaedia of Islam, ed. 2, ed. H. A. R. Gibbs et al., 11 vols. Leiden: Brill, 1960-2005.

Ephraim Syrus. *Commentary on Genesis,* trans. Edward G. Mathews Jr. and Joseph P. Amar, in *St. Ephrem the Syrian: Selected Prose Works,* ed. Kathleen McVey, 67-220. Washington: Catholic University of America Press, 1994.

Erdmann, Carl. "Das ottonische Reich als Imperium Romanum," *Deutsches Archiv für Erforschung des Mittelalters* 6 (1943), 412-41.

Esmeijer, Anna C. *Divina Quaternitas. A Preliminary Study in the Method and Application of Visual Exegesis.* Amsterdam; Assen: Van Gorcum, 1978.

Esposito, Mario. "An Irish Teacher at the Carolingian Court: Dicuil," *Studies* 3 (1914), 651-76.

Estey, F. N. "Charlemagne's Silver Celestial Table," *Speculum* 18 (1943), 112-17.

Flint, Valerie. "Monsters and Antipodes in the Early Middle Ages and Enlightenment," *Viator* 15 (1984), 65-80.

Fontaine, Jacques. *Isidore de Séville, Traité de la Nature.* Bordeaux: Féret, 1960.

Fox-Friedman, Jeanne. "Sacred and Secular: Modena Cathedral and Monumental World Maps," *Arte Medievale,* 2nd ser., 10 (1996), 39-55.

Fox-Friedman, Jeanne. "Vision of the World: Romanesque Art of Northern Italy and the Hereford Mappamundi," in *The Hereford World Map: Medieval World Maps and their Context,* ed. Paul D. A. Harvey, 137-51. London: The British Library, 2006.

Friedman, John B., and Kristen M. Figg, eds. *Trade, Travel, and Exploration in the Middle Ages: an Encyclopedia.* New York and London: Garland, 2000.

Frolow, A. "Deux églises byzantines d'après des sermons peu connus de Léon VI le sage," *Études Byzantines* 3 (1945), 43-91.

Frugoni, Chiara. "La figurazione basso-medioevale dell' imago mundi," in *"Imago Mundi": La conoscenza scientifica nel pensiero bassomedioevale*, Convegni del Centro di Studi sulla Spiritualità Medievale 22, 225-69. Todi: presso l'Accademia Tudertina, 1983.

Gallazzi, Claudio, and Barbel Krämer, "Artemidor in Zeichensaal. Eine Papyrusrolle mit Text, Landkarte und Skizzenbüchern aus späthellenistischer Zeit," *Archiv für Papyrusforschung* 44 (1998), 189-208.

Gallazzi, Claudio, and Salvatore Settis, eds. *Le Tre Vite del Papiro di Artemidoro: Voci e Sguardi dall' Egitto Greco-Romano*. Milan: Electa, 2006.

Garrison, Eliza. "Henry II's *Renovatio* in the Pericope Book and Regensburg Sacramentary," in *The White Mantle of Churches: Architecture, Liturgy, and Art Around the Millenium*, ed. Nigel Hiscock, 57-79. Turnhout: Brepols, 2003.

Gautier Dalché, Patrick. "Les diagrammes topographiques dans les manuscrits des classiques latins," *Bibliologia* 20 (1983), 291-306.

Gautier Dalché, Patrick. "Tradition et renouvellement dans la représentation de l'espace géographique au IXe siècle," *Studi Medievali*, 3rd ser. 24 (1983), 121-65.

Gautier Dalché, Patrick. "Les 'quatre sages' de Jules César et la 'mesure du monde' selon Julius Honorius: réalité antique et tradition médiévale, II: la tradition médiévale," *Journal des Savants* (1987), 184-209.

Gautier Dalché, Patrick. *La "Descriptio Mappe Mundi" de Hugues de Saint Victor*. Paris: Études Augustiniennes, 1988.

Gautier Dalché, Patrick. "L'espace de l'histoire: le role de la géographie dans les chroniques universelles," in *L'Historiographie médiévale en Europe*, ed. Jean-Philippe Genet, 287-300. Paris: CNRS, 1991.

Gautier Dalché, Patrick. "De la glose à la contemplation: Place et fonction de la carte dans les manuscrits du Haut Moyen Âge," in *Testo e Immagine nell'Alto Medioevo*, Settimane di Studio del Centro Italiano di Studi sull' Alto Medioevo, XLI, 693-771. Spoleto, Presso la Sede del Centro, 1994.

Gautier Dalché, Patrick. "Notes sur la 'carte de Théodose II' et sur la 'mappemonde de Théodulf d'Orléans'," *Geographia Antiqua* 3 (1994), 91-108.

Gautier Dalché, Patrick. *Géographie et culture. La représentation de l'espace du VIe au XIIe siècle*. Variorum, Collected Studies Series. Aldershot: Ashgate, 1997.

Gautier Dalché, Patrick. "*Mappae mundi* antérieures au XIIIe siècle dans les manuscrits de la Bibliothèque nationale de France," *Scriptorium* 52 (1998), 102-62.

Gautier Dalché, Patrick. "Le souvenir de la *Géographie* de Ptolémée dans le monde latin médiéval (VIe-XIVe siècles)," *Euphrosyne* 27 (1999), 79-106.

Gautier Dalché, Patrick. "Les diagrammes topographiques dans les manuscrits des classiques latin (Lucain, Solin, Salluste)," in *La tradition vive. Mélanges d'histoire des textes en l'honneur de Louis Holtz*. Turnhout: Brepols, 2002, 291-306.

Gautier Dalché, Patrick. "Principes et modes de la représentation de l'espace géographique durant le haut moyen âge," in *Uomo e spazio nell'alto medioevo*, Settimane di Studio del Centro Italiano di Studi sull'Alto Medioevo 50 (Spoleto, 2003), 117-50.

Gautier Dalché, Patrick. "La trasmissione medievale e rinascimentale della *Tabula Peutingeriana*," in *Tabula Peutingeriana: Le Antiche Vie del Mondo*, ed. Francesco Prontera, 43-52. Florence: L. S. Olschki, 2003.

Gautier Dalché, Patrick. "Du nouveau sur la transmission et la découverte de la *Tabula Peutingeriana*: la '*cosmographia vetustissima*' de Pellegrino Prisciani (†1518)," *Geographia antiqua* 13 (2004), 71-81.

Gautier Dalché, Patrick. *Du Yorkshire à l'Inde. Une "géographie" urbaine et maritime de la fin du XIIe siècle (Roger de Howden ?)*. Geneva: Droz, 2005.

Gautier Dalché, Patrick. "Maps in words: the descriptive logic of medieval geography, from the eighth to the twelfth century," in *The Hereford World Map: Medieval World Maps and their Context*, ed. Paul D. A. Harvey, 223-42. London: The British Library, 2006.

Gautier Dalché, Patrick, and Jean-Yves Tilliette. "Un nouveau document sur la tradition du poème de Baudri de Bourgueil à la comtesse Adèle," *Bibliothèque de l'École des Chartes*, 144/2 (1986), 241-57.

Gehl, Paul F. "Latin Readers in Fourteenth-Century Florence," *Scrittura e civiltà* 13 (1989), 387-440.

Gehl, Paul F. *A Moral Art: Grammar, Society, and Culture in Trecento Florence.* Ithaca: Cornell University Press, 1993.

Geus, K. "Space and geography," in *A Companion to the Hellenistic World*, ed. Andrew Erskine, 232-45. Oxford: Blackwell, 2003.

Görich, Knut. *Otto III., Romanus, Saxonicus, et Italicus: Kaiserliche Rompolitik und Sachsische Historiographie.* Sigmaringen: Thorbecke, 1993.

Goldthwaite, Richard A. *Wealth and the Demand for Art in Italy, 1300-1600.* Baltimore: Johns Hopkins University Press, 1993.

Gormley, C., M.A. and R.H. Rouse. "The medieval circulation of the *De chorographia* of Pomponius Mela," *Medieval Studies* 46 (1984), 266-320.

Graf, Arturo. *Roma nella memoria e nelle immaginazioni del medio evo.* 2 vols. Turin: Ermanno Loescher, 1882-1883.

Gregory the Great. *Dialogues.* Ed. Adalbert de Vogüé and trans. Paul Antin. 3 vols. Paris: Les Éditions du Cerf, 1978-1980.

Grendler, Paul F. *Culture and Censorship in Late Renaissance Italy and France.* London: Variorum, 1981.

Grendler, Paul F. "What Zuanne Read in School: Vernacular Texts in Sixteenth-Century Venetian Schools," *Sixteenth Century Journal* 13 (1982), 41-53.

Grendler, Paul F. *Schooling in Renaissance Italy: Literacy and Learning, 1300-1600*, Johns Hopkins University Studies in Historical and Political Science; 107th Ser., 1. Baltimore: Johns Hopkins University Press, 1989.

Grendler, Paul F. *Books and Schools in the Italian Renaissance.* Aldershot: Variorum, 1995.

Grendler, Paul F. *The Universities of the Italian Renaissance.* Baltimore: Johns Hopkins University Press, 2002.

Gribaudi, Pietro. "Di un mosaico cosmografico medioevale scoperto a Torino," *Bollettino della Società Geografica Italiana*, 4th ser., 12 (1911), 619-37.

Gross, H. *Zur Entstehungsgeschichte der Tabula Peutingeriana.* PhD diss., Universität Berlin, 1913.

Grossman, Avraham. "Marginal Notes and Addenda of R. Shemaiah and the Text of Rashi's Commentary," *Tarbiz* 60 (1990-91), 67-98 [in Hebrew].

Grossman, Avraham. "Eretz-Israel in Rashi's Teaching," *Shalem* 7 (2002), 15-31 [in Hebrew].

Grossman, Avraham. *Rashi: R. Shlomo Yitzhaqi.* Jerusalem: Merkaz Zalman Shazar le-toldot Yiśra'el, 2006 [in Hebrew].

Grossman, Avraham and Benjamin Z. Kedar. "Rashi's Maps of the Land of Israel and their Historical Significance," *Newsletter of the Israel Academy of Sciences and Humanities* 25 (November 2003), 26-29 [in Hebrew].

Gruber, Mayer I. "What Happened to Rashi's Pictures ?," *The Bodleian Record* 14 (1992), 111-24.

Gruber, Mayer I. "Light on Rashi's Diagrams from the Asher Library of Spertus College of Judaica," *The Solomon Goldman Lectures* 6 (1993), 73-85.

Gruber, Mayer I. "The Sources of Rashi's Cartography," in *Letters and Texts of Jewish History*, ed. Norman Simms. Hamilton, New Zealand: Outrigger Publ., 1998.

Harley, J. Brian, and David Woodward, eds. *The History of Cartography*, vol. 1, *Cartography in Prehistoric, Ancient, and Medieval Europe and the Mediterranean.* Chicago: University of Chicago Press, 1987.

Harley, J. Brian. "Text and Contexts in the Interpretation of Early Maps," in *From Sea Charts to Satellite Images: Interpreting North American History through Maps,* ed. David Buisseret, 3-15. Chicago: University of Chicago Press, 1990.

Harris, William V. *Ancient Literacy.* Cambridge: Harvard University Press, 1989.

Harvey, Paul D.A. *Mappa Mundi: The Hereford World Map.* London: British Library, 1996.

Harvey, Paul D.A. "The Sawley map and other world maps in twelfth-century England," *Imago Mundi* 49 (1997), 33-42.

Hedeman, Anne D. *The Royal Image. Illustrations of the Grandes Chroniques de France 1274-1422.* Berkeley: University of California Press, 1991.

Herren, Michael W. "The "De imagine Tetrici" of Walahfrid Strabo: Edition and Translation," *Journal of Medieval Latin* 1 (1991), 118-39.

Hiatt, A. "Mapping the ends of empire," in *Postcolonial Approaches to the European Middle Ages: Translating Cultures,* ed. A.J. Kabir and D. Williams, 48-76. Cambridge: Cambridge University Press, 2005.

Hindman, Sandra, Michael Camille, Nina Rowe, and Rowan Watson. *Manuscript Illumination in the Modern Age: Recovery and Reconstruction.* Evanston: Mary and Leigh Block Museum of Art, Northwestern University, 2001.

Hoffmann, Hartmut. *Bamberger Handschriften des 10. und des 11. Jahrhunderts.* Hannover: Hahn, 1995.

Hoffmann, Konrad. "Das Herrscherbild im "Evangeliar Ottos III." (clm 4453)," *Frühmittelalterliche Studien* 7 (1973), 324-41.

Honorius Augustodunensis. *Gemma Animae.* In *Patrologia Latina* 172, ed. J.-P. Migne, 541-738. Paris: Migne, 1895.

Honorius Augustodunensis. *De Imagine Mundi.* In *Patrologia Latina* 172, ed. J.-P. Migne, 115-88. Paris: Migne, 1895.

Honorius Augustodunensis. *Imago Mundi*, ed. Valerie I.J. Flint. *Archives d'histoire littéraire et doctrinale du Moyen Age* 49 (1983), 7-153.

Houts, Elisabeth M. C. van. "Latin Poetry and the Anglo-Norman Court 1066-1135: The *Carmen de Hastingae Proelio,*" *Journal of Medieval History* 15 (1989), 39-62.

Huot, Sylvia. "Authors, Scribes, Remanieurs: A Note on the Textual History of the Romance of the Rose," in *Rethinking the Romance of the Rose. Text, Image, Reception,* ed. Kevin Brownlee and Sylvia Huot, 203-33. Philadelphia: University of Pennsylvania Press, 1992.

Husslein, G. "Konrad Miller," *Orbis Terrarum* 1 (1995), 213-33.

Inglebert, H. *Interpretatio christiana. Les mutations des savoirs (cosmographie, géographie, ethnographie, histoire) dans l'Antiquité chrétienne 30-630 après J.-C.* Paris: Inst. d'Études Augustiniennes, 2001.

Isidore. *Isidori Hispalensis Episcopi, Etymologiarum sive Originum.* Ed. W. M. Lindsay. 2 vols. Oxford: Clarendon Press, 1911.

Ivo of Chartres. *Sermo V, sive opusculum de convenientia veteris et novi sacrificii.* In *Patrologia Latina* 162, ed. J.-P. Migne, 535-62. Paris: Migne, 1889.

Jacob, Christian. *La description de la terre habitée de Denys d'Alexandrie.* Paris: Michel, 1990.

Janni, Pietro. *La Mappa e il Periplo: Cartografia Antica e Spazio Odologico.* Rome: Bretschneider, 1984.

Janvier, Yves. *Le Géographie d'Orose.* Collection d'études anciennes. Paris: Belles Lettres, 1982.

Jerome. *Commentarium in Esaiam.* Ed. Marc Adriaen. *S. Hieronymi Presbyteri Opera,* 1.2-2a, Corpus Christianorum Series Latina 73-73A. 2 vols. Turnhout: Brepols, 1963.

Johns, Jeremy, and Emilie Savage-Smith. "*The Book of Curiosities*: A Newly Discovered Series of Islamic Maps," *Imago Mundi* 55 (2003), 7-24.

Jonas of Bobbio. *Vita Columbani,* I, 27 (56), *M. G. H., Scriptores rerum Germ. in usum schol.,* ed. G. Krush. Hanover: Hahn, 1905.

Jones, R. M. "Posidonius and the flight of the mind through the universe," *Classical Philology* 21 (1926), 97-113.

Julius Honorius. *Cosmographia.* In *Geographi Latini Minores.* Ed. A. Riese, 24-53. Heilbronn: apud Henningeros fratres, 1878.

Jungmann, Joseph A. *The Mass of the Roman Rite: Its Origins and Development.* 2 vols. Trans. Francis A. Brunner. Blackrock: Four Courts Press, 1986.

Kahsnitz, Rainer. "Herrscherbilder der Ottonen," in *Krönungen: Könige in Aachen —Geschichte und Mythos: Katalog der Ausstellung in zwei Bänden,* ed. Mario Kramp, vol. 1, 283-93. Mainz: von Zabern, 2000.

Kedar, Benjamin Z. "Some Reflections on Maps, Crusading and Logistics," in *Logistics of Warfare in the Age of the Crusades,* ed. John H. Pryor, 159-84. Aldershot: Ashgate, 2006.

Kehr, Paul Fridolin. *Italia Pontificia.* Vol. 6, *Liguria sive Provincia Mediolanensis,* 2, *Pedemontium, Liguria Maritima.* Berlin: apud Weidmannos, 1914.

Kennedy, E.S. "Suhrāb and the world-map of Ma'mūn," in *From Ancient Omens to Statistical Mechanics: Essays on the Exact Sciences Presented to Asger Aaboe,* ed. J. Lennart Berggren and B. R. Goldstein, 113–19. Acta Historica Scientiarum Naturalium et Medicinalium, 39. Copenhagen: University Library, 1987.

Kennedy, E.S., and M. H. Kennedy, *Geographical Coordinates of Localities from Islamic Sources.* Veröffentlichungen des Instituts für Geschichte der Arabisch-Islamischen Wissenschaften, Reihe A: Texte und Studien, 2. Frankfurt am Main: Institut für Geschichte der Arabisch-Islamischen Wissenschaften an der Johann Wolfgang Goethe-Universität, 1987.

al-Khwārazmī, *Das Kitāb sūrat al-arḍ des Abū Ǧaʿfar Muḥammad ibn Mūsā al-Huwārizm,* ed. Hans van Mžik. Bibliothek Arabischer Historiker und Geographen, 3. Leipzig: Harrassowitz, 1926.

King, David A. *World-Maps for Finding the Direction and Distance to Mecca: Innovation and Tradition in Islamic Science.* Islamic Philosophy, Theology and Science: Texts and Studies, 3. London: al-Furqān Foundation; Leiden: Brill, 1999.

Kitzinger, Ernst. "World Map and Fortune's Wheel: A Medieval Mosaic Floor in Turin," *Proceedings of the American Philosophical Society* 117, no. 5 (1973), 344-73.

Koelsch, W.A. "Squinting back at Strabo," *The Geographical Review* 94 (2004), 502-18.

Koller, D., and M. Levoy. "Computer-aided Reconstruction and New Matches in the *Forma Urbis Romae,*" in *Formae Urbis Romae: Nuovi Frammenti di Piante Marmoree dallo Scavo dei Fori Imperiali. Bullettino della Commissione Archeologica Comunale di Roma* Suppl. 15, ed. R. Meneghini and R. Santangeli Valenziani. Rome: Bretschneider, 2006: 103-25.

Koller, D., J. Trimble, T. Najbjerg, N. Gelfand and M. Levoy, "Fragments of the City: Stanford's Digital Forma Urbis Romae Project," in *Imaging Ancient Rome:*

Documentation—Visualization—Imagination, ed. L. Haselberger and J. Humphrey, 237-52. *Journal of Roman Archaeology Suppl.* 61, 2006.

Kominko, Maja. "The Map of Cosmas, The Albi Map and the Tradition of Ancient Geography," *Mediterranean Historical Review* 20 (2005), 163-86.

Kordosis, Michael. "The Limits of the Known Land (Ecumene) in the East according to Cosmas Indicopleustes. Tzinista (China) and the Ocean," *Byzantion* 69 (1999), 99-106.

Kramer, Barbel. "The earliest known map of Spain (?) and the Geography of Artemidorus of Ephesus on papyrus," *Imago Mundi* 53 (2001), 115-20.

Kretschmer, Konrad. *Die Italienischen Portolane des Mittelalters: Ein Beitrag zur Geschichte der Kartographie und Nautik.* Berlin: E. S. Mittler, 1909.

Kretschmer, Konrad. "Die mittelalterliche Weltkarte nach Anlage und Herkunft," *Petermanns Mitteilungen,* Erg. Heft 209 (Gotha, 1930), 59-64.

Kuder, Ulrich. "Die Ottonen in der ottonischen Buchmalerei: Identification und Ikonographie," in *Herrschaftsrepräsentation im ottonischen Sachsen*, ed. Gerd Althoff and Ernst Schubert, 137-218. Sigmaringen: Thorbecke, 1998.

Kupfer, Marcia. "Medieval World Maps: Embedded Images, Interpretive Frames," *Word and Image* 10 (1994), 262-88.

Kupfer, Marcia. "The Lost Wheel Map of Ambrogio Lorenzetti," *The Art Bulletin* 78-2 (1996), 286-310.

La Rocca, Eugenio. "L'affresco con veduto di città del colle Oppio," in *Romanization and the City: Creation, Transformations, and Failures*, ed. E. Fentress, 57-71. *Journal of Roman Archaeology Suppl.* 38, 2000.

La Rocca, Eugenio. "La Nuova Immagine dei Fori Imperiali," *Mitteilungen des Deutschen Archäologischen Instituts, Römische Abteilung* 108 (2001), 171-213.

Ladner, Gerhart. *L' immagine dell' imperatore Ottone III.* Rome: Unione Internazionale degli Istituti di Archeologia, Storia e Storia dell'Arte, 1988.

Lecoq, Danielle. "La mappemonde du *Liber Floridus* ou la vision du monde de Lambert de Saint-Omer," *Imago Mundi* 39 (1987), 9-49.

Lecoq, Danielle. "La Mappemonde d'Henri de Mayence, ou l'Image du Monde au XIIe Siècle," *in Iconographie Médiévale: Image, Texte, Contexte*, ed. Gaston Duchet-Suchaux, 155-207. Paris: Éditions du Centre National de la Recherche, 1990.

Legrand, Émile, ed. *Version Grecque du 'Liber Insularum Archipelagi' c. 1420 par un anonyme; Publiée d'après le manuscrit du Sérail, avec une traduction en Français; Suivie de la 'Descriptio Insule Candie' et de la 'Descriptio Cretae' et précédée d'une préface historique et critique*, Publications Ecole des Langues Orientales (4th Series). Paris: E. Leroux, 1897.

Legras, B. "L'horizon géographique de la jeunesse grecque d'Égypte (IIIe s. avant notre ère-VIe siècle de notre ère)," in *Proceedings of the 20th international Congress of papyrologists*, ed. Adam Bülow-Jacobsen, 165-76. Copenhagen: Museum Tusculanum Press, 1994.

Lehmann, P. *Die mittelalterliche Dombibliothek zu Speyer*, in *Erforschung des Mittelalters*, II. Stuttgart: Hiersemann, 1959.

Lenfant, D., ed. *Ctésias de Cnide*, La Perse; L' Inde; autres fragments. Paris: Budé, 2004.

Leonardi, C. "Nota introduttiva per un indagine sull fortuna di Marziano Capella nel Medioevo," *Bulletino dell'Istituto storico italiano per il medio evo e Archivio Muratoriano* 67 (1955), 265-88.

Leonardi, C. "I codici di Marziano Capella," *Aevum* 33 (1959), 459-62.

Levi, A. and M. *Itineraria picta. Contributo allo studio della Tabula Peutingeriana.* Rome: Bretschneider, 1967.

Levi, A. and M. "The medieval map of Rome in the Ambrosian library manuscript of Solinus (C.246 inf)," *Proceedings of the American Philosophical Society* 118 (1974), 567-94.

Lewis, Michael J.T. *Surveying Instruments of Greece and Rome.* Cambridge: Cambridge University Press, 2001.

Liber numerorum qui in sanctis scripturis occurrunt. In *Patrologia Latina* 83, ed. J.-P. Migne, 179-200. Paris: Migne, 1862.

Lieb, Hans. "Zur Herkunft der Tabula Peutingeriana," in *Die Abtei Reichenau: Neue Beiträge zur Geschichte und Kultur des Inselklosters*, ed. Helmut Maurer, 31-34. Sigmaringen: Thorbecke, 1974.

Lilienthal, [Max]. "Bibliographische Notizen über die hebräischen Manuscripte der Königlichen Bibliothek zu München," *Allgemeine Zeitung des Judenthums* 2 (1838), 49-52.

Lindgren, Uta. *Gerbert von Aurillac und das Quadrivium: Untersuchungen zur Bildung im Zeitalter der Ottonen.* Wiesbaden: Steiner, 1976.

Little, A., ed. *Arrian*, Periplus Ponti Euxini. London: Bristol Classical Press, 2003.

Lowe, E. A., ed. *Codices Latini Antiquiores: A Palaeographical Guide to Latin Manuscripts prior to the Ninth Century.* Oxford: Clarendon Press, 1956.

Lozovsky, Natalia. "Roman Geography and Ethnography in the Carolingian Empire," *Speculum* 81 (2006), 325-64.

Lugli, P.M. "Considerazioni urbanistiche sulla Pianta Marmorea del Foro della Pace," *Bollettino di Archeologia* 16-18 (1992), 19-31.

MacCormack, Sabine. "Latin Prose Panegyrics," in *Empire and Aftermath: Silver Latin II*, ed. Thomas Alan Dorey, 143-205. London: Routledge and Kegan Paul, 1975.

Macrobius. *Commentary on the Dream of Scipio.* Trans. William Harris Stahl. New York: Columbia University Press, 1952 and 1990.

Mallett, Michael Edward, and Luca degli Albizzi. *The Florentine Galleys in the Fifteenth Century.* Oxford: Clarendon Press, 1967.

Manacorda, D. *Archeologia Urbana a Roma: Il Progetto della Crypta Balbi*, 1. Florence: all' Insegna del Giglio, 1982.

Manacorda, D. et al. *Museo Nazionale Romano: Crypta Balbi.* Rome: Electa, 2000.

Mango, Cyril Alexander, ed. *The Art of the Byzantine Empire 312-1453: Sources and Documents.* Englewood Cliffs, N. J.: Prentice-Hall, 1972.

Marcotte, D., ed. *Géographes grecs* 1. Paris: Belles Lettres, 2000.

Markus, R. A. "How on earth could places become holy ? Origin of the Christian idea of holy places," *Journal of Early Christian Studies* 2 (1994), 257-71.

Marrou, H.-I. *Saint-Augustin et la fin de la culture antique.* Paris: Boccard, 1958.

Martianus Capella. *De nuptiis Philologiae et Mercurii.* Ed. J. Willis. Leipzig: Teubner, 1983. Trans. William H. Stahl, Richard Johnson, and E. L. Burge, in *Martianus Capella and the Seven Liberal Arts*, vol. 2. New York: Columbia University Press, 1977.

al-Masʿūdī, *Kitāb al-Tanbīh wa-al-ishrāf.* Ed. M. J. de Goeje. Bibliotheca Geographorum Arabicorum, 8. Leiden: Brill, 1894.

Mattern, Susan P. *Rome and the Enemy: Imperial Strategy in the Principate.* Berkeley: University of California Press, 1999.

Maurmann, Barbara. *Die Himmelsrichtungen im Weltbild des Mittelalters: Hildegard von Bingen, Honorius Augustodunensis und Andere Autoren.* Munich: Wilhelm Fink Verlag, 1976.

Mayerson, P. "A Confusion of Indias: Asian India and African India in the Byzantine Sources," *Journal of the American Oriental Society* 113 (1993), 169-74.

Mayr-Harting, Henry. *Ottonian Book Illumination: An Historical Study*, 2 vols. London and New York: Harvey Miller, 1991.

McKitterick, Rosamond. *The Carolingians and the Written Word*. Cambridge: Cambridge University Press, 1989.

McVey, Kathleen. "The Domed Church as Microcosm: Literary Roots of an Architectural Symbol," *Dumbarton Oaks Papers* 37 (1983), 91-121.

Meiss, Millard. *French Painting in the Time of Jean of Berry. The Limbourg Brothers and Their Contemporaries*. New York: George Braziller, 1974.

Meneghini, R., and R. Santangeli Valenziani, eds. *Formae Urbis Romae: Nuovi Frammenti di Piante Marmoree dallo Scavo dei Fori Imperiali. Bullettino della Commissione Archeologica Comunale di Roma Suppl.* 15. Rome: Bretschneider, 2006.

Merrills, Andrew H. *History and Geography in Late Antiquity*. Cambridge: Cambridge University Press, 2005.

Metzger, Thérèse. "Le manuscrit illuminé Cod. Hebr. 5 de la Bibliothèque d'Etat à Munich," in *Études de civilisation médiévale (IXe-XIIe siècles). Mélanges offerts à Edmond-René Labande*, 537-52. Poitiers: CESCM, 1974.

Miller, Konrad. *Die Weltkarte des Castorius genannt die Peutingersche Tafel*. Ravensburg: Maier, 1887/1888.

Miller, Konrad. *Mappaemundi: Die Ältesten Weltkarten*. 6 vols. Stuttgart: Roth, 1895-1898.

Minguzzi, Simonetta. *I Mosaici Pavimentali della Cattedrale di Novara dal Tardoantico al Medioevo*. Ravenna: Edizioni del Girasole, 1995.

Moralejo, Serafín. "Las Islas del Sol. Sobre el Mapamundi del Beato del Burgo de Osma (1086)," in *A Imaagem do Mundo na Idade Média. Actas do Colóquio Internacional, 1989*. Lisbon: Ministério da Educaçao, Instituto de Cultura e Língua Portuguesa, 1992.

Mortet, Victor, ed. *Recueil de textes relatifs à l'histoire de l'architecture et à la condition des architectes en France au moyen age, XIe—XIIe siècles*. 2 vols. Paris: Alphonse Picard, 1911.

Muhammad Ibn Hauqal. *Configuration de la Terre (Kitab surat al-ard)*, trans. J.H. Kramers and G. Wiet, 2 vols. Beirut and Paris: Aubenas, 1964.

Mütherich, Florentine. "Das Kaiserbild," in *Das Evangeliar Ottos III.: Clm 4453 Bayerischen Staatsbibliothek München*, ed. Florentine Mütherich and Karl Dachs, 31-35. Munich: Prestel, 2001.

Murphy, Trevor M. *Pliny the Elder's* Natural History*: The Empire in the Encyclopedia*. Oxford: Oxford University Press, 2004.

Mynors, Roger A.B., ed. *XII Panegyrici Latini*. Oxford: Clarendon Press, 1964.

Najbjerg, Tina, and Jennifer Trimble. "Ancient Maps and Mapping in and around Rome," *Journal of Roman Archaeology* 17 (2004), 577-83.

Najbjerg, Tina, and Jennifer Trimble. "The Severan Marble Plan since 1960," in *Formae Urbis Romae: Nuovi Frammenti di Piante Marmoree dallo Scavo dei Fori Imperiali*, ed. R. Meneghini and R. Santangeli Valenzani, 75-101. *Bullettino della Commissione Archeologica Comunale di Roma Suppl.* 15, 2006.

Nallino, Carlo Alfonso. "Al-Ḫuwârismî e il suo rifacimento della Geografia di Tolomeo," in *Raccolta di scritti editi e inediti*, ed. Maria Nallino, 6 vols, 5:458-532. Rome: Istituto per l' Oriente, 1939-1948.

Narkiss, Bezalel. "Rashi's Maps," in *Zev Vilnay's Jubilee Volume*, ed. Ely Schiller, 435-49. Jerusalem: Ariel, 1984.

Nelson, Janet L. "The Frankish Kingdoms, 814-898: the West," in *New Cambridge Medieval History*, vol. 2, c. 700-c. 900, ed. Rosamond McKitterick, 110-41. Cambridge: Cambridge University Press, 1995.

Nicolet, Claude. "Les quatre sages de Jules César et la 'mesure du monde' selon Julius Honorius: réalité antique et tradition médiévale. I. La réalité antique," *Journal des savants* (1987), 157-83.

Nicolet, Claude. *L'inventaire du monde. Géographie et politique aux origines de l'Empire romain.* Paris: Fayard, 1988.

Nicolet, Claude. *Space, Geography, and Politics in the Early Roman Empire.* Ann Arbor: University of Michigan Press, 1991.

Nilgen, Ursula. "Blonde Roma? Zum Sinn des Blondhaars in der Buchmalerei der Reichenau," *Zeitschrift für Kunstgeschichte* 66 (2003), 19-32.

Nimmo Smith, J. "The River Alpheus in Greek Christian and Byzantine Thought," *Byzantion* 74 (2004), 416-32.

Nixon, C. Edward V. and Barbara S. Rodgers. *In Praise of Later Roman Emperors: The Panegyrici Latini.* Berkeley: University of California Press, 1995.

Norbye, Marigold Anne. "The King's blood: royal genealogies, dynastic rivalries and historical culture in the Hundred Years War. A case study of *A tous nobles qui aiment beaux faits et bonnes histories.*" PhD diss., University College London, 2004.

Nordenskiöld, A. E. "Dei disegni marginali negli antichi manoscritti della *Sfera* del Dati," *Bibliofilia* 3 (1901), 49-55.

Ordo Ambrosianus ad Consecrandam Ecclesiam et Altaria, in *Antiche Reliquie Liturgiche Ambrosiane e Romane, con un excursus sui frammenti dogmatici Ariani del Mai*, Studi e Testi 7, ed. Giovanni Mercati, 1-32. Rome: Tipografia Vaticana, 1902.

O'Reilly, Jennifer. "Patristic and insular traditions of the evangelists: exegesis and iconography," in *Le Isole Britanniche e Roma in Età Romanobarbarica*, ed. A. M. Luiselli Fadda and É. Ó. Carragáin, 49-94. Rome: Herder, 1998.

O'Reilly, Jennifer. "Islands and idols at the ends of the earth: exegesis and conversion in Bede's *Historia Ecclesiastica,*" in *Bède le Vénérable entre tradition et postérité / The Venerable Bede. Tradition and posterity*, ed. Stéphane Lebecq, Michel Perrin and Olivier Szerwiniack, 119-45. Université Charles-de-Gaulle, Lille 3, 2005.

Orosius, Paulus. *Seven Books of History Against the Pagans.* Trans. Roy J. Defferari. Washington, D. C.: Catholic University Press, 1964.

Ott, Joachim. "*regi nostro se subdit Roma benigno*—Die Stiftung des Perikopenbuche Heinrichs II. (Clm 4452) für den Bamberger Dom vor dem Hintergrund der bevorstehenden Kaiserkrönung," *Jahrbuch für fränkische Landesforschung* 54 (1994), 347-70.

Otto of Freising. *Gesta Friderici I. Imperatoris.* Ed. B. von Simson and G. Waitz. Ed. 3. *Scriptores Rerum Germanicarum ex Monumentis Germaniae historicis recusi* 46. Hannover, Leipzig: impensis bibliopolii Hahniani, 1912.

Patetta, Federico. "A proposito del mosaico medioevale scoperto a Torino nel Marzo del 1909," *Atti della Società Piemontese di Archeologia e Belle Arti* 8 (1910/17), 318-40.

Pedroni, L. "Per una lettura verticale della Forma Urbis Marmorea," *Ostraka* 1.2, (1992), 223-30.

Peeters, Paul. "Observations sur la vie syriaque de Mar Aba, catholicos de l'église Perse (540-552)," *Recherches d'Histoire et de Philologie Orientales, Subsidia Hagiographica* 27 (1951), 109-61.

Pejrani Baricco, Luisella. "La Basilica del Salvatore e la Cattedrale di Torino: Considerazioni su uno scavo in corso," in *Archeologia in Piemonte.* Vol. 3, *Il Medioevo*, ed. L. Mercando and E. Micheletto, 132-49. Turin: Allemandi, 1998.

Pemberton, Steven et al., *XHTML™ 1.0 The Extensible HyperText Markup Language (Second Edition): A Reformulation of HTML 4 in XML 1.0*, W3C Recommendation 26 January 2000, revised August 1, 2002. http://www.w3.org/TR/xhtml1.

Perinetti, Renato. "I mosaici medievali di Aosta," in *Atti del VI Colloquio dell'Associazione Italiana per lo Studio e la Conservazione del Mosaico, Venezia, 20-23 gennaio 1999*, ed. Federico Guidobaldi and Andrea Paribeni, 161-74. Ravenna: Edizioni del Girasole, 2000.

Pertz, K. *De Cosmographia Aethici libri tres.* Berlin, 1853.

Petrarca, Francesco, T. J. Cachey, and Biblioteca statale di Cremona. *Petrarch's Guide to the Holy Land : Itinerary to the Sepulcher of Our Lord Jesus Christ = Itinerarium Ad Sepulchrum Domini Nostri Yehsu Christi: Facsimile Edition of Cremona, Biblioteca Statale, Deposito Libreria Civica, Manuscript Bb.1.2.5.* Notre Dame: University of Notre Dame Press, 2002.

Pianea, Elena. "I mosaici pavimentali," in *Piemonte Romanico*, ed. Giovanni Romano, 393-420. Turin: Cassa di Risparmio di Torino, 1994.

Piccirillo, M., and E. Alliata, eds. *The Madaba Map Centenary 1897-1997. Travelling through the Byzantine Umayyad Period.* Jerusalem: Studium Biblicum Franciscanum, 1998.

Pinto, Karen C. "Ways of Seeing. 3: Scenarios of the World in the Medieval Islamic Cartographic Imagination." PhD diss., Columbia University, 2002.

Pontificale Romano-Germanicum. Ed. Cyrille Vogel and Reinhard Elze. *Le Pontifical Romano-Germanique du Dixième Siècle.* 3 vols. Studi e Testi 226-27, 269. Rome: Biblioteca Apostolica Vaticana, 1963-1972.

Praepositinus of Cremona. *Tractatus de officiis.* Ed. James A. Corbett. *Publications in Mediaeval Studies 21.* Notre Dame: University of Notre Dame Press, 1969.

Prontera, Francesco, ed. *Tabula Peutingeriana: Le antiche vie del mondo.* Florence: Olschki, 2003.

Pseudo-Aethicus, *Cosmographia.* In *Geographi Latini Minores*, ed. A. Riese, 71-103. Heilbronn: apud Henningeros fratres, 1878.

Pseudo-Alcuin, *De Divinis Officiis Liber.* In *Patrologia Latina* 101, ed. J.-P. Migne, 1173-286. Paris: Migne, 1863.

Ptolemy. *Cosmographia.* Trans. Simon Knight. Wigston: Magna Books, 1990.

Quintavalle, Arturo Carlo. *Wiligelmo e Matilde: L' officina romanica.* Milan: Electa, 1991.

Rapoport, Yossef, and Emilie Savage-Smith. "Medieval Islamic View of the Cosmos: The Newly Discovered *Book of Curiosities*," *The Cartographic Journal* 41 (2004), 253-59.

Ratkowitsch, Christine. *Descriptio picturae. Die literarische Funktion der Beschreibung von Kunstwerken in der lateinischen Großdichtung des 12. Jahrhunderts.* Vienna: Verlag der Österreichischen Akademie der Wissenschaften, 1991.

Rees, Roger. *Layers of Loyalty in Latin Panegyric*, AD 289-307. Oxford: Oxford University Press, 2002.

Remak-Honnef, Elisabeth, and Hermann Hauke. *Katalog der lateinischen Handschriften der Bayerischen Staatsbibliothek München, Clm 10001-10930.* Wiesbaden: Harrossowitz, 1991.

Repsher, Brian. *The Rite of Church Dedication in the Early Medieval Era.* Lewiston: Edwin Mellen Press, 1998.

Revelli, P. "Figurazione cartografiche dell'età imperiale in un codice ambrosiano di Solino," in *Raccolta di studi in onore di F. Ramorino*, 615-26. Milan: Vita e Pensiero, 1927.

Reynaud, Nicole. *Jean Fouquet.* Paris: Réunion des Musées Nationaux, 1981.

Reynolds, David West. *Forma Urbis Romae: The Severan Marble Plan and the Urban Form of Ancient Rome.* PhD diss., University of Michigan, 1996.

Reynolds, L. D., ed. *Texts and transmission. A survey of Latin classics.* Oxford: Clarendon Press, 1983.

Richardson, L. "The Approach to the Temple of Saturn," *American Journal of Archaeology* 84 (1980), 51-62.

Riese, Alexander, ed. *Geographi Latini Minores.* Heilbronn: apud Henningeros fratres, 1878.

Robert le Moine, *Historiens des Croisades,* III, *Historiens occidentaux.* Paris, 1866.

Roberts, C.H., and E.G. Turner, eds. *Catalogue of the Greek and Latin Papyri in the John Rylands Library Manchester* IV. Manchester: Manchester University Press, 1952.

Rodríguez-Almeida, E. *Forma Urbis Marmorea. Aggiornamento Generale 1980.* Rome: Quasar, 1981.

Rodríguez-Almeida, E. "Un Frammento di Una Nuova Pianta Marmorea di Roma," *Journal of Roman Archaeology* 1 (1988), 120-31.

Rodríguez-Almeida, E. "Euristica Materiale e Forma Marmorea. Alcuni Falsi Problemi," *Atti della Pontificia Accademia Romana di Archeologia: Rendiconti* 68 (1995-96), 3-20.

Rodríguez-Almeida, E. *Formae Urbis Antiquae: Le Mappe Marmoree di Roma tra la Repubblica e Settimio Severo.* Collection de l'Ecole Française de Rome 305, 2002.

Rom als Idee. Ed. Bernhard Kytzler. Darmstadt: Wissenschaftliche Buchgesellschaft, 1993.

Romm, James S. *The Edges of the Earth in Ancient Thought: Geography, Exploration and Fiction.* Princeton: Princeton University Press, 1992.

Salway, R.W. Benet. "Travel, *itineraria* and *tabellaria,*" in *Travel and Geography in the Roman Empire,* ed. C. Adams and R. Laurence, 22-66. London and New York: Routledge, 2001.

Salway, R.W. Benet. "Sea and river travel in the Roman itinerary literature," in *Space in the Roman World: its Perception and Presentation,* ed. R. Talbert and K. Brodersen, 43-96. Münster: LIT, 2004.

Salway, R. W. Benet. "The Nature and Genesis of the Peutinger Map," *Imago Mundi* 57 (2005), 119-35.

Santangeli Valenzani, R.. "Distruzione e Dispersione della *Forma Urbis* Severiana alla Luce dei Dati Archeologici," in *Formae Urbis Romae: Nuovi Frammenti di Piante Marmoree dallo Scavo dei Fori Imperiali,* ed. R. Meneghini and R. Santangeli Valenziani, 53-59. *Bullettino della Commissione Archeologica Comunale di Roma Suppl.* 15. Rome: Bretschneider, 2006.

Scafi, Alessandro. "Mapping Eden: Cartographies of the Earthly Paradise," in *Mappings,* ed. Denis Cosgrove, 50-70. London: Reaktion Books, 1999.

Scafi, Alessandro. *Mapping Paradise. A History of Heaven on Earth.* Chicago: University of Chicago Press, 2006.

Scetti, Aurelio, and Luigi Monga. *The Journal of Aurelio Scetti: A Florentine Galley Slave at Lepanto (1565-1577).* Tempe: Arizona Center for Medieval and Renaissance Studies, 2004.

Schnabel, P. "Die Weltkarte des Agrippa als wissenschaftliche Mittelglied zwischen Hipparch und Ptolemaeus," *Philologus* 90 (1935), 405-40.

Schneider, Fedor. *Rom und Romgedanke im Mittelalter: die geistige Grundlagen der Renaissance.* Munich: Drei Masken Verlag, 1926.

Schnetz, J., ed. *Ravennatis Anonymi Cosmographia et Guidonis Geographica.* Leipzig: B. G. Teubner, 1940; reissued with index by M. Zumschlinge, Stuttgart: Teubner, 1990.

Schramm, Percy Ernst. "Das Herrscherbild in der Kunst des frühen Mittelalters," *Vorträge der Bibliothek Warburg* 2/1 (1922-23), 145-224.

Schramm, Percy Ernst. *Kaiser, Rom und Renovatio: Studien und Texte zur Geschichte des römischen Erneuerungsgedankens vom Ende des karolingischen Reiches bis zum Investiturstreit.* Leipzig: Teubner, 1929.

Schramm, Percy Ernst. *Sphaira, Globus, Reichsapfel.* Stuttgart: Hiersemann, 1958.

Schramm, Percy Ernst. *Die deutschen Kaiser und Könige in Bildern ihrer Zeit 751-1190.* Ed. Florentine Mütherich. Munich: Prestel, 1983.

E. Schweder, "Über eine Weltkarte des achten Jahrhunderts," *Hermes* 24 (1889), 586-604.

Scott, James M. *Geography in Early Judaism and Christianity:* The Book of Jubilees. Cambridge: Cambridge University Press, 2002.

Sears, Elizabeth. *The Ages of Man: Medieval Interpretations of the Life Cycle.* Princeton: Princeton University Press, 1986.

Sed-Rajna, Gabrielle. "Some Further Data on Rashi's Diagrams to his Commentary on the Bible." *Jewish Studies Quarterly* 1 (1993-94), 149-53.

Sed-Rajna, Gabrielle. *Les manuscrits hébreux enluminés des bibliothèques de France.* Leuven and Paris: Peeters, 1994.

Segatto, Filiberto. "Un'immagine quattrocentesca del mondo: la *Sfera* del Dati," *Memorie, Accademia nazionale dei Lincei: Classe di scienze morali, storiche e filologiche,* 8th series, 27 (1983), 169-70.

Serchuk, Camille. "Picturing France in the Fifteenth Century: The Map in BnF MS Fr 4991," *Imago Mundi* 58 (2006), 133-49.

Seyfarth, W. *Der Codex Fuldensis und der Codex E des Ammianus Marcellinus.* Berlin: Akademie-Verlag, 1962.

Sezgin, Fuat. *Mathematische Geographie und Kartographie im Islam und ihr Fortleben im Abendland. Historische Darstellung,* Teil I and Kartenband. Geschichte des arabischen Schrifttums, 10 and 12. Frankfurt am Main: Institut für Geschichte der Arabisch-Islamischen Wissenschaften an der Johann Wolfgang Goethe-Universität, 2000.

Sicard, P. *Diagrammes médiévaux et exégèse visuelle. Le "Libellus de formatione arche" de Hugues de Saint-Victor.* Paris and Turnhout: Brepols, 1993.

Silberman, A., ed. *Pomponius Mela,* Chorographie. Paris: Belles Lettres, 1988.

Silberman, A., ed. *Arrien,* Périple du Pont-Euxin. Paris: Belles Lettres, 1995.

Simar, Théophile. "La géographie de l'Afrique Centrale dans l'antiquité et au Moyen-Age," *Revue Congolaise* 3 (1912-13), 1-23, 81-102, 145-69, 225-52, 289-310, 440-41.

Smith, R. R. R. "*Simulacra Gentium*: The *Ethne* from the Sebasteion at Aphrodisias," *Journal of Roman Studies* 78 (1988), 50-77.

Sorabji, R. "John Philoponus," in *Philoponus and the Rejection of Aristotelian Science,* ed. R. Sorabji, 5-18. London: Duckworth, 1987.

Southern, Richard. *Medieval Humanism and Other Studies.* Oxford: Basil Blackwell, 1970.

Stanford Digital Forma Urbis Romae Project. http://formaurbis.stanford.edu.

Stebbins, F.A. "A Roman sun-dial," *Journal of the Royal Astronomical Society of Canada* 52 (1958), 250-54.

Steinby, Eva Margareta. "Il Frammento 18a della *Forma Urbis Romae,*" in *Lacus Iuturnae* I. Rome: De Luca, 1989, 24-33.

Steinschneider, Moritz. *Die hebräischen Handschriften der K. Hof- und Staatsbibliothek in München.* Munich, 1875. Ed. 2, Munich: Palm in Comm., 1895.

Stevens, Wesley. "The Figure of the Earth in Isidore's 'De Natura Rerum'," *Isis* 71 (June 1980), 268-77.

Strayer, J. R. "France: The Holy Land, the Chosen People, and the Most Christian King," in *Medieval Statecraft and the Perspectives of History*, 300-14. Princeton: Princeton University Press, 1971.

Struve, Tilman. "Kaisertum und Romgedanke in salischer Zeit," *Deutsches Archiv für Erforschung des Mittelalters* 44 (1988), 424-54.

Suhrāb, *Das Kitāb ʿağāʾib al-akālim as-sabʿa*. Ed. Hans von Mžik. Bibliothek arabischer Historiker und Geographen, 5. Leipzig: Harrassowitz, 1930.

Sullivan, Richard E. "What Was Carolingian Monasticism ? The Plan of St. Gall and the History of Monasticism," in *After Rome's Fall: Narrators and Sources of Early Medieval History. Essays presented to Walter Goffart*, ed. Alexander Callander Murray, 251-87. Toronto: University of Toronto Press, 1988.

Talbert, Richard J.A. Review of O.A.W. Dilke, *Greek and Roman Maps, Journal of Roman Studies* 77 (1987), 210-12.

Talbert, Richard J.A., ed. *Barrington Atlas of the Greek and Roman World*. Princeton and Oxford: Princeton University Press, 2000.

Talbert, Richard J.A. "Cartography and taste in Peutinger's Roman map," in *Space in the Roman World: its Perception and Presentation*, ed. R. Talbert and K. Brodersen, 113-41. Münster: LIT, 2004.

Talbert, Richard J.A. "Rome's provinces as framework for world-view," in *Roman Rule and Civic Life: Local and Regional Perspectives*, ed. L. De Ligt et al., 21-37. Amsterdam: Gieben, 2004.

Talbert, Richard J.A. "Rome's Marble Plan and Peutinger's Map: continuity in cartographic design," *Althistorisch-Epigraphische Studien* (Österreichischen Gesellschaft für Archäologie) 5 (2005), 627-34.

Talbert, Richard J.A. "Author, audience and the Roman empire in the *Antonine Itinerary*," in *Herrschen und Verwalten: Der Alltag der römischen Administration in der Hohen Kaiserzeit*, ed. R. Haensch and J. Heinrichs, 256-70. Köln, Weimar, Wien: Böhlau, 2007.

Thomson, J. O. *History of Ancient Geography*. Cambridge: Cambridge University Press, 1948.

Thorndike, Lynn, Joannes de Sacro Bosco, and Anglicus Robertus. *The Sphere of Sacrobosco and Its Commentators*. Chicago: University of Chicago Press, 1949.

Tibbetts, Gerald R. *A Study of the Arabic Texts Containing Material on South-East Asia*. Leiden: Brill and London: Royal Asiatic Society, 1979.

Toesca, Pietro. "Vicende di un' antica chiesa di Torino: Scavi e scoperte," *Bollettino d'Arte* 4 (1910), 1-16.

Tolias, George. *The Greek Portolan Charts: 15th-17th Centuries. A Contribution to the Mediterranean Cartography of the Modern Period*. Trans. Geoffrey Cox and John Solman. Athens: Olkos, 1999.

Trimble, Jennifer. "Rome as Souvenir: the Septizodium and the Severan Marble Plan," in *Common Ground: Archaeology, Art, Science and Humanities. Proceedings of the XVI International Congress of Classical Archaeology, Boston, August 23-26, 2003*, ed. C. Mattusch, A.A. Donohue, and A. Brauer, 106-109. Oxford: Oxbow Books, 2006.

Tsafrir, Y. "The maps used by Theodosius: On the Pilgrim Maps of the Holy Land and Jerusalem in the Sixth Century," *Dumbarton Oaks Papers* 40 (1986), 129-45.

Tucci, P.L. "Dov'erano il Tempio di Nettuno e la Nave di Enea," *Bullettino della Commissione Archeologica Comunale di Roma* 98 (1997), 15-42.

Tucci, P.L. "Eight Fragments of the Marble Plan of Rome Shedding New Light on the Transtiberim," *Papers of the British School at Rome* 72 (2004), 185-202.

Tucci, P.L. "L'Arx Capitolina: tra Mito e Realtà," in *Imaging Ancient Rome: Documentation—Visualization—Imagination*, ed. L. Haselberger and J. Humphrey, 63-73. *Journal of Roman Archaeology Suppl. 61*, 2006.
Uhden, Richard. "Zur Herkunft und Systematik der mittelalterlichen Weltkarten," *Geographische Zeitschrift* 37 (1931), 321-40.
Uhden, Richard. "Die Weltkarte des Isidorus von Sevilla," *Mnemosyne, Bibliotheca classica Batava* 3 (1936), 1-28.
Uhden, Richard. "Die Weltkarte des Martianus Capella," *Mnemosyne. Bibliotheca classica Batava* 3 (1936), 97-124.
Unger, Richard. *The Ship in the Medieval Economy 600-1600*. Toronto: McGill-Queen's University Press, 1980.
Vasari, Giorgio. *The Lives of the Artists*. Trans. Julia Conaway Bondanella and Peter Bondanella. Oxford: Oxford University Press, 1991.
Vidier, Alexandre. "La mappemonde de Théodulphe et la mappemonde de Ripoll (IXe-XIe siècles)," *Bulletin de géographie historique et descriptive* 26 (1911), 285-313.
Vilnay, Zev. *The Hebrew Maps of the Holy Land: A Research in Hebrew Cartography*. Ed. 1, Jerusalem, 1944; ed. 2, Jerusalem, 1968 [in Hebrew].
Wajntraub, Eva and Gimpel. *Hebrew Maps of the Holy Land.* Vienna: Hollinek, 1992.
Wallis, Faith E. "MS Oxford, St John's 17: a Medieval Manuscript in its Context." Ph.D. diss., University of Toronto, 1985.
Weber, Ekkehard. *Tabula Peutingeriana: Codex Vindobonensis 324, Kommentar*. Graz: Akademische Druck- und Verlagsanstalt, 1976.
Weber, Ekkehard. "Zur Datierung der Tabula Peutingeriana," in *Labor omnibus unus: Gerold Walser zum 70. Geburtstag*, ed. Heinz E. Herzig and Regula Frei-Stolba, 113-17. Stuttgart: Steiner-Verl.-Wiesbaden-GmbH, 1989.
Weizsäcker, W. "Imperator und huldigende Frauen," in *Festschrift für Karl Gottfried Hugelmann zum 80. Geburtstag am 26. September 1959 dargebracht von Freunden, Kollegen und Schülern*, vol. 2, 815-31. Aalen: Scientia Verlag, 1959.
Werckmeister, O. K. "The Political Ideology of the Bayeux Tapestry," *Studi medievali*, 3rd ser. 17.2 (1976), 535-95.
Westrem, Scott D. *The Hereford Map: A Transcription and Translation of the Legends with Commentary*. Turnhout: Brepols, 2001.
Whittaker, C. R. *Frontiers of the Roman Empire. A Social and Economic Study*. Baltimore: John Hopkins University Press, 1994.
Whittaker, C. R. "Itinerari romani e spazio in età antica," in *Itinerari medievali e identità europea, Atti del Congresso internazionale Parma, 27-28 febbraio 1998,* ed. R. Greci. Bologna: CLUEB, 1999.
Whittaker, C. R. "Mental maps: seeing like a Roman," in *Thinking Like a Lawyer: Essays on Legal History and General History for John Crook on his Eightieth Birthday*, ed. Paul McKechnie, 81-112. Leiden: Brill, 2002.
Whittaker, C. R. *Rome and its Frontiers: the Dynamics of Empire*. London and New York: Routlege, 2004.
Williams, John, ed. *The Illustrated Beatus: A Corpus of the Illustrations of the Commentary on the Apocalypse*. 5 vols. London: Harvey Miller Publishers, 1994-2003.
Williams, John. "Isidore, Orosius, and the Beatus Map," *Imago Mundi* 49 (1997), 7-32.
Winstedt, Eric O. *The Christian Topography of Cosmas Indicopleustes*. Cambridge: Cambridge University Press, 1909.
Wolska-Conus, Wanda. *La Topographie Chrétienne de Cosmas Indicopleustes. Théologie et Science au VIe siècle*. Bibliothèque Byzantine, ed. Paul Lemerle. Paris: PUF, 1962.

Wolska-Conus, Wanda. *Cosmas Indicoplèustes. Topographie Chrétienne, Sources Chrétiennes.* Paris: Éditions du Cerf, 1968, 1970, 1973.

Wolska-Conus, Wanda. "Deux contributions à l'histoire de la géographie. II. La carte de Theodose II: sa destination ?," *Centre de recherche d'histoire et civilization de Byzance. Travaux et memoirs* 5 (1973), 274-79.

Wood, Denis. "Pleasure in the Idea/the Atlas as Narrative Form," *Cartographica* 14, no. 1 (1987), 24-45.

Woodward, David. "Reality, Symbolism, Time and Space in the Medieval World Map," *Annals of the Association of American Geographers* 75 (1985), 510-21.

Wright, John Kirtland. *The Geographical Lore of the Time of the Crusades: A Study in the History of Medieval Science and Tradition in Western Europe.* Corr. ed. New York: Dover Publications, 1965.

Xu-Shan, Zhang. "The Name of China and its Geography in Cosmas Indicopleustes," *Byzantion* 74, fasc. 2 (2004), 452-62.

Zale, Sanford C. "Unofficial Histories of France in the late Middle Ages." PhD diss., Ohio State University, 1994.

Zuchetti, G. *Codice topografico della città di Roma*, vol. 3. Rome: Tipografia del Senato, 1946.

INDEX

COLOUR PLATES

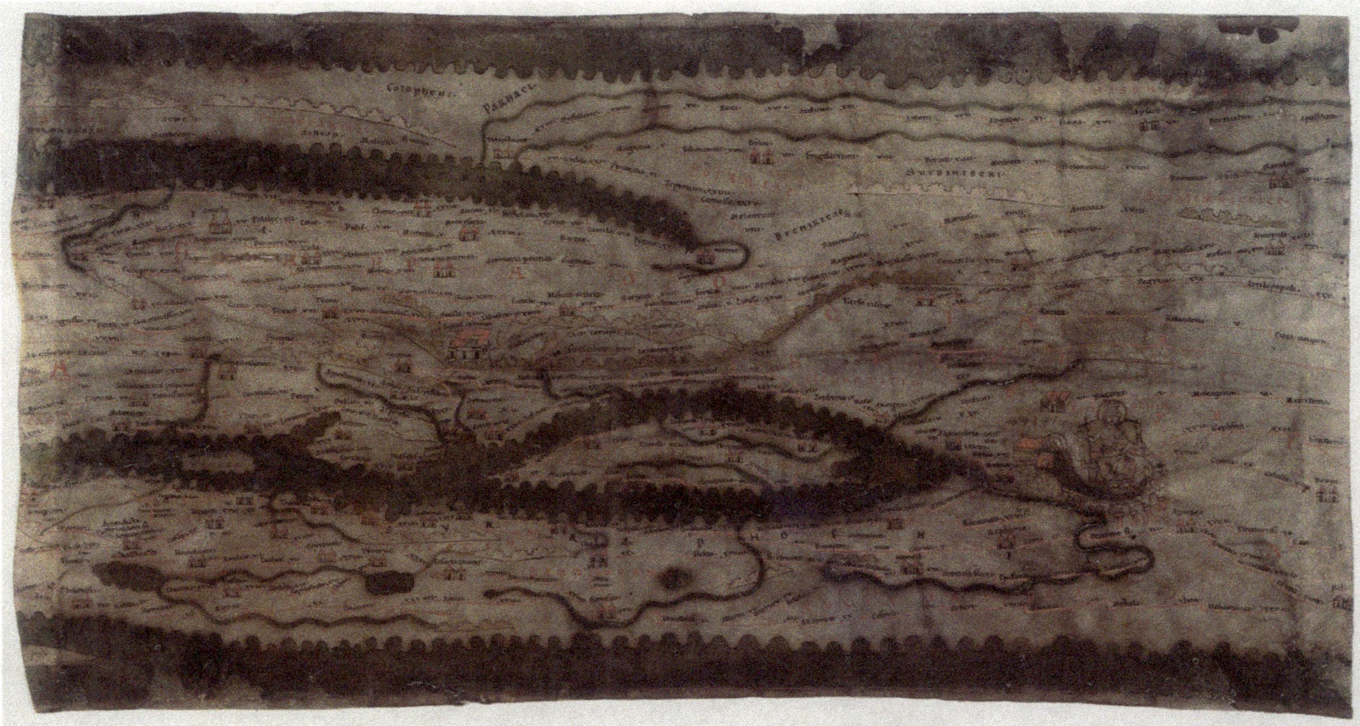

Plate I. The Peutinger Map. Detail, with Cappadocia stretching along its centre. Vienna. Österreichische Nationalbibliothek, Codex Vindobonensis 324, segment 9 (of 11). Reproduced with permission from the Austrian National Library Vienna, Picture Archive.

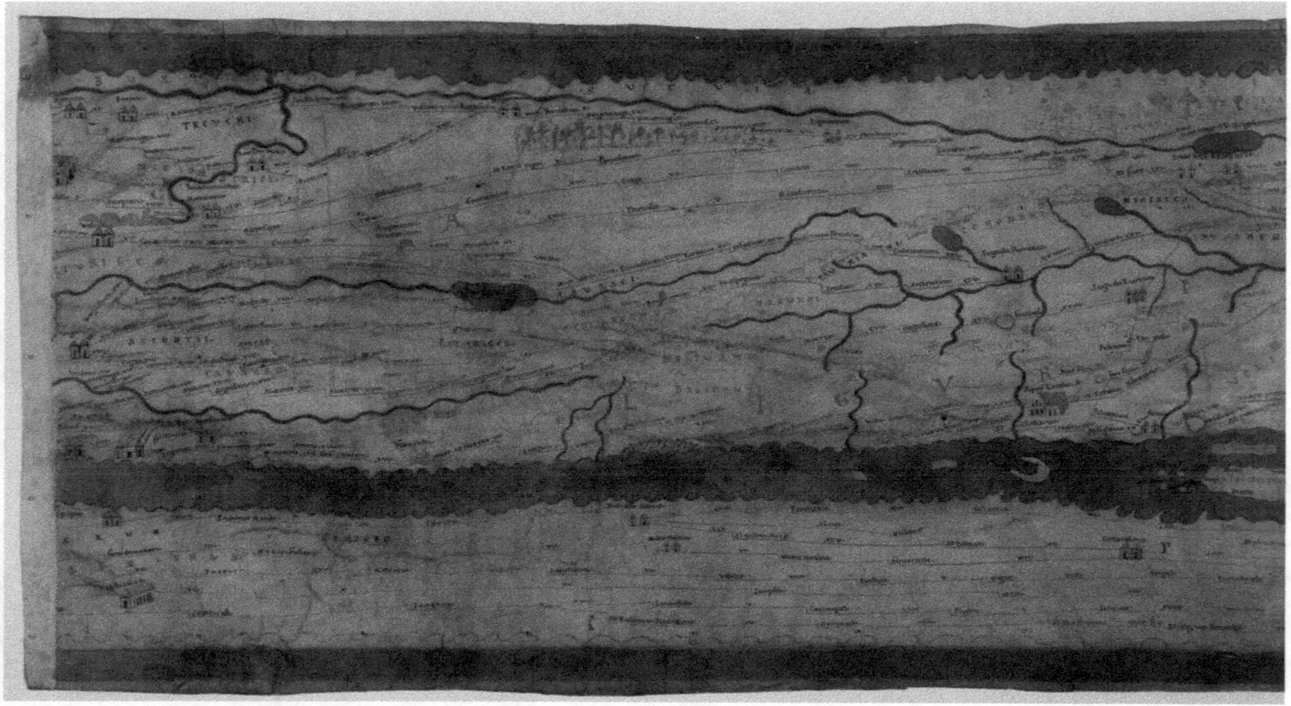

Plate II (over two pages). The Peutinger Map. Detail showing the Black Forest (*silua Marciana*) and, beneath it, the Rhine flowing into Lake Constance. Vienna. Österreichische Nationalbibliothek, Codex Vindobonensis 324, segments 2 and 3. Reproduced with permission from the Austrian National Library Vienna, Picture Archive.

Plate IIb.

Plate III. The Plan of St Gall. Cod. 1092. Reproduced with permission from Stiftsbibliothek St. Gallen.

Plate IV. The rectangular world map from the anonymous *Books of Curiosities* compiled about 1020-1050. Oxford, Bodleian Library, MS arab. C. 90, fols. 23b-24a. Undated, c1200 (?). Reproduced with permission from the Bodleian Library.

Plate V. World map by the 'Balkhī school', from a Persian translation of *Kitāb al-Masālik wa-al-mamālik* (Book of Routes and Provinces) by al-Iṣṭakhrī (d. *c*951). Oxford, Bodleian Library, MS Ouseley 373, fols. 3b-4a. Copied 1297 (696 H). Reproduced with permission from the Bodleian Library.

Plate VI. Map of the River Nile from the anonymous *Book of Curiosities* compiled about 1020-1050. Oxford, Bodleian Library, MS arab. c. 90, fol. 42a. Undated, *c* 1200 (?). Reproduced with permission from the Bodleian Library.

Plate VII. *Gospels of Otto III*, Reichenau, late tenth century. Munich, Bayerische Staatsbibliothek, Clm. 4453, Reichenau, c. 1000, fol. 23v-24r.

Plate VIIb.

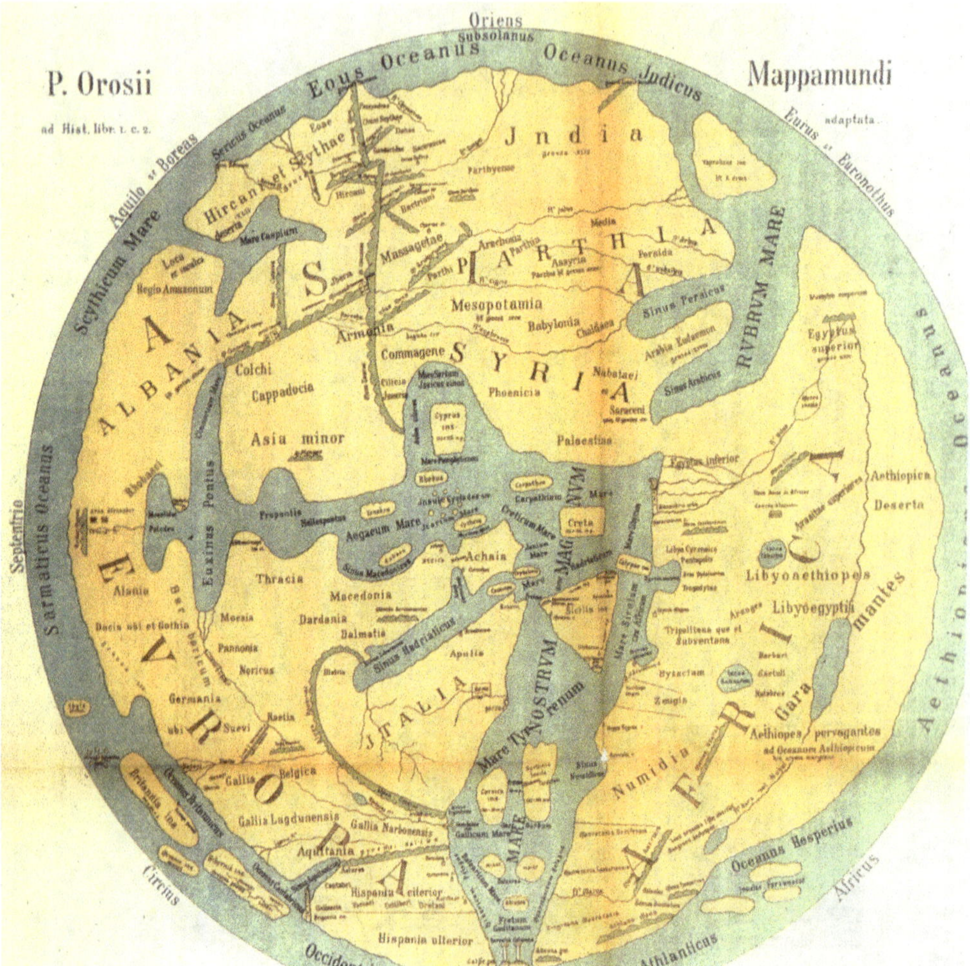

Plate VIII. Konrad Miller, *Mappaemundi: Die Ältesten Weltkarten* (Stuttgart: Roth, 1895-1898), 6, Taf. 3 (Orosius).

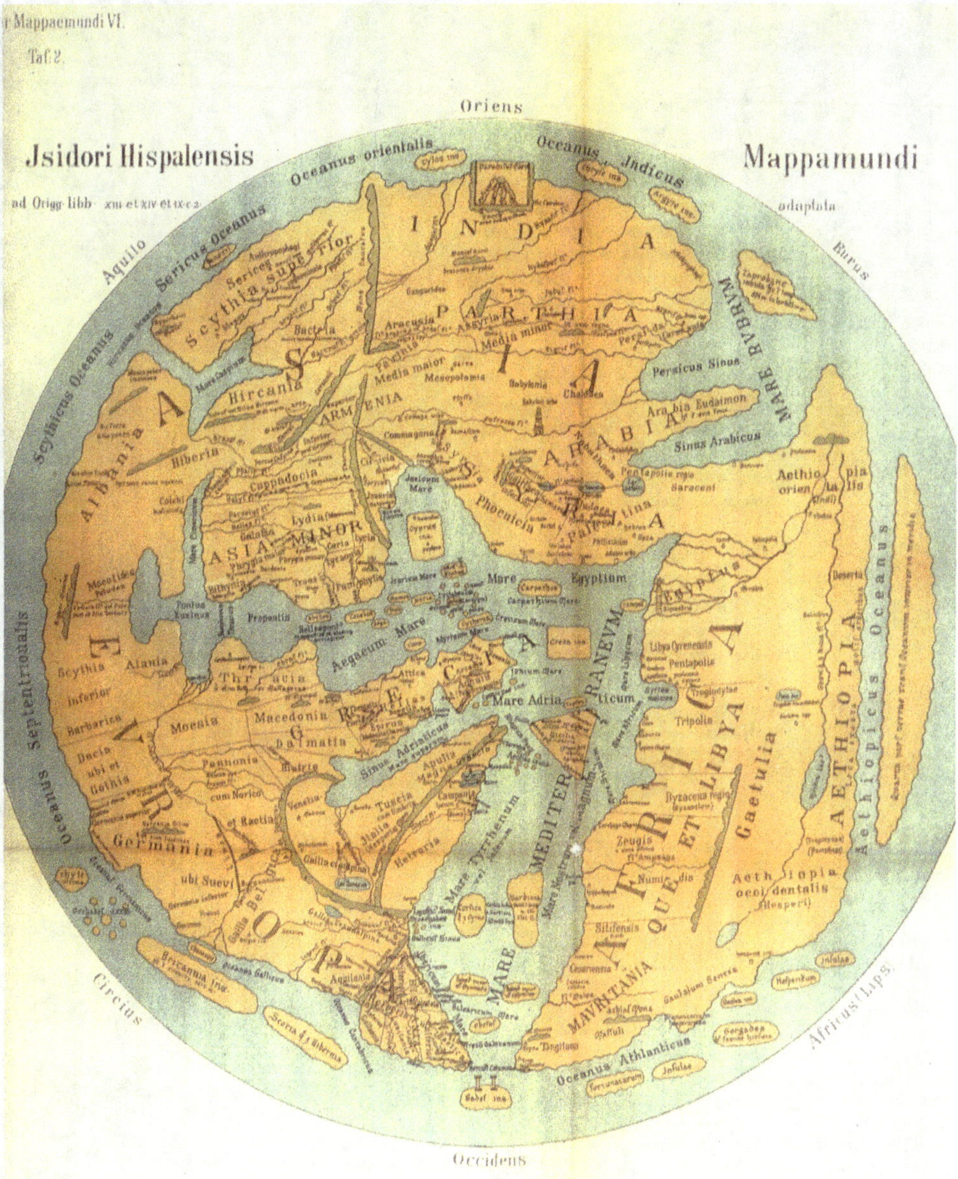

Plate IX. Konrad Miller, *Mappaemundi*, 6, Taf. 2 (Isidore).

Plate X. Gregorio Dati, *Sfera*. Newberry Ayer MS Map 1, f. 19r.

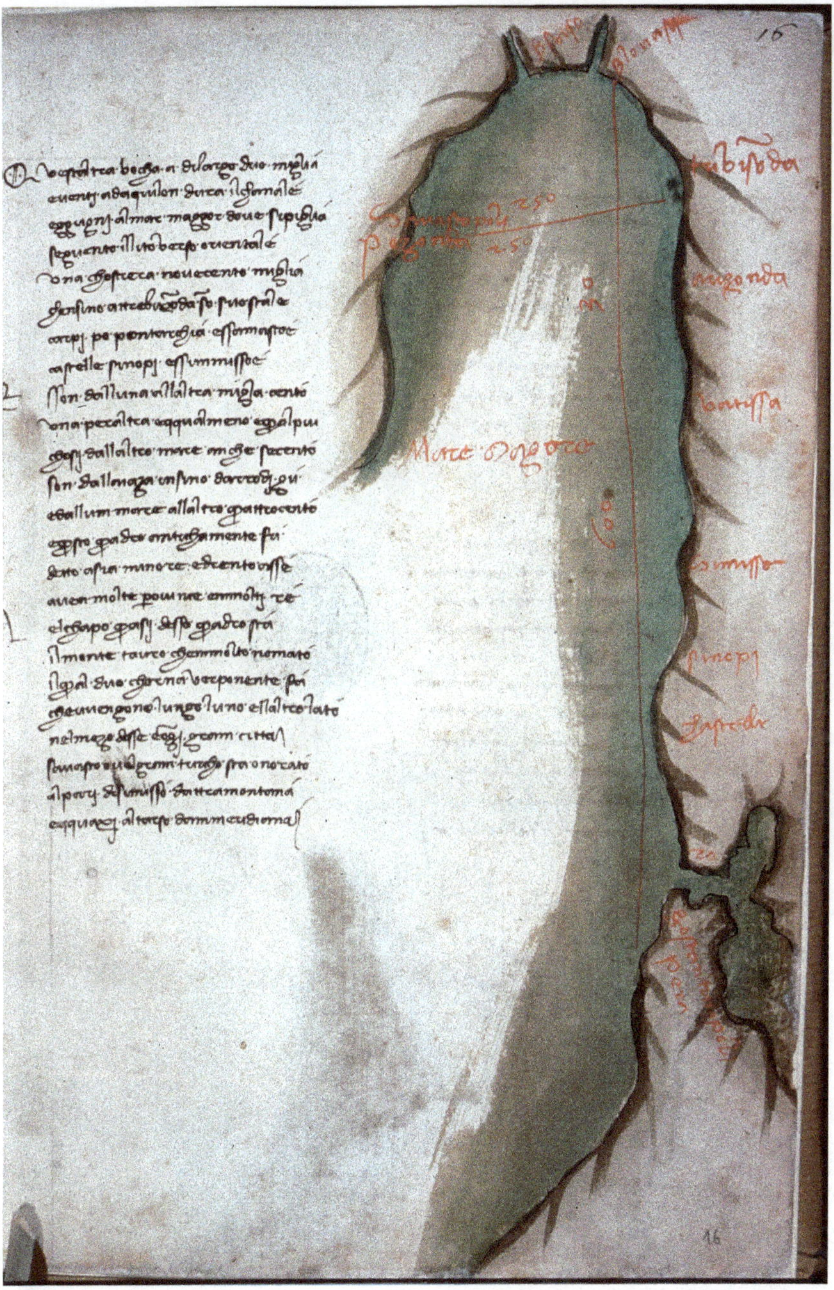

Plate XI. Gregorio Dati, *Sfera*. Laurenziana Mediceo-Palatino MS 89, f. 16r.

Plate XII. Gregorio Dati, *Sfera*. Newberry Ayer MS Map 1, f. 15v-16r.

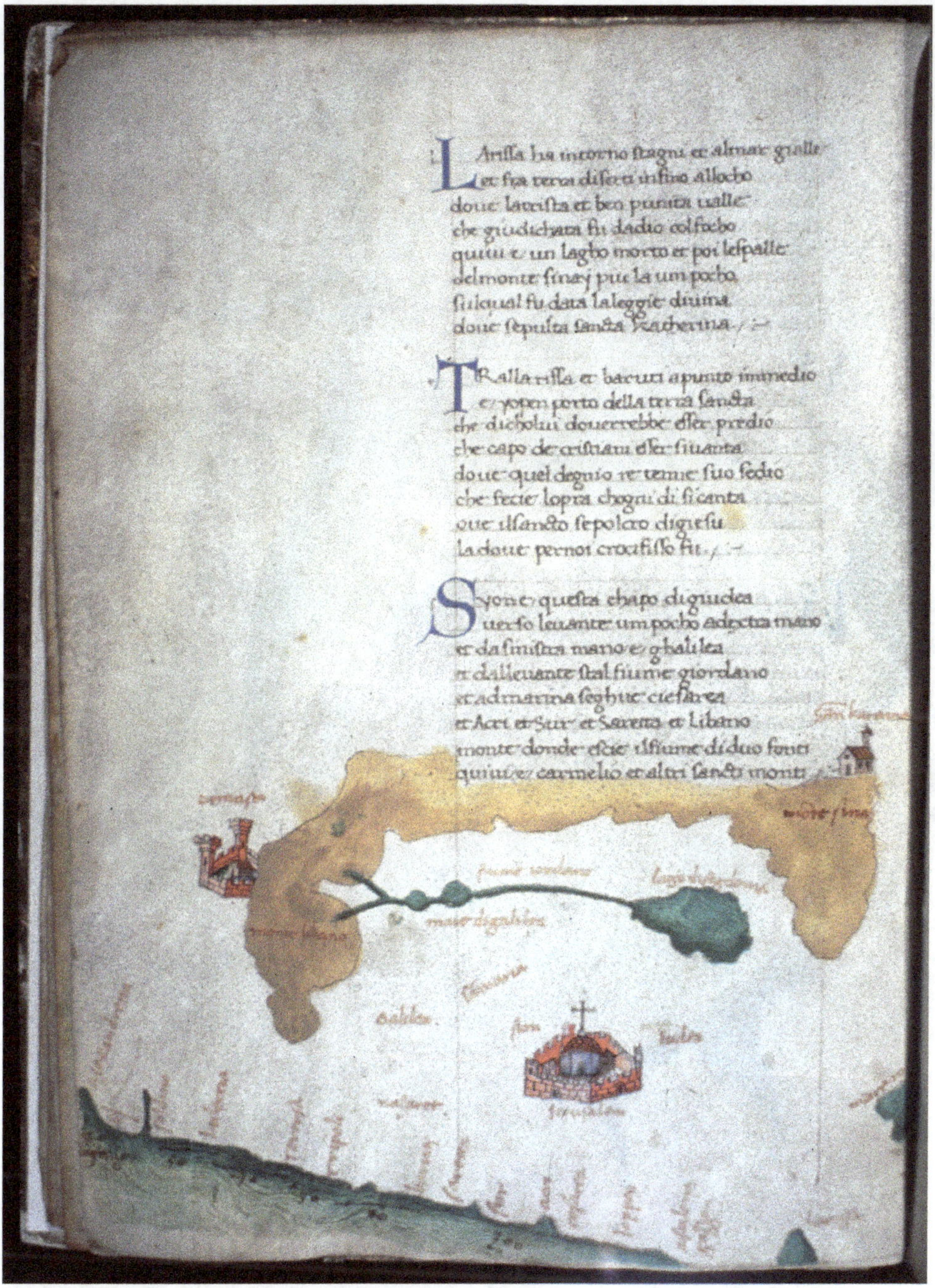

Larissa ha intorno stagni et almar gialle
et fra terra diserti infino allocho
doue latrista et ben punita ualle
che giudichata fu dadio colfocho
quiui e un lagho morto et poi lespalle
delmonte sinaj piu la um pocho
sulqual fu data laleggie diuina
doue sepulta sancta katherina

Tralla rissa et baruti apunto inmedio
e yoppen porto della terra sancta
che dicholui douerrebbe esser predio
che capo de cristiani esser si uanta
doue quel degnio re tenne suo sedio
che fecie sopra chogni di si canta
oue ilsancto sepolcro digiesu
ladoue pernoi crocifisso fu

Syone e questa chapo digiudea
uerso leuante um pocho adextra mano
et da sinistra mano e ghalilea
et dallevante sta il fiume giordano
et admarina seghue ciesarea
et Acri et Sur et Sarepta et Libano
monte donde escie ilfiume di duo fonti
quiui e carmelio et altri sancti monti

Plate XIII. Gregorio Dati, *Sfera*. Newberry Ayer MS Map 1, f. 14v.

Plate XIV. Paris, Bibliothèque nationale de France Ms Fr 4991, f. 5v. Map of France.

www.ingramcontent.com/pod-product-compliance
Lightning Source LLC
LaVergne TN
LVHW020633300926
849161LV00004B/686